Lecture Notes in Bioinformatics 7262

Edited by S. Istrail, P. Pevzner, and M. Waterman

Subseries of Lecture Notes in Computer Science

Benny Chor (Ed.)

Research in Computational Molecular Biology

16th Annual International Conference, RECOMB 2012
Barcelona, Spain, April 21-24, 2012
Proceedings

 Springer

Series Editors

Sorin Istrail, Brown University, Providence, RI, USA
Pavel Pevzner, University of California, San Diego, CA, USA
Michael Waterman, University of Southern California, Los Angeles, CA, USA

Volume Editor

Benny Chor
Tel-Aviv University
School of Computer Science
69978 Tel-Aviv, Israel
E-mail: benny@cs.tau.ac.il

ISSN 0302-9743   e-ISSN 1611-3349
ISBN 978-3-642-29626-0   e-ISBN 978-3-642-29627-7
DOI 10.1007/978-3-642-29627-7
Springer Heidelberg Dordrecht London New York

Library of Congress Control Number: 2012935493

CR Subject Classification (1998): F.2.2, H.2.8, I.6, F.2, J.3

LNCS Sublibrary: SL 8 – Bioinformatics

Typesetting: Camera-ready by author, data conversion by Scientific Publishing Services, Chennai, India

Printed on acid-free paper

Springer is part of Springer Science+Business Media (www.springer.com)

Preface

This volume contains the papers presented at RECOMB 2012, the 16th Annual International Conference on Research in Computational Molecular Biology, held in Barcelona during April 21–24, 2012. The RECOMB conference series was started in 1997 by Sorin Istrail, Pavel Pevzner, and Michael Waterman. RECOMB 2012 was hosted by the Centre for Genomic Regulation (CRG), and took place at the Fira Palace hotel, Barcelona. This year, 31 papers were accepted for presentation out of 200 submissions. The papers presented were selected by the Program Committee (PC), with the help of many external reviewers. Each paper was reviewed by at least three members of the PC, or by external reviewers. Following the initial reviews, there was a vivid and extensive e-mail-based discussion over a period of two weeks, leading to the final decisions. Accepted papers were also invited for submission to a special issue of the *Journal of Computational Biology*. The highlights track, first introduced during RECOMB 2010, continued. Works published or accepted for publication in a journal during the last 15 months were eligible for submission. There were 42 submissions, and 6 were selected for oral presentation.

In addition to the contributed and highlight talks, RECOMB 2012 featured five keynote addresses on a broad theme of life science and computational biology. The keynote speakers were Richard Durbin (The Wellcome Trust Sanger Institute, UK), Eileen Furlong (Genome Biology Unit, EMBL, Germany), Thomas Gingeras (Cold Spring Harbor Laboratory, USA), Alfonso Valencia (National Cancer Research Centre, Spain), and Ada E. Yonath (Department of Structural Biology, Weizmann Institute of Science, Israel). Finally, RECOMB 2012 featured an NSF-sponsored special session on emerging areas and challenges in computational biology. This session was initiated by Mitra Basu (NSF, USA) and Mona Singh (Princeton University, USA). Five speakers were invited to give talks in this session and participate in a follow-up panel. They were Dirk Evers (UK), Christina Leslie (Memorial Sloan-Kettering Cancer Center, USA), Laxmi Parida (IBM, USA), Teresa Przytycka (NIH–NLM–NCBI, USA), and Amos Tanay (Weizmann Institute of Science, Israel).

RECOMB 2012 would not have been possible without the dedication, hard work, and substantial investment of many individuals and organizations. We thank the PC and external reviewers who helped form a high-quality conference program, the Organizing Committee, coordinated by Roderic Guigó, the Organization and Finance Chair, and Blanka Wysocka, the event coordinator, for hosting the conference and providing the administrative, logistic, and financial support. We also thank our sponsors, including the CRG (Centre for Genomic Regulation), the MICINN (Spanish Ministry of Science and Innovation), the ISCB (International Society for Computational Biology), BioMed Central, CRC Press, Cambridge University Press, the INB (Spanish Institute of Bioinformatics), and the NSF (USA National Science Foundation). Without them the conference would not have been financially viable. We thank all the authors who contributed papers and posters, as well as the attendees of the conference for their enthusiastic participation. Finally, I would like to personally thank Roderic Guigó and Blanka Wysocka for the smooth cooperation we had regarding various organization details, and Michal Linial (Steering Committee), Martin Vingron (Chair of the Steering Committee), Bonnie Berger, and Vineet Bafna (Program Committee Chairs of RECOMB 2010 and 2011, respectively), for numerous discussions, helpful tips, and advice.

February 2012   Benny Chor

Conference Organization

Program Committee

Tatsuya Akutsu	Kyoto University
Rolf Backofen	University of Freiburg, Germany
Vineet Bafna	UC San Diego, USA
Nuno Bandeira	UC San Diego, USA
Vikas Bansal	Scripps Translational Science Institute, USA
Bonnie Berger	Massachusetts Institute of Technology, USA
Mathieu Blanchette	McGill University, Canada
Elhanan Borenstein	University of Washington, USA
Michael Brudno	University of Toronto, Canada
Benny Chor (Chair)	Tel Aviv University, Israel
Lenore Cowen	Tufts University, USA
Eleazar Eskin	UC Los Angeles, USA
Nir Friedman	The Hebrew University of Jerusalem, Israel
David Gifford	Massachusetts Institute of Technology, USA
Bjarni Halldorsson	Reykjavik University, Iceland
Eran Halperin	Tel Aviv University, Israel, and ICSI, Berkeley, USA
Alexander Hartemink	Duke University, USA
Barbara Holland	University of Tasmania, Australia
Katharina Huber	University of East Anglia, UK
Daniel Huson	University of Tübingen, Germany
Sorin Istrail	Brown University, USA
Mehmet Koyuturk	Case Western Reserve University, USA
Jens Lagergren	KTH Royal Institute of Technology, Sweden
Thomas Lengauer	Max Planck Institute for Informatics, Germany
Michal Linial	The Hebrew University of Jerusalem, Israel
Navodit Misra	Max Planck Institute for Molecular Genetics, Germany
Satoru Miyano	University of Tokyo, Japan
Luay Nakhleh	Rice University, USA
William Stafford Noble	University of Washington, USA
Nadia Pisanti	Università di Pisa, Italy
Teresa Przytycka	NIH, USA
Mark A. Ragan	The University of Queensland, Australia
Ben Raphael	Brown University, USA
Knut Reinert	Freie Universität Berlin, Germany
Marie-France Sagot	Université de Lyon, France

S. Cenk Sahinalp	Simon Fraser University, Canada
David Sankoff	University of Ottawa, Canada
Russell Schwartz	Carnegie Mellon University, USA
Charles Semple	University of Canterbury, New Zealand
Ron Shamir	Tel Aviv University, Israel
Mona Singh	Princeton University, USA
Sagi Snir	University of Haifa, Israel
Jens Stoye	Bielefeld University, Germany
Fengzhu Sun	University of Southern California, USA
Amos Tanay	The Weizmann Institute of Science, Israel
Haixu Tang	Indiana University, USA
Anna Tramontano	Sapienza University, Italy
Martin Vingron	Max Planck Institute for Molecular Genetics, Germany
Jerome Waldispuhl	McGill University, Canada
Yufeng Wu	University of Connecticut, USA
Jinbo Xu	Toyota Tech. Inst. at Chicago, USA
Zohar Yakhini	Agilent Technologies and the Technion, Israel
Michal Ziv-Ukelson	Ben-Gurion University, Israel

Steering Committee

Vineet Bafna	UC San Diego, USA
Serafim Batzoglou	Stanford University, USA
Bonnie Berger	Massachusetts Institute of Technology, USA
Sorin Istrail	Brown University, USA
Michal Linial	The Hebrew University of Jerusalem, Israel
Martin Vingron (Chair)	Max Planck Institute for Molecular Genetics, Germany

Organizing Committee

Josep F Abril	Universitat de Barcelona, Spain
Francisco Cámara	CRG, Spain
Eduardo Eyras	UPF, Spain
Pedro Ferreira	CRG, Spain
Toni Gabaldón	CRG, Spain
Roderic Guigó (Chair)	CRG, Spain
Cédric Nortedame	CRG, Spain
Gabriel Valiente	Univ. Politècnica de Catalunya, Spain
Blanka Wysocka	CRG, Spain

Previous RECOMB Meetings

Dates	Hosting Institution	Program Chair	Conference Chair
January 20-23, 1997 Santa Fe, NM, USA	Sandia National Lab	Michael Waterman	Sorin Istrail
March 22-25, 1998 New York, NY, USA	Mt. Sinai School of Medicine	Pavel Pevzner	Gary Benson
April 22-25, 1999 Lyon, France	INRIA	Sorin Istrail	Mireille Regnier
April 8-11, 2000 Tokyo, Japan	University of Tokyo	Ron Shamir	Satoru Miyano
April 22-25, 2001 Montreal, Canada	Université de Montreal	Thomas Lengauer	David Sankoff
April 18-21, 2002 Washington, DC, USA	Celera	Gene Myers	Sridhar Hannenhalli
April 10-13, 2003 Berlin, Germany	German Federal Ministry for Education and Research	Webb Miller	Martin Vingron
March 27-31, 2004 San Diego, USA	University of California, San Diego	Dan Gusfield	Philip E. Bourne
May 14-18, 2005 Boston, MA, USA	Broad Institute of MIT and Harvard	Satoru Miyano	Jill P. Mesirov and Simon Kasif
April 2-5, 2006 Venice, Italy	University of Padova	Alberto Apostolico	Concettina Guerra
April 21-25, 2007 San Francisco, CA, USA	QB3	Terry Speed	Sandrine Dudoit
March 30-April 2, 2008 Singapore, Singapore	National University of Singapore	Martin Vingron	Limsoon Wong
May 18-21, 2009 Tucson, AZ, USA	University of Arizona	Serafim Batzoglou	John Kececioglu
August 12-15, 2010 Lisbon, Portugal	INESC-ID and Instituto Superior Técnico	Bonnie Berger	Arlindo Oliveira
March 28-31, 2011 Vancouver, Canada	Lab for Computational Biology, Simon Fraser University	Vineet Bafna	S. Cenk Sahinalp

External Reviewers

Akey, Josh
Alkan, Can
Allali, Julien
Altmann, Andre
Ander, Christina
Andreotti, Sandro
Arndt, Peter
Arvestad, Lars
Aviran, Sharon
Ay, Ferhat
Bachmat, Eitan
Backoffen, Rolf
Bader, Gary
Bansal, Mukul
Bar, Amir
Baumbach, Jan
Beiko, Robert
Ben Bassat, Ilan
Ben-Dor, Amir

Ben-Hur, Asa
Benyamini, Hadar
Bernstein, Lars
Berry, Vincent
Birmele, Etienne
Bliven, Spencer
Böcker, Sebastian
Caglar, Mehmet Umut
Capra, Tony
Carmel, Liran
Carr, Rogan
Catanzaro, Daniele
Chen, Quan
Chen, Yang-Ho
Chinappi, Mauro
Chiu, Hsuan-Chao
Cho, Dongyeon
Cho, Sungje
Chowdhury, Salim

Cloonan, Nicole
Cohen, Reuven
Colbourn, Charles
Comin, Matteo
Costa, Fabrizio
Da Fonseca, P.G.S.
Dao, Phuong
Dasgupta, Bhaskar
Daskalakis, Constantinos
David, Matei
Davis, Melissa
Denise, Alain
Diallo, Abdoulaye
Dinitz, Yefim
Donmez, Nilgun
Du, Xiangjun
Dörr, Daniel
Edwards, Matthew
El-Mabrouk, Nadia
Emde, Anne-Katrin
Erten, Sinan
Farahani, Hossein
Ferreira, Rui
Fertin, Guillaume
Frånberg, Mattias
Furlotte, Nicholas
Furman, Ora
Gaspin, Christine
Geiger, Dan
Giegerich, Robert
Goeke, Jonathan
Golan, David
Gophna, Uri
Gorecki, Pawel
Gottlieb, Assaf
Greenblum, Sharon
Grossi, Roberto
Guidoni, Leonardo
Guindon, Stephane
Guthals, Adrian
Habin, Naomi
Hach, Faraz
Hajirasouliha, Iman
Halloran, John
Hamilton, Nicholas

Han, Buhm
Hashimoto, Tatsunori
Hayashida, Morihiro
He, Dan
Hogan, James
Hoinka, Jan
Holtgrewe, Manuel
Hormozdiari, Farhad
Hormozdiari, Fereydoun
Howbert, Jeff
Hu, Jialu
Huang, Yang
Imoto, Seiya
Jaenicke, Sebastian
Joo, Jong Wha
Kamburov, Atanas
Kang, Eun Yong
Kang, John
Kaplan, Tommy
Kato, Yuki
Kayano, Mitsunori
Keasar, Chen
Kenigsberg, Effi
Kim, Sangtae
Kim, Yoo-Ah
Kinsella, Marcus
Kirkpatrick, Bonnie
Klaere, Steffen
Kostem, Emrah
Krause, Roland
Kundu, Kousik
Laframboise, Thomas
Landan, Gilad
Lange, Sita
Lasserre, Julia
Lee, Heewook
Lee, Hyunju
Lee, Su-In
Lempel, Noa
Levin, Alex
Levitt, Michael
Levy, Roie
Li, Jing
Lingjaerde, Ole-Christian
Linial, Nathan

Linz, Simone
Lipson, Doron
Liu, Kevin
Liu, Zhenqiu
Loytynoja, Ari
Lu, Bingwen
Lubling, Yaniv
Luccio, Fabrizio
Mahmoody, Ahmad
Mahony, Shaun
Mandoiu, Ion
Margelevicius, Mindaugas
Maticzka, Daniel
Mayrose, Itay
Mcilwain, Sean
Mckenney, Mark
Mcpherson, Andrew
Medvedev, Paul
Melsted, Pall
Mendelson-Cohen, Netta
Mendes, Nuno
Mezlini, Aziz
Milazzo, Paolo
Milo, Nimrod
Misra, Navodit
Moll, Mark
Mysickova, Alena
Nachman, Iftach
Nair, Nishanth
Navarro, Gonzalo
Navon, Roy
Niida, Atsushi
Numanagic, Ibrahim
O'Donnell, Charles
Oesper, Layla
Oostenbrink, Chris
Ophir, Ron
Orenstein, Yaron
Oshlack, Alicia
Pagli, Linda
Park, Hyun Jung
Parrish, Nathaniel
Pasaniuc, Bogdan
Peng, Jian
Peterlongo, Pierre

Pfeifer, Nico
Picard, Franck
Pinhas, Tamar
Preisner, Robert
Press, Maximilian
Privman, Eyal
Przulj, Natasa
Qi, Yanjun
Rahnenführer, Jörg
Raimondo, Domenico
Rawitz, Droro
Renda, M. Elena
Roch, Sebastien
Rohs, Remo
Rose, Dominic
Roshan, Usman
Rosset, Saharon
Rozov, Roye
Rubinstein, Nimrod
Ruffalo, Matthew
Ruths, Troy
Safer, Hershel
Sankararaman, Sriram
Savel, Daniel
Schlicker, Andreas
Schnall-Levin, Michael
Schoenhuth, Alexander
Scornavacca, Celine
Serin, Akdes
Shatkay, Hagit
Sheridan, Paul
Shibuya, Tetsuo
Shimamura, Teppei
Shomron, Noam
Sindi, Suzanne
Singer, Meromit
Singh, Ajit
Sjöstrand, Joel
Smoly, Ilan
Snedecor, June
Sommer, Ingolf
Souaiaia, Tade
Spivak, Marina
Stacho, Juraj
Steel, Mike

Table of Contents

Protein Structure by Semidefinite Facial Reduction

Babak Alipanahi[1], Nathan Krislock[2], Ali Ghodsi[3], Henry Wolkowicz[4], Logan Donaldson[5], and Ming Li[1,*]

[1] David R. Cheriton School of Computer Science, University of Waterloo, Waterloo, Ontario, Canada
[2] INRIA Grenoble Rhône-Alpes, France
[3] Department of Statistics and Actuarial Science, University of Waterloo, Waterloo, Ontario, Canada
[4] Department of Combinatorics and Optimization, University of Waterloo, Waterloo, Ontario, Canada
[5] Department of Biology, York University, Toronto, Ontario, Canada
mli@uwaterloo.ca

Abstract. All practical contemporary protein NMR structure determination methods use molecular dynamics coupled with a simulated annealing schedule. The objective of these methods is to minimize the error of deviating from the NOE distance constraints. However, this objective function is highly nonconvex and, consequently, difficult to optimize. Euclidean distance geometry methods based on semidefinite programming (SDP) provide a natural formulation for this problem. However, complexity of SDP solvers and ambiguous distance constraints are major challenges to this approach. The contribution of this paper is to provide a new SDP formulation of this problem that overcomes these two issues for the first time. We model the protein as a set of intersecting two- and three-dimensional cliques, then we adapt and extend a technique called semidefinite facial reduction to reduce the SDP problem size to approximately one quarter of the size of the original problem. The reduced SDP problem can not only be solved approximately 100 times faster, but is also resistant to numerical problems from having erroneous and inexact distance bounds.

Keywords: Molecular structural biology, nuclear magnetic resonance.

1 Introduction

Computing three-dimensional protein structures from their amino acid sequences has been one of the most widely studied problems in bioinformatics because knowing the structure of protein structure is key to understanding its physical, chemical, and biological properties. The protein nuclear magnetic resonance (NMR) method is fundamentally different from the X-ray method: It is not a

---

* Corresponding author.

B. Chor (Ed.): RECOMB 2012, LNBI 7262, pp. 1–11, 2012.

"microscope with atomic resolution"; rather it provides a network of distance measurements between spatially proximate hydrogen atoms [12]. As a result, the NMR method relies heavily on complex computational algorithms. The existing methods for protein NMR can be categorized into four major groups: (*i*) methods based on Euclidean distance matrix completion (EDMC) [6,15,5,20], (*ii*) methods based on molecular dynamics and simulated annealing [24,8,28,14,13], (*iii*) methods based on local/global optimization [7,23,34], and (*iv*) methods originating from sequence-based protein structure prediction algorithms [29,25,2].

1.1   Gram Matrix Methods

Using the Gram matrix, or the matrix of inner products, has many advantages: (*i*) The Gram matrix and Euclidean distance matrix (EDM) are linearly related to each other. (*ii*) Instead of enforcing all of the triangle inequality constraints, it is sufficient to enforce that the Gram matrix is positive semidefinite. (*iii*) The embedding dimension and the rank of the Gram matrix are directly related. Semidefinite programming (SDP) is a natural choice for formulating the problem using the Gram matrix. However, the major obstacle is the computational complexity of SDP solvers.

1.2   Contributions of the Proposed SPROS Method

Most of the existing methods make some of the following assumptions: (*i*) assuming to know the (nearly) exact distances between atoms, (*ii*) assuming to have the distances between any type of nuclei (not just hydrogens), (*iii*) ignoring the fact that not all hydrogens can be uniquely assigned, and (*iv*) overlooking the ambiguity in the NOE cross-peak assignments. In order to automate the NMR protein structure determination process, we need a robust structure calculation method that tolerates more errors. We give a new SDP formulation that does not assume (*i–iv*) above. Moreover, the new method, called "SPROS" (Semidefinite Programming-based Protein structure determination), models the protein molecule as a set of intersecting two- and three-dimension cliques. We adapt and extend a technique called semidefinite facial reduction which makes the SDP problem strictly feasible and reduces its size to approximately one quarter the size of the original problem. The reduced problem is more numerically stable to solve and can be solved nearly 100 times faster.

2   The SPROS Method

2.1   Euclidean Distance Geometry

Euclidean Distance Matrix. A symmetric matrix D is called a Euclidean Distance Matrix (EDM) if there exists a set of points $\{x_1, \ldots, x_n\}$, $x_i \in \mathbb{R}^r$ such that:
$$D_{ij} = \|x_i - x_j\|^2, \quad \forall i, j. \tag{1}$$
The smallest value of r is called the *embedding dimension* of D, and is denoted **embdim**$(D) = r$. The space of all $n \times n$ EDMs is denoted $\mathcal{E}^n$.

The Gram Matrix. If we define $X := [\boldsymbol{x}_1, \ldots, \boldsymbol{x}_n] \in \mathbb{R}^{r \times n}$, then the matrix of inner-products, or *Gram Matrix*, is given by $G := X^\top X$. It immediately follows that $G \in \mathcal{S}_+^n$, where $\mathcal{S}_+^n$ is the set of symmetric positive semidefinite $n \times n$ matrices. The Gram matrix and the Euclidean distance matrix are linearly related:

$$D = \mathsf{K}(G) := \mathbf{diag}(G) \cdot \mathbf{1}^\top + \mathbf{1} \cdot \mathbf{diag}(G)^\top - 2G, \tag{2}$$

where $\mathbf{1}$ is the all-ones vector of the appropriate size. To go from the EDM to the Gram matrix, we use the $\mathsf{K}^\dagger \colon \mathcal{S}^n \to \mathcal{S}^n$ linear map:

$$G = \mathsf{K}^\dagger(D) := -\tfrac{1}{2} HDH, \quad D \in \mathcal{S}_H^n, \tag{3}$$

where $H = I - \tfrac{1}{n} \mathbf{1}\mathbf{1}^\top$ is the *centering* matrix, $\mathcal{S}^n$ is the space of symmetric $n \times n$ matrices, and $\mathcal{S}_H^n := \{A \in \mathcal{S}^n : \mathbf{diag}(A) = \mathbf{0}\}$, is the set of symmetric matrices with zero diagonal.

Schoenberg's Theorem. Given a matrix D, we can determine if it is an EDM with the following well-known theorem [27]:

Theorem 1. *A matrix $D \in \mathcal{S}_H^n$ is a Euclidean distance matrix if and only if $\mathsf{K}^\dagger(D)$ is positive semidefinite. Moreover,* $\mathbf{embdim}(D) = \mathbf{rank}(\mathsf{K}^\dagger(D))$ *for all* $D \in \mathcal{E}^n$.

2.2 The SDP Formulation

Semidefinite optimization or, more commonly, semidefinite programming is a class of convex optimization problems that has attracted much attention in the optimization community and has found numerous applications in different science and engineering fields. Notably, several diverse convex optimization problems can be formulated as SDP problems [31]. Current state-of-the-art SDP solvers are based on *primal-dual interior-point* methods.

Preliminary Problem Formulation. There are three types of constraints in our formulation: (*i*) equality constraints, which are the union of equality constraints preserving bond lengths ($\mathcal{B}$), bond angles ($\mathcal{A}$), and planarity of the coplanar atoms ($\mathcal{P}$), giving $\mathcal{E} = \mathcal{E}_\mathcal{B} \cup \mathcal{E}_\mathcal{A} \cup \mathcal{E}_\mathcal{P}$; (*ii*) upper bounds, which are the union of NOE-derived ($\mathcal{N}$), hydrogen bonds ($\mathcal{H}$), disulfide and salt bridges ($\mathcal{D}$), and torsion angle ($\mathcal{T}$) upper bounds, giving $\mathcal{U} = \mathcal{U}_\mathcal{N} \cup \mathcal{U}_\mathcal{H} \cup \mathcal{U}_\mathcal{D} \cup \mathcal{U}_\mathcal{T}$; (*iii*) lower bounds, which are the union of steric or van der Waals ($\mathcal{W}$) and torsion angle ($\mathcal{T}$) lower bounds, giving $\mathcal{L} = \mathcal{L}_\mathcal{W} \cup \mathcal{L}_\mathcal{T}$. We assume the target protein has n atoms, $a_1, \ldots, a_n$. The preliminary problem formulation is given by:

$$\begin{aligned}
\text{minimize} \quad & \gamma \langle I, K \rangle + \sum_{ij} w_{ij} \xi_{ij} + \sum_{ij} w'_{ij} \zeta_{ij} \tag{4}\\
\text{subject to} \quad & \langle A_{ij}, K \rangle = e_{ij}, \ (i,j) \in \mathcal{E}\\
& \langle A_{ij}, K \rangle \leq u_{ij} + \xi_{ij}, \ (i,j) \in \mathcal{U}\\
& \langle A_{ij}, K \rangle \geq l_{ij} - \zeta_{ij}, \ (i,j) \in \mathcal{L}\\
& \xi_{ij} \in \mathbb{R}_+, \ (i,j) \in \mathcal{U}, \quad \zeta_{ij} \in \mathbb{R}_+, \ (i,j) \in \mathcal{L}\\
& K\mathbf{1} = \mathbf{0}, \quad K \in \mathcal{S}_+^n,
\end{aligned}$$

where $A_{ij} = (e_i - e_j)(e_i - e_j)^\top$ and e_i is the ith column of the identity matrix. The *centering* constraint $K\mathbf{1} = \mathbf{0}$, ensures that the embedding of K is centered at the origin. Since both upper bounds and lower bounds may be inaccurate and noisy, non-negative penalized slacks, ζ_{ij}'s and ξ_{ij}'s, are included to prevent infeasibility and manage ambiguous upper bounds. The heuristic rank reduction term, $\gamma\langle I, K\rangle$, with $\gamma < 0$, in the objective function, produces lower-rank solutions [33].

Challenges in Solving the SDP Problem. Solving the optimization problem in (4) can be challenging: For small to medium sized proteins, the number of atoms, n, is 1,000-3,500, and current primal-dual interior-point SDP solvers cannot solve problems with $n > 2,000$ efficiently. Moreover, the optimization problem in (4) does not satisfy *strict feasibility*, causing numerical problems; see [32].

It can be observed that the protein contains many small intersecting cliques. For example, peptide planes or aromatic rings, are 2D cliques, and tetrahedral carbons form 3D cliques. As we show later, whenever there is a clique in the protein, the corresponding Gram matrix, K, can never be full-rank, which violates strict feasibility. By adapting and extending a technique called *semidefinite facial reduction*, not only do we obtain an equivalent problem that satisfies strict feasibility, but we also significantly reduce the SDP problem size.

2.3   Cliques in a Protein Molecule

A protein molecule with ℓ amino acid residues has $\ell + 1$ planes in its backbone. Moreover, each amino acid has a different side chain with a different structure; therefore, the number of cliques in each side chain varies. We assume that the i-th residue, r_i, has s_i cliques in its side chain, denoted by $\mathcal{S}_i^{(1)}, \ldots, \mathcal{S}_i^{(s_i)}$. For all amino acids (except glycine and proline), the first side chain clique is formed around the tetrahedral carbon CA, $\mathcal{S}_i^{(1)} = \{N_i, CA_i, HA_i, CB_i, C_i\}$, which intersects with two peptide planes $\mathcal{P}_{i-1}$ and $\mathcal{P}_i$ in two atoms: $\mathcal{S}_i^{(1)} \cap \mathcal{P}_{i-1} = \{N_i, CA_i\}$ and $\mathcal{S}_i^{(1)} \cap \mathcal{P}_i = \{CA_i, C_i\}$. There is a total of $q = \ell + 1 + \sum_{i=1}^{\ell} s_i$ cliques in the distance matrix of any protein. To simplify, let $\mathcal{C}_i = \mathcal{P}_{i-1}$, $1 \le i \le \ell + 1$, and $\mathcal{C}_{\ell+2} = \mathcal{S}_1^{(1)}$, $\mathcal{C}_{\ell+2} = \mathcal{S}_1^{(2)}$, $\ldots$, $\mathcal{C}_q = \mathcal{S}_\ell^{(s_\ell)}$.

2.4   Algorithm for Finding the Face of the Structure

For $t < n$ and $U \in \mathbb{R}^{n \times t}$, the set of matrices $U\mathcal{S}_+^t U^\top$ is a face of $\mathcal{S}_+^n$ (in fact every face of $\mathcal{S}_+^n$ can be described in this way); see, e.g., [26]. We let $\mathbf{face}(\mathcal{F})$ represent the smallest face containing a subset $\mathcal{F}$ of $\mathcal{S}_+^n$; then we have the important property that $\mathbf{face}(\mathcal{F}) = U\mathcal{S}_+^t U^\top$ if and only if there exists $Z \in \mathcal{S}_{++}^t$ such that $UZU^\top \in \mathcal{F}$. Furthermore, in this case, we have that every $Y \in \mathcal{F}$ can be decomposed as $Y = UZU^\top$, for some $Z \in \mathcal{S}_+^t$, and the reduced feasible set $\{Z \in \mathcal{S}_+^t : UZU^\top \in \mathcal{F}\}$ has a strictly feasible point, giving us a problem that is more numerically stable to solve (problems that are not strictly feasible have a dual optimal set that is unbounded and therefore can be difficult to solve numerically; for more information, see [32]). Moreover, if $t \ll n$, this results in a significant reduction in the matrix size.

The Face of a Single Clique. Here, we solve the SINGLE CLIQUE problem, which is defined as follows: Let D be a partial EDM of a protein. Suppose the first n_1 points form a clique in the protein, such that for $C_1 = \{1, \ldots, n_1\}$, all distances are known. That is, the matrix $D_1 = D[C_1]$ is completely specified, where for an *index set* $\mathcal{I} \subseteq \{1, \ldots, n\}$, $B = A[\mathcal{I}]$ is the $|\mathcal{I}| \times |\mathcal{I}|$ matrix formed by rows and columns of A indexed by $\mathcal{I}$. Moreover, let $r_1 = \mathbf{embdim}(D_1)$. We now show how to compute the smallest face containing the feasible set:

$$\mathcal{F}_1 := \{K \in \mathcal{S}_+^n : \mathsf{K}(K[C_1]) = D_1\}. \tag{5}$$

Theorem 2 (SINGLE CLIQUE, [17]). *Let $U_1 \in \mathbb{R}^{n \times (n-n_1+r_1+1)}$ be defined as follows:*

- *let $V_1 \in \mathbb{R}^{n_1 \times r_1}$ be a full column rank matrix such that* $\mathbf{range}(V_1)$ $= \mathbf{range}(\mathsf{K}^\dagger(D_1))$;

- *let $\bar{U}_1 := \begin{bmatrix} V_1 & 1 \end{bmatrix}$ and* $U_1 := \begin{matrix} \\ n_1 \\ n-n_1 \end{matrix} \begin{matrix} \overset{r_1+1\ \ n-n_1}{} \\ \begin{bmatrix} \bar{U}_1 & 0 \\ 0 & I \end{bmatrix} \end{matrix} \in \mathbb{R}^{n \times (n-n_1+r_1+1)}.$

Then U_1 has full column rank, $1 \in \mathbf{range}(U)$, and

$$\mathbf{face}\{K \in \mathcal{S}_+^n : \mathsf{K}(K[C_1]) = D[C_1]\} = U_1 \mathcal{S}_+^{n-n_1+r_1+1} U_1^\top.$$

Computing the V_1 Matrix. In Theorem 2, we can find V_1 by computing the eigendecomposition of $\mathsf{K}^\dagger(D[C_1])$ as follows:

$$\mathsf{K}^\dagger(D[C_1]) = V_1 \Lambda_1 V_1^\top, \quad V_1 \in \mathbb{R}^{n_1 \times r_1}, \ \Lambda_1 \in \mathcal{S}_{++}^{r_1}. \tag{6}$$

It can be seen that V_1 has full column rank (columns are orthonormal) and also $\mathbf{range}(\mathsf{K}^\dagger(D[C_1])) = \mathbf{range}(V_1)$.

2.5 The Face of a Protein Molecule

The protein molecule is made of q cliques, $\{C_1, \ldots, C_q\}$, such that $D[C_l]$ is known, and we have $r_l = \mathbf{embdim}(D[C_l])$, and $n_l = |C_l|$. Let $\mathcal{F}$ be the feasible set of the SDP problem. If for each clique C_l, we define $\mathcal{F}_l := \{K \in \mathcal{S}_+^n : \mathsf{K}(K[C_l]) = D[C_l]\}$, then

$$\mathcal{F} \subseteq \left(\bigcap_{l=1}^q \mathcal{F}_l \right) \cap \mathcal{S}_C^n, \tag{7}$$

where $\mathcal{S}_C^n = \{K \in \mathcal{S}^n : K\mathbf{1} = \mathbf{0}\}$. For $l = 1, \ldots, q$, let $F_l := \mathbf{face}(\mathcal{F}_l) = U_l \mathcal{S}_+^{n-n_l+r_l+1} U_l^\top$, where U_l is computed as in Theorem 2. We have [17]:

$$\left(\bigcap_{l=1}^q F_l \right) \cap \mathcal{S}_C^n \subseteq \left(\bigcap_{l=1}^q U_l \mathcal{S}_+^{n-n_l+r_l+1} U_l^\top \right) \cap \mathcal{S}_C^n = (U \mathcal{S}_+^k U^\top) \cap \mathcal{S}_C^n, \tag{8}$$

where $U \in \mathbb{R}^{n \times k}$ is a full column rank matrix that satisfies:

$$\mathbf{range}(U) = \bigcap_{l=1}^q \mathbf{range}(U_l). \tag{9}$$

We now have an efficient method for computing the face of the feasible set $\mathcal{F}$.

After computing U, we can decompose the Gram matrix as $K = UZU^\top$, for $Z \in \mathcal{S}_+^k$. However, by exploiting the centering constraint, $K\mathbf{1} = \mathbf{0}$, we can reduce the matrix size one more. If $V \in \mathbb{R}^{k \times (k-1)}$ has full column rank and satisfies $\mathbf{range}(V) = \mathbf{null}(\mathbf{1}^\top U)$, then we have [17]:

$$\mathcal{F} \subseteq (UV)\mathcal{S}_+^{k-1}(UV)^\top. \tag{10}$$

For more details on facial reduction for Euclidean distance matrix completion problems, see [16].

Constraints for Preserving the Structure of Cliques. If we find a *base* set of points $\mathcal{B}_l$ in each clique $\mathcal{C}_l$ such that $\mathbf{embdim}(D[\mathcal{B}_l]) = r_l$, then by fixing the distances between points in the base set and fixing the distances between points in $\mathcal{C}_l \setminus \mathcal{B}_l$ and points in $\mathcal{B}_l$, the entire clique is kept rigid.

Therefore, we need to fix *only* the distances between base points [3], resulting in a three- to four-fold reduction in the number of equality constraints. We call the reduced set of equality constraints $\mathcal{E}_{\mathrm{FR}}$.

2.6   Solving and Refining the Reduced SDP Problem

From equation (10), we can formulate the reduced SDP problem as follows:

$$
\begin{aligned}
\text{minimize} \quad & \gamma\langle I, Z\rangle + \sum_{ij} w_{ij}\xi_{ij} + \sum_{ij} w'_{ij}\zeta_{ij} \\
\text{subject to} \quad & \langle A'_{ij}, Z\rangle = e_{ij}, \ (i,j) \in \mathcal{E}_{\mathrm{FR}} \\
& \langle A'_{ij}, Z\rangle \leq u_{ij} + \xi_{ij}, \ (i,j) \in \mathcal{U} \\
& \langle A'_{ij}, Z\rangle \geq l_{ij} - \zeta_{ij}, \ (i,j) \in \mathcal{L} \\
& \xi_{ij} \in \mathbb{R}_+, \ (i,j) \in \mathcal{U}, \quad \zeta_{ij} \in \mathbb{R}_+, \ (i,j) \in \mathcal{L} \\
& Z \in \mathcal{S}_+^{k-1},
\end{aligned}
\tag{11}
$$

where $A'_{ij} = (UV)^\top A_{ij}(UV)$.

Post-Processing. We perform a refinement on the raw structure determined by the SDP solver. For this refinement we use a BFGS-based quasi-Newton method [21] that only requires the value of the objective function and its gradient at each point.

Letting $X^{(0)} = X_{\mathrm{SDP}}$, we iteratively minimize the following objective function:

$$
\phi(X) = w_\mathcal{E} \sum_{(i,j)\in\mathcal{E}} (\|\boldsymbol{x}_i - \boldsymbol{x}_j\| - e_{ij})^2 + w_\mathcal{U} \sum_{(i,j)\in\mathcal{U}} f(\|\boldsymbol{x}_i - \boldsymbol{x}_j\| - u_{ij})^2
$$

$$
+ w_\mathcal{L} \sum_{(i,j)\in\mathcal{L}} g(\|\boldsymbol{x}_i - \boldsymbol{x}_j\| - l_{ij})^2 + w_r \sum_{i=1}^n \|\boldsymbol{x}_i\|^2, \tag{12}
$$

where $f(\alpha) = \max(0, \alpha)$ and $g(\alpha) = \min(0, -\alpha)$, and w_r is the weight of the regularization term. We also employ a hybrid protocol from XPLOR-NIH that incorporates thin-layer water refinement [22] and a multidimensional torsion angle database [18,19].

3   Results

We tested the performance of SPROS on 18 proteins: 15 protein data sets from the DOCR database in the NMR Restraints Grid [10,11] and three protein data sets from Donaldson's laboratory at York University. We chose proteins with different sizes and topologies, as listed in Table 1. Finally, the input to the SPROS method is exactly the same as the input to the widely-used CYANA method.

3.1   Implementation

The SPROS method has been implemented and tested in MATLAB 7.13 (apart from the water refinement, which is done by XPLOR-NIH). For solving the SDP problem, we used the SDPT3 method [30]. For minimizing the post-processing objective function (12), we used the BFGS-based quasi-Newton method implementation by Lewis and Overton [21]. All the experiments were carried out on an Ubuntu 11.04 Linux PC with a 2.8 GHz Intel Core i7 Quad-Core processor and 8 GB of memory.

3.2   Determined Structures

From the 18 test proteins, 9 of them were calculated with backbone RMSDs less than or equal to 1 Å, and 16 have backbone RMSDs less than 1.5 Å. Detailed analysis of calculated structures is listed in Table 2. The superimposition of the SPROS and reference structures for three of the proteins are depicted in Fig. 1. More detailed information about the determined structures can be found in [4].

To further assess the performance of SPROS, we compared the SPROS and reference structures for 1G6J, Ubiquitin, and 2GJY, PTB domain of Tensin, with their corresponding X-ray structures, 1UBQ and 1WVH, respectively. For 1G6J, the backbone (heavy atoms) RMSDs for SPROS and the reference structures are 0.42 Å (0.57 Å) and 0.73±0.04 Å (0.98±0.04 Å), respectively. For 2GJY, the backbone (heavy atoms) RMSDs for SPROS and the reference structures are 0.88 Å (1.15 Å) and 0.89 ± 0.08 Å (1.21 ± 0.06 Å), respectively.

3.3   Discussion

The SPROS method was tested on 18 experimentally derived protein NMR data sets of sequence lengths ranging from 76 to 307 (weights ranging from 8 to 35 KDa). Calculation times were in the order of a few minutes per structure. Accurate results were obtained for all of the data sets, although with some variability in precision. The best attribute of the SPROS method is its tolerance for, and efficiency at, managing many incorrect distance constraints (that are typically defined as upper bounds).

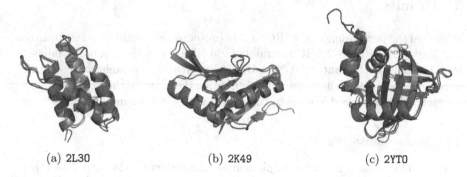

(a) 2L30   (b) 2K49   (c) 2YTO

Fig. 1. Superimposition of structures determined by SPROS in blue and the reference structures in red

Table 1. Information about the proteins used in testing SPROS. The second, third, and fourth columns, list the topologies, sequence lengths, and molecular weight of the proteins, the fifth and sixth columns, n and n', list the original and reduced SDP matrix sizes, respectively. The seventh column lists the number of cliques in the protein. The eights and ninth columns, $m_\mathcal{E}$ and $m'_\mathcal{E}$, list the number of equality constraints in the original and reduced problems, respectively. The 10th column, $m_\mathcal{U}$, lists the total number of upper bounds for each protein. The 11th column, bound types, lists intra-residue, $|i - j| = 0$, sequential, $|i - j| = 1$, medium range, $1 < |i - j| \leq 4$, and long range, $|i-j| > 4$, respectively, in percentile. The 12th column, $\bar{m}_\mathcal{U} \pm s_\mathcal{U}$, lists the average number of upper bounds per residue, together with the standard deviation. The 13th column, $m_\mathcal{N}$, lists the number of NOE-inferred upper bounds. The 14th column, $p_\mathcal{U}$, lists the fraction of pseudo-atoms in the upper bounds in percentile. The last two columns, $m_\mathcal{T}$ and $m_\mathcal{H}$, list the number of upper bounds inferred from torsion angle restraints, and hydrogen bonds, disulfide and salt bridges, respectively.

ID	topo.	len.	weight	n	n'	cliques (2D/3D)	$m_\mathcal{E}$	$m'_\mathcal{E}$	$m_\mathcal{U}$	bound types	$\bar{m}_\mathcal{U} \pm s_\mathcal{U}$	$m_\mathcal{N}$	$p_\mathcal{U}$	$m_\mathcal{T}$	$m_\mathcal{H}$
1G6J	a+b	76	8.58	1434	405	304 (201/103)	5543	1167	1354	21/ 29/ 17/ 33	31.9±15.3	1291	32	63	0
1B4R	B	80	7.96	1281	346	248 (145/103)	4887	1027	787	26/ 25/ 6 / 43	17.1±10.8	687	30	22	78
2E80	A	103	11.40	1523	419	317 (212/105)	5846	1214	3157	19/ 29/ 26/ 26	71.4±35.4	3070	24	87	0
1CN7	a/b	104	11.30	1927	532	393 (253/140)	7399	1540	1560	46/ 24/ 12/ 18	23.1±13.4	1418	31	80	62
2KTS	a/b	117	12.85	2075	593	448 (299/149)	7968	1719	2279	22/ 28/ 14/ 36	34.6±17.4	2276	25	0	3
2K49	a+b	118	13.10	2017	574	433 (291/142)	7710	1657	2612	22/ 27/ 18/ 38	40.9±21.1	2374	27	146	92
2K62	B	125	15.10	2328	655	492 (327/165)	8943	1886	2367	21/ 32/ 15/ 32	33.9±18.6	2187	32	180	0
2L30	A	127	14.30	1867	512	393 (269/124)	7143	1492	1270	24/ 38/ 20/ 18	22.5±12.7	1055	25	156	59
2GJY	a+b	144	15.67	2337	639	474 (302/172)	8919	1875	1710	7 / 30/ 19/ 44	25.0±16.6	1536	29	98	76
2KTE	a/b	152	17.21	2576	717	542 (360/182)	9861	2089	1899	17/ 31/ 22/ 30	24.3±20.8	1669	30	124	106
1XPW	B	153	17.44	2578	723	541 (355/186)	9837	2081	1206	0 / 31/ 11/ 58	17.0±10.8	934	37	210	62
2K7H	a/b	157	16.66	2710	756	563 (363/200)	10452	2196	2768	29/ 33/ 13/ 25	30.3±11.3	2481	19	239	48
2KVP	A	165	17.28	2533	722	535 (344/191)	9703	2094	5204	31/ 26/ 23/ 20	59.2±25.0	4972	22	232	0
2YTO	a+b	176	19.17	2940	828	627 (419/208)	11210	2404	3357	23/ 28/ 14/ 35	34.9±22.3	3237	30	120	0
2L7B	A	307	35.30	5603	1567	1205 (836/369)	21421	4521	4355	10/ 30/ 44/ 16	27.6±14.4	3459	23	408	488
1Z1V	A	80	9.31	1259	362	272 (181/91)	4836	1046	1261	46/ 24/ 18/ 13	28.6±16.3	1189	15	0	72
HACS1	B	87	9.63	1150	315	237 (156/81)	4401	923	828	46/ 21/ 5 / 27	20.2±14.2	828	20	0	36
2LJG	a+b	153	17.03	2343	662	495 (327/168)	9009	1909	1347	40/ 29/ 8 / 22	16.4±11.9	1065	28	204	78

Table 2. Information about determined structures of the test proteins. The second, third, and fourth columns list SDP time, water refinement time, and total time, respectively. For the backbone and heavy atom RMSD columns, the mean and standard deviation between the determined structure and the reference structures is reported (backbone RMSDs less than 1.5 Å are shown in bold). The seventh column, CBd, lists the number of residues with "CB deviations" larger than 0.25 Å computed by MolProbity, as defined by [9]. The eighth and ninth columns list the percentage of upper bound violations larger than 0.1 Å and 1.0 Å, respectively (the numbers for the reference structures are in parentheses). The last three columns, list the percentage of residues with favorable and allowed backbone torsion angles and outliers, respectively.

ID	t_s	t_w	t_t	RMSD backbone	heavy atoms	CBd.	violations 0.1 Å	1.0 Å	Ramachandran fav.	alw.	out.
1G6J	44.5	175.5	241.0	**0.68±0.05**	0.90±0.05	0	4.96 (0.08±0.07)	0.85 (0)	100	100	0
1B4R	21.4	138.0	179.0	**0.85±0.06**	1.06±0.06	0	20.92 (13.87±0.62)	6.14 (2.28±0.21)	80.8	93.6	6.4
2E80	129.8	181.3	340.9	**0.58±0.02**	0.68±0.01	0	31.33 (31.93±0.14)	9.98 (10.75±0.13)	96.2	100	0
1CN7	75.0	230.1	339.7	1.53±0.11	1.80±0.10	0	10.27 (7.63±0.80)	3.18 (2.11±0.52)	96.1	99.0	1.0
2KTS	116.7	231.0	398.5	**0.92±0.06**	1.13±0.06	0	25.36 (27.44±0.58)	6.49 (10.36±0.68)	86.1	95.7	4.3
2K49	140.7	240.7	422.7	**0.99±0.14**	1.24±0.16	0	13.75 (15.79±0.67)	2.80 (4.94±0.46)	93.8	97.3	2.7
2K62	156.1	259.0	464.2	**1.40±0.08**	1.72±0.08	1	33.74 (42.92±0.95)	10.79 (21.20±1.20)	87.8	95.9	4.1
2L30	61.7	212.0	310.0	**1.28±0.15**	1.59±0.15	0	21.53 (19.81±0.58)	7.33 (7.61±0.31)	80.4	92.8	7.2
2GJY	113.7	285.9	455.7	**0.99±0.07**	1.29±0.09	0	11.67 (8.36±0.59)	0.36 (0.49±0.12)	85.4	92.3	7.7
2KTE	139.9	297.7	503.2	**1.39±0.17**	1.85±0.16	1	35.55 (31.97±0.46)	11.94 (11.96±0.40)	79.4	90.8	9.2
1XPW	124.8	297.1	489.7	**1.30±0.10**	1.68±0.10	0	9.74 (0.17±0.09)	1.20 (0.01±0.02)	87.9	97.9	2.1
2K7H	211.7	312.0	591.0	**1.24±0.07**	1.49±0.07	0	17.60 (16.45±0.30)	4.39 (4.92±0.35)	92.3	96.1	3.9
2KVP	462.0	282.4	814.8	**0.94±0.08**	1.05±0.09	0	15.15 (17.43±0.29)	4.01 (5.62±0.21)	96.6	100	0
2YTO	292.1	421.5	800.1	**0.79±0.05**	1.04±0.06	1	29.04 (28.9±0.36)	6.64 (6.60±0.30)	90.5	97.6	2.4
2L7B	1101.1	593.0	1992.1	2.15±0.11	2.55±0.11	3	19.15 (21.72±0.36)	4.23 (4.73±0.23)	79.2	91.6	8.4
1Z1V	30.6	158.8	209.2	**1.44±0.17**	1.74±0.15	0	3.89 (2.00±0.25)	0.62 (0)	90.9	98.5	1.5
HACS1	17.4	145.0	176.1	**1.00±0.07**	1.39±0.10	0	20.29 (15.68±0.43)	4.95 (3.73±0.33)	83.6	96.7	3.3
2LJG	94.7	280.4	426.3	**1.24±0.09**	1.70±0.10	1	28.35 (25.3±0.51)	10.76 (8.91±0.49)	80.6	90.7	9.3

Our final goal is a fully automated system for NMR protein structure determination, from peak picking [1] to resonance assignment [2], to protein structure determination. An automated system, without the laborious human intervention will have to tolerate more errors than usual. This was the initial motivation of designing SPROS. The key is to tolerate more errors. Thus, we are working towards incorporating an adaptive violation weight mechanism to identify the most significant outliers in the set of distance restraints automatically.

Acknowledgments. This work was supported in part by NSERC Grant OGP0046506, Canada Research Chair program, an NSERC Collaborative Grant, David R. Cheriton Graduate Scholarship, OCRiT, Premier's Discovery Award, and Killam Prize.

References

1. Alipanahi, B., Gao, X., Karakoc, E., Donaldson, L., Li, M.: PICKY: a novel SVD-based NMR spectra peak picking method. Bioinformatics 25(12), i268–i275 (2009)
2. Alipanahi, B., Gao, X., Karakoc, E., Li, S., Balbach, F., Feng, G., Donaldson, L., Li, M.: Error tolerant NMR backbone resonance assignment and automated structure generation. Journal of Bioinformatics and Computational Biology 0(1), 1–26 (2011)
3. Alipanahi, B., Krislock, N., Ghodsi, A.: Manifold learning by semidefinite facial reduction (2011) (unpublished manuscript) (in preparation)
4. Alipanahi, B.: New Approaches to Protein NMR Automation. Ph.D. thesis, University of Waterloo (2011)
5. Biswas, P., Toh, K.C., Ye, Y.: A distributed SDP approach for large-scale noisy anchor-free graph realization with applications to molecular conformation. SIAM J. Sci. Comput. 30, 1251–1277 (2008)
6. Braun, W., Bösch, C., Brown, L.R., Go, N., Wüthrich, K.: Combined use of proton-proton overhauser enhancements and a distance geometry algorithm for determination of polypeptide conformations. application to micelle-bound glucagon. Biochimica et Biophysica Acta 667(2), 377–396 (1981)
7. Braun, W., Go, N.: Calculation of protein conformations by proton-proton distance constraints. a new efficient algorithm. Journal of Molecular Biology 186(3), 611–626 (1985)
8. Brünger, A.T.: X-PLOR Version 3.1: A System for X-ray Crystallography and NMR. Yale University Press (1993)
9. Chen, V.B., Arendall, W.B., Headd, J.J., Keedy, D.A., Immormino, R.M., Kapral, G.J., Murray, L.W., Richardson, J.S., Richardson, D.C.: MolProbity: all-atom structure validation for macromolecular crystallography. Acta Crystallographica. Section D, Biological Crystallography 66(pt.1), 12–21 (2010)
10. Doreleijers, J.F., Mading, S., Maziuk, D., Sojourner, K., Yin, L., Zhu, J., Markley, J.L., Ulrich, E.L.: BioMagResBank database with sets of experimental NMR constraints corresponding to the structures of over 1400 biomolecules deposited in the protein data bank. Journal of Biomolecular NMR 26(2), 139–146 (2003)
11. Doreleijers, J.F., Nederveen, A.J., Vranken, W., Lin, J., Bonvin, A.M., Kaptein, R., Markley, J.L., Ulrich, E.L.: BioMagResBank databases DOCR and FRED containing converted and filtered sets of experimental NMR restraints and coordinates from over 500 protein PDB structures. Journal of Biomolecular NMR 32(1), 1–12 (2005)
12. Güntert, P.: Structure calculation of biological macromolecules from NMR data. Quarterly Reviews of Biophysics 31(2), 145–237 (1998)
13. Güntert, P.: Automated NMR structure calculation with CYANA. Methods in Molecular Biology 278, 353–378 (2004)
14. Güntert, P., Mumenthaler, C., Wüthrich, K.: Torsion angle dynamics for NMR structure calculation with the new program DYANA. Journal of Molecular Biology 273, 283–298 (1997)
15. Havel, T.F., Wüthrich, K.: A Distance Geometry Program for Determining the Structures of Small Proteins and Other Macromolecules From Nuclear Magnetic Resonance Measurements of Intramolecular H-H Proxmities in Solution. Bulletin of Mathematical Biology 46(4), 673–698 (1984)
16. Krislock, N.: Semidefinite Facial Reduction for Low-Rank Euclidean Distance Matrix Completion. Ph.D. thesis, University of Waterloo (2010)

17. Krislock, N., Wolkowicz, H.: Explicit sensor network localization using semidefinite representations and facial reductions. SIAM J. Optimiz. 20, 2679–2708 (2010)
18. Kuszewski, J., Gronenborn, A.M., Clore, G.M.: Improving the quality of NMR and crystallographic protein structures by means of a conformational database potential derived from structure databases. Protein Sci. 5(6), 1067–1080 (1996)
19. Kuszewski, J., Gronenborn, A.M., Clore, G.M.: Improvements and extensions in the conformational database potential for the refinement of NMR and x-ray structures of proteins and nucleic acids. Journal of Magnetic Resonance 125(1), 171–177 (1997)
20. Leung, N.H.Z., Toh, K.C.: An SDP-based divide-and-conquer algorithm for large-scale noisy anchor-free graph realization. SIAM J. Sci. Comput. 31, 4351–4372 (2009)
21. Lewis, A., Overton, M.: Nonsmooth optimization via BFGS. Submitted to SIAM J. Optimiz. (2009)
22. Linge, J.P., Habeck, M., Rieping, W., Nilges, M.: ARIA: automated NOE assignment and NMR structure calculation. Bioinformatics 19(2), 315–316 (2003)
23. Moré, J.J., Wu, Z.: Global continuation for distance geometry problems. SIAM J. Optimiz. 7, 814–836 (1997)
24. Nilges, M., Clore, G.M., Gronenborn, A.M.: Determination of three-dimensional structures of proteins from interproton distance data by hybrid distance geometry-dynamical simulated annealing calculations. FEBS Letters 229(2), 317–324 (1988)
25. Raman, S., Lange, O.F., Rossi, P., Tyka, M., Wang, X., Aramini, J., Liu, G., Ramelot, T.A., Eletsky, A., Szyperski, T., Kennedy, M.A., Prestegard, J., Montelione, G.T., Baker, D.: NMR structure determination for larger proteins using Backbone-Only data. Science 327(5968), 1014–1018 (2010)
26. Ramana, M.V., Tunçel, L., Wolkowicz, H.: Strong duality for semidefinite programming. SIAM J. Optimiz. 7(3), 641–662 (1997)
27. Schoenberg, I.J.: Remarks to Maurice Fréchet's article Sur la définition axiomatique d'une classe d'espace distanciés vectoriellement applicable sur l'espace de Hilbert. Ann. of Math. 36(3), 724–732 (1935)
28. Schwieters, C., Kuszewski, J., Tjandra, N., Clore, G.: The Xplor-NIH NMR molecular structure determination package. Journal of Magnetic Resonance 160, 65–73 (2003)
29. Shen, Y., Lange, O., Delaglio, F., Rossi, P., Aramini, J.M., Liu, G., Eletsky, A., Wu, Y., Singarapu, K.K., Lemak, A., Ignatchenko, A., Arrowsmith, C.H., Szyperski, T., Montelione, G.T., Baker, D., Bax, A.: Consistent blind protein structure generation from NMR chemical shift data. Proceedings of the National Academy of Sciences of the United States of America 105(12), 4685–4690 (2008)
30. Tütüncü, R., Toh, K., Todd, M.: Solving semidefinite-quadratic-linear programs using SDPT3. Math. Program. 95(2, ser. B), 189–217 (2003)
31. Vandenberghe, L., Boyd, S.: Semidefinite programming. SIAM Review 38(1), 49–95 (1996)
32. Wei, H., Wolkowicz, H.: Generating and measuring instances of hard semidefinite programs. Mathematical Programming 125, 31–45 (2010)
33. Weinberger, K.Q., Saul, L.K.: Unsupervised learning of image manifolds by semidefinite programming. In: IEEE Computer Society Conference on Computer Vision and Pattern Recognition, vol. 2, pp. 988–995 (2004)
34. Williams, G.A., Dugan, J.M., Altman, R.B.: Constrained global optimization for estimating molecular structure from atomic distances. Journal of Computational Biology 8(5), 523–547 (2001)

Ancestry Inference in Complex Admixtures via Variable-Length Markov Chain Linkage Models

Sivan Bercovici[1,*], Jesse M. Rodriguez[1,2,*,**],
Megan Elmore[1], and Serafim Batzoglou[1]

[1] Department of Computer Science, Stanford University
[2] Biomedical Informatics Program, Stanford University
jesserod@cs.stanford.edu

Abstract. Inferring the ancestral origin of chromosomal segments in ad-mixed individuals is key for genetic applications, ranging from analyzing population demographics and history, to mapping disease genes. Previous methods addressed ancestry inference by using either weak models of linkage disequilibrium, or large models that make explicit use of ancestral haplotypes. In this paper we introduce ALLOY, an efficient method that incorporates generalized, but highly expressive, linkage disequilibrium models. ALLOY applies a factorial hidden Markov model to capture the parallel process producing the maternal and paternal admixed haplotypes, and models the background linkage disequilibrium in the ancestral populations via an inhomogeneous variable-length Markov chain. We test ALLOY in a broad range of scenarios ranging from recent to ancient admixtures with up to four ancestral populations. We show that ALLOY outperforms the previous state of the art, and is robust to uncertainties in model parameters.

Keywords: Population genetics, ancestry inference, VLMC, FHMM.

1 Introduction

Determining the ancestral origin of chromosomal segments in admixed individuals is a problem that has been addressed by several methods [13, 21, 20, 14, 11, 4]. The development of these methods was motivated by various applications such as studying population migration patterns [9, 7], increasing the statistical power of association studies by accounting for population structure [12], and enhancing admixture-mapping [23, 19] for both disease-gene mapping as well as personalized drug therapy applications [3]. The ability to accurately infer ancestry is important in genome-wide association studies (GWAS). These studies are based on the premise that a homogenous population sample was collected. Population stratification, however, poses a significant challenge in association studies; the existence of different sub-populations within the examined cases and controls can

---

* These authors contributed equally to this work.
** Corresponding author.

B. Chor (Ed.): RECOMB 2012, LNBI 7262, pp. 12–28, 2012.
© Springer-Verlag Berlin Heidelberg 2012

yield many spurious associations originating from the population sub-structure rather than the disease status. Inferred sub-structure within the population enables the correction for this effect, consequently improving the statistical power of these studies. A second disease-gene mapping technique that benefits from an accurate inference of ancestry is admixture-mapping [23]. This statistically powerful and efficient method identifies genomic regions containing disease susceptibility genes in recently admixed populations, which are populations formed from the merging of several distinct ancestral populations (e.g., African Americans). The statistical power of admixture-mapping increases as the disease prevalence exhibits a greater difference between the ancestral populations from which the admixed population was formed. Admixed individuals carrying such a disease are expected to show an elevated frequency of the ancestral population with the higher disease risk near the disease gene loci. Hence, the effectiveness of this method relies on the ability to accurately infer the ancestry along the chromosomes of admixed individuals.

The problem of ancestry inference is commonly viewed at one of two levels: (a) at the global scale, predicting an individual's single origin out of several possible homogenous ancestries, or determining an individual's ancestral genomic composition; (b) at the finer local scale, labeling the different ancestries along the chromosomes of an admixed individual. In the context of local ancestry inference, most previous methods are based on hidden Markov models (HMM), where the hidden states correspond to ancestral populations and generate the observed genotypes. The work of Patterson et al. [13] employed such an HMM, integrated into a Markov chain Monte Carlo (MCMC), for estimating ancestry along the genome. The method accounted for uncertainties in model parameters such as number of generation since admixture, admixture proportions, and ancestral allele frequencies. For simplicity, the work assumed that, given the ancestry, the sampled markers are in linkage equilibrium (i.e., independent). This assumption was then relaxed in the work by Tang et al. [21], applying a Markov hidden Markov model (MHMM) to account for the dependencies between neighboring markers as exhibited within the ancestral populations. While the modeled first-order Markovian dependencies accounted for some of the linkage disequilibrium (LD) between markers, the complex nature of the linkage patterns presented an opportunity for more accurate LD models that would yield better performance in inferring local ancestry. The explicit use of ancestral haplotypes, in methods such as HAPAA [20] and HAPMIX [14], enabled a more comprehensive account for background LD (i.e., LD within the ancestral population) over longer segments. In these methods, the hidden states corresponded to specific ancestral haplotypes, and the transition between the states corresponded to intra-population mixture and inter-populations admixture processes. While efficient inference algorithms were applied, the model size grew linearly with the number of parental individuals, and the time complexity grew quadratically with the numbers of parental individuals for the case of genotype-based analysis. The time complexity of such an analyses became prohibitively high with more than a modest number of model individuals.

Other work explored window-based techniques, in which a simple ancestral composition was assumed to occur within a window (i.e., at most a single admixture event within an examined segment). LAMP [18], and its extension WIN-POP [11], used a naïve Bayes approach, assuming markers within a window are independent given ancestry, applying the inference over a sliding window. Although LD was not modeled, the methods demonstrated an accuracy superior to methods that did account for background LD. An additional window-based framework was developed in [4], decoupling the admixture process from the background LD model. Chromosomal ancestral profiles were efficiently enumerated using a dynamic-programming (DP) technique, enabling the instantiation of various LD models for the single-ancestry segments from which a profile was composed. Multiple LD models were studied within the framework, showing that higher-order LD models yield an increase in inference accuracy.

In this work we describe ALLOY, a novel local ancestry inference method that enables the incorporation of complex models for linkage disequilibrium in the ancestral populations. ALLOY applies a factorial hidden Markov model (FHMM) to capture the parallel process producing the maternal and paternal admixed haplotypes. We model background LD in ancestral populations via an inhomogeneous variable-length Markov chain (VLMC). The states in our model correspond to ancestral haplotype clusters, which are groups of haplotypes that share local structure within a chromosomal region, as in [5]. In our method, each ancestral population is described by a separate LD model that locally fits the varying LD complexities along the genome. We provide an inference algorithm that is sub-cubic in the maximal number of haplotype clusters at any position. This allows ALLOY to scale well when analyzing admixtures of more than two populations, or incorporating more elaborate LD models.

We demonstrate through simulations that ALLOY accurately infers the position-specific ancestry in a wide range of complex and ancient admixtures. For instance, ALLOY achieves 87% accuracy on a 3-population admixture between individuals sampled from Yoruba in Ibadan, Nigeria, Maasai in Kinyawa, Kenya, and northern and western Europe. Our results represent substantial improvements over previous state of the art. Further, we explore the landscape of background LD models, and find that the highest performance is achieved by LD models that lie between models that assume independence of markers and models that explicitly use the reference haplotypes. Finally, our results demonstrate that as more samples representing the ancestral populations become available, our LD models improve and enable more accurate local ancestry inference.

2   Methods

We consider the problem of local ancestry inference, defined as labeling each genotyped position along the genome of an admixed individual with its ancestry. Here, admixture is assumed to follow the Hybrid-Isolated model [10], in which a single past admixture event mixing K ancestral populations with proportions $\pi = (\pi_1, ..., \pi_K)$ is followed by g generations of consecutive random mating. For clarity, we assume that a set of L bi-allelic SNPs was observed along the genome of an individual; we relax the bi-allelic marker assumption in the Discussion section. Furthermore, at each position, we define a state space of haplotype clusters A_l, each of which represents a collection of ancestral haplotypes that share a common local structure (i.e., allelic sequence surrounding a particular location). It immediately follows that each such haplotype cluster $a_l \in A_l$ at location l is mapped to a single allele, denoted by $e(a_l) \in \{0, 1\}$. In our model, each of the K populations is represented by a separate mutually-exclusive subset of haplotype clusters. We denote by $anc(a_l)$ the ancestry, out of K, of a particular haplotype cluster $a_l \in A_l$. We denote by H_l^m, H_l^p the (hidden) haplotype cluster membership drawn from A_l on the maternal and paternal haplotype at position l, respectively, and by $G_l \in \{0, 1, 2\}$ the genotype observed at th/e same marker position, representing the minor allele count. The vectors of haplotype cluster memberships and genotypes across all L marker positions are denoted by $H^{\{m,p\}} = (H_1^{\{m,p\}}, H_2^{\{m,p\}}, ..., H_L^{\{m,p\}})$ and $G = (G_1, G_2, ..., G_L)$, respectively.

We use a factorial hidden Markov model (FHMM) [6] to statistically model the dual mosaic ancestral pattern along the genome of an admixed individual, as depicted in Figure 1. In factorial HMMs, which are equivalent in expressive power to hidden Markov models (HMM) [15], the single chain of hidden variables is replaced by a chain of a hidden vector of independent factors. In our application, the FHMM representation allows us to naturally decouple the state space into two parallel dynamic processes generating H^m and H^p, pertaining to the presumably independent maternal and paternal admixture processes, and producing the single composed admixed offspring G. The decomposition of the state space into independent processes allows efficient inference by leveraging the structure in the compound state transition probabilities. In our model, the values of $H_l = (H_l^m, H_l^p)$ at specific position l are drawn from the Cartesian product $A_l \times A_l$, corresponding to the alleles within specific ancestral haplotypes originating from a restricted prior set of K hypothesized ancestral populations. Note that A_l extends the notion of an allele, which is simply a binary variable, to an allele within an ancestral haplotype; for position l, multiple states in A_l may correspond to the same allele.

To infer local ancestry, we first compute the posterior marginals given the sampled genotypes $P(H_l^m, H_l^m | G)$ by applying the forward-backward algorithm

$$P(H_l^m = a_l, H_l^p = a_l' | G) \propto \alpha_l(a_l, a_l') \cdot \beta_l(a_l, a_l') \tag{1}$$

where $\alpha_l(a_l, a_l') = P(G_1, ..., G_l, H_l^m = a_l, H_l^p = a_l')$ and $\beta_l(a_l, a_l') = P(G_{l+1}, ..., G_L | H_l^m = a_l, H_l^p = a_l')$. A naive recursive computation of α and β yields $O(|A_1|^2 + \sum_{l=2}^{L} |A_{l-1}|^2 \cdot |A_l|^2)$ time complexity as the transition from each pair of haplotype cluster memberships to each consecutive pair of haplotype cluster memberships is explicitly assessed. However, the dependency structure

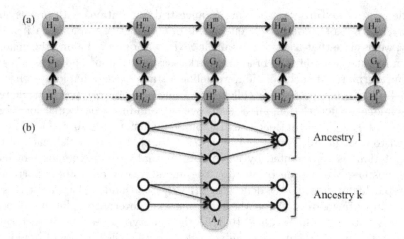

Fig. 1. A factorial hidden Markov model capturing the parallel admixture processes generating the maternal and paternal haplotypes, and giving rise to the sampled genotypes of the admixed offspring. (a) A graphical model depicting the conditional independencies in our model. Each variable in the hidden chains $H_l^{\{m,p\}}$ corresponds to a haplotype cluster membership, and G_l corresponds to the observed genotype at location l. (b) The state space A_l for a particular location l along the genome. Each ancestry is modeled by an independent set of haplotype cluster membership states, and each such state can emit a single allele. Edges in the illustration connecting states correspond to intra-population observed transitions, namely local haplotypic sequences that were frequent in the corresponding ancestral population. Edges corresponding to admixture transitions, connecting states of different ancestries, are omitted from this illustration for clarity.

of FHMMs allows for a more efficient recursive computation of α and β, as described in [6]. Specifically, α is computed in the forward direction in three steps as follows

$$\alpha_{l-1}^m(a_l, a'_{l-1}) = \sum_{a_{l-1} \in A_{l-1}} \alpha_{l-1}(a_{l-1}, a'_{l-1}) \cdot P(H_l^m = a_l | H_{l-1}^m = a_{l-1}) \quad (2)$$

$$\alpha_{l-1}^p(a_l, a'_l) = \sum_{a'_{l-1} \in A_{l-1}} \alpha_{l-1}^m(a_l, a'_{l-1}) \cdot P(H_l^p = a'_l | H_{l-1}^p = a'_{l-1})$$

$$\alpha_l(a_l, a'_l) = \alpha_{l-1}^p(a_l, a'_l) \cdot P(G_l | H_l^m = a_l, H_l^p = a'_l),$$

namely, advancing on the maternal track, followed by advancing on the paternal track, and finally incorporating the local observation by multiplying by the emission probability $P(G_l | H_l^m = a_l, H_l^p = a'_l)$. Similarly, β is computed in a backward recursion as

$$\beta_l^e(a_l, a_l') = \beta_l(a_l, a_l') \cdot P(G_l | H_l^m = a_l, H_l^p = a_l') \tag{3}$$

$$\beta_l^m(a_{l-1}, a_l') = \sum_{a_l \in A_l} P(H_l^m = a_l | H_{l-1}^m = a_{l-1}) \cdot \beta_l^e(a_l, a_l')$$

$$\beta_{l-1}(a_{l-1}, a_{l-1}') = \sum_{a_l' \in A_l} P(H_l^p = a_l' | H_{l-1}^p = a_{l-1}') \cdot \beta_l^m(a_{l-1}, a_l').$$

To complete the description, we define $\alpha_1(a_1, a_1') = P(H_1^m = a_1) \cdot P(H_1^p = a_1') \cdot P(G_1 | H_1^m = a_1, H_1^p = a_1')$ and $\beta_L(a_L, a_L') = 1$. When computing β, advancing on the maternal track takes $(|A_{l-1}| \cdot |A_l|) \cdot |A_l|$ time, while advancing on the paternal track takes $(|A_{l-1}| \cdot |A_{l-1}|) \cdot |A_l|$ time, as determined by the size of the corresponding composite state space. Similarly, a single forward step α_l is computed in $|A_l| \cdot |A_{l-1}| \cdot (|A_l| + |A_{l-1}|)$ time. Hence, the time complexity is now reduced to $O(|A_1|^2 + \sum_{l=2}^{L} |A_l| \cdot |A_{l-1}| \cdot (|A_l| + |A_{l-1}|))$.

To model genotyping error, the emission probability $P(G_l | H_l^m, H_l^p)$ used in Equations 2 and 3 is defined as follows

$$P(G_l | H_l^m = a_l, H_l^p = a_l') = \begin{cases} 1 - 2\epsilon, & e(a_l) + e(a_l') = G_l \\ \epsilon, & \text{otherwise} \end{cases} \tag{4}$$

where ϵ corresponds to the genotyping error rate.

To increase the numerical stability in the forward-backward computation, scaling is applied. Specifically, α_l and β_l are scaled by $s_l = \sum_{a_l, a_l'} \alpha_l(a_l, a_l')$ as follows

$$\alpha_l^*(a_l, a_l') = \frac{\alpha_l(a_l, a_l')}{s_l} \quad \beta_l^*(a_l, a_l') = \frac{\beta_l(a_l, a_l')}{s_l}. \tag{5}$$

Next, the unordered ancestry pair $\{Z_l^1, Z_l^2\}$ at location l is called by determining the maximal *a posteriori* assignment

$$\{\hat{Z}_l^1, \hat{Z}_l^2\} = \arg\max_{Z_l^1, Z_l^2} \sum_{\substack{a_l, a_l' \text{ s.t.} \\ \{anc(a_l), anc(a_l')\} = \{Z_l^1, Z_l^2\}}} \alpha_l^*(a_l, a_l') \cdot \beta_l^*(a_l, a_l') \tag{6}$$

where, for each $\{Z_l^1, Z_l^2\}$ pair, we sum over all (a_l, a_l') haplotype cluster membership pairs that are consistent in their ancestry with the unordered ancestry pair $\{Z_l^1, Z_l^2\}$.

We proceed by describing the transition probabilities $P(H_l | H_{l-1})$. Let R_l be defined as the event in which at least one post-admixture recombination occurred between position $l - 1$ and position l since the first population admixture event, and let $\bar{R}_l$ be defined as the complementary event. The transition probability $P(H_l | H_{l-1})$, which captures the process in which an admixed haplotype is generated, mixes the event of intra-ancestral-population transition, $P(H_l | H_{l-1}, \bar{R}_l)$, with the event corresponding to the introduction of a new ancestral haplotype, $P(H_l|H_{l-1}, R_l)$, as described by

$$P(H_l|H_{l-1}) = P(R_l) \cdot P(H_l|H_{l-1}, R_l) + P(\bar{R}_l) \cdot P(H_l|H_{l-1}, \bar{R}_l) \qquad (7)$$
$$= P(R_l) \cdot P_{anc(H_l)}(H_l) + P(\bar{R}_l) \cdot P_{anc(H_l)}(H_l|H_{l-1})$$

where $P_{anc(H_l)}(H_l)$ is the position specific ancestral haplotype cluster prior, and $P_{anc(H_l)}(H_l|H_{l-1})$ models the transition within the ancestral population $anc(H_l)$, capturing the background population-specific LD. Namely, if a post-admixture recombination was introduced ($P(R_l)$), a haplotype H_l is sampled based on the local ancestry prior $P_{anc(H_l)}(H_l)$; if no post-admixture recombination was introduced ($P(\bar{R}_l)$), the next marker is sampled based on the haplotypic structure within population $anc(H_l)$, as defined by $P_{anc(H_l)}(H_l|H_{l-1})$. Assuming the Hybrid-Isolated model, the probability of post-admixture recombination $P(R_l)$ is approximated via the Halden function [8]

$$P(R_l) = 1 - e^{-\phi(g \cdot d_l)} \qquad (8)$$

where d_l is the genetic distance, in Morgans (M), between marker $l - 1$ and l, and $g + 1$ generations are assumed to have passed since the first admixture event. We note that $\phi(z)$ is defined as a function of the recombination rate $g \cdot d_l$ to enable smoothing; the number of false ancestry changes can be reduced by controlling the probability for recombination (e.g., $\phi(z) = \frac{z}{10}$), overcoming local inaccuracies in ancestry inference due to an imperfect ancestral linkage model. The prior probability of the ancestral haplotype cluster is governed by the mixture proportions π and the intra-population haplotype cluster prior $P_{anc(H_l)}(H_l)$, as given by

$$P(H_l) = \pi_{anc(H_l)} \cdot P_{anc(H_l)}(H_l). \qquad (9)$$

Finally, we describe the background model we use to capture the ancestral linkage disequilibrium between markers. The range of explored background LD models is illustrated in Figure 2. The most basic models used for ancestry inference assume markers are independent given their corresponding ancestry assignment. An immediate extension which can be captured by our FHMM model incorporates first-order Markovian dependencies to model LD between neighboring markers. However, the model is not limited to first-order dependencies; to capture longer range dependencies between ancestral alleles, the state space A_l from which H_l is drawn, can be enriched so as to track ancestral haplotype clusters over a longer range. Specifically, longer range dependencies are effectively translated to additional states that map to specific ancestral local haplotype clusters. Moreover, a different number of states can be introduced at each position, fitting the local ancestral haplotypic complexity. The higher the local complexity is, the more states are used to track dependencies reaching further away. In essence, the model is equivalent to an inhomogeneous VLMC in which regions exhibiting complex LD structures are modeled using longer dependencies (i.e., edges connecting distant nodes in the underlying graphical model). At one extreme,

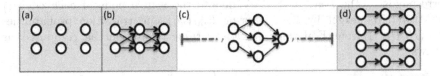

Fig. 2. The state space A_l over three consecutive locations in different background LD models, pertaining to the marker dependencies exhibited within a single ancestral population. (a) Markers are independent given ancestry. The model contain two states per location, each emitting one of the two possible alleles matching the marginal distribution observed in the ancestral population. (b) First order Markovian dependency between adjacent markers. The transition between the neighboring states, which correspond to alleles at specific positions, is derived from the conditional probability estimated from the ancestral population sample. (c) Generalized linkage model via haplotype clusters. The number of states at each position correspond to the number of haplotype clusters, each emitting an allele. The local transition probabilities correspond to the Markovian property by which haplotype cluster membership at a given location l is determined by the cluster membership at the previous location $l-1$. (d) Explicit use of ancestral haplotypes. For each position, the number of states equals the number of training haplotypes, each emitting a single allele observed in the corresponding haplotype.

the state space A_l can be constructed assuming a zero-order Markov model (i.e., markers are independent), while at the other extreme, A_l can be extended to have one state per ancestral haplotype instance used in the training phase.

An algorithm for fitting inhomogeneous VLMCs was described by Ron et al. [16], and extended by [5] to model haplotypes. We apply Beagle, an implementation of this procedure, to empirically model the local haplotypic structure. Specifically, we determine both the state space of A_l as well as the transition probability through the use of a localized haplotype cluster model described in [5]. Briefly, given a set of training haplotypes from a single ancestry, the algorithm processes the markers in chromosomal order. With each additional marker considered, nodes, representing some history of allele sequences, are split by considering the subsequent alleles for each such node. Then, nodes at location l are merged based on a Markov criterion roughly guaranteeing that given the cluster membership at position l, prior cluster memberships are irrelevant for the prediction of subsequent cluster memberships. Namely, given some parameter t, two clusters at position l are merged if the probabilities of allele sequences at markers $l+1, l+2, ..., l+t$ resemble each other. For each population anc, the procedure yields a weighted directed acyclic graph (DAG), where edges are labeled by alleles, and each training haplotype traces a path through the graph from a root node to a terminal node, defining the weights. For each edge e_l^i at location l, the weight w_l^i is defined as the number of haplotypes in the ancestral population sample that pass through the ith cluster. In our model, the state space $A_l^{anc} \subset A_l$ for population anc at location l is defined so that each edge e_l^i in the weighted DAG corresponds to the state $a_l^{anc,i}$. We denote the source node of each edge e_l^i by s_l^i, and its target by t_l^i. The prior $P_{anc}(H_l)$ and transition probabilities $P_{anc}(H_l|H_{l-1})$ from Equations 7 and 9, respectively, can be computed as follows

$$P_{anc}(H_l = a_l^{anc,i}|H_{l-1} = a_{l-1}^{anc,j}) = \begin{cases} \dfrac{w_l^i}{\sum\limits_{k \text{ s.t. } t_{l-1}^k = s_l^i} w_{l-1}^k}, & \text{if } t_{l-1}^j = s_l^i \\ 0, & \text{otherwise} \end{cases} \quad (10)$$

$$P_{anc}(H_l = a_l^{anc,i}) = \frac{w_l^i}{\sum_j w_{l-1}^j}. \quad (11)$$

The process is repeated for each ancestry separately, producing the population specific $P_{anc(a_l)}(H_l = a_l)$ and $P_{anc(a_l)}(H_l = a_l|H_{l-1} = a_{l-1})$.

3   Results

Simulation of Admixed Individuals and Training the Background LD Models. We evaluated the performance of ALLOY for local ancestry inference. In our experiments, we simulated admixed individuals and trained ALLOY's background model using data from six HapMap [2] populations: individuals from the Centre d'Etude du Polymorphisme Humain collected in Utah, USA, with ancestry from northern and western Europe (CEU); Han Chinese in Beijing, China (CHB); Japanese in Tokyo, Japan (JPT); Yoruba in Ibadan, Nigeria (YRI); Maasai in Kinyawa, Kenya (MKK); and Tuscans in Italy (Toscani in Italia, TSI). All SNPs present in the HapMap Phase III panel on the first arm of Chromosome 1 were used to expedite the results. We partitioned the HapMap data such that 100 individuals from each population were used as training data, and the remainder were used as test data to evaluate the performance of our method. We used haplotypes from the test set to simulate admixed individuals for 6 different combinations of ancestral populations: YRI-MKK (YM), CHB-JPT (CJ), YRI-MKK-CEU (YMC), CHB-JPT-CEU, (CJC) YRI-MKK-CEU-CHB (YMCC), CHB-JPT-CEU-YRI (CJCY). Each test data set contained 100 simulated admixed individuals. In this section, we use g_{sim} and π_{sim} to denote the parameters used for simulation, and g and π, to denote the parameters used for inference. Each simulated admixed individual was generated by traversing the set of markers in chromosomal order, generating a pair of admixed maternal and paternal haplotypes in parallel. The initial pair of ancestries and alleles, corresponding to the first marker, was randomly selected based on the prior ancestral admixture proportions π_{sim}. Alleles were then copied from the ancestral reference haplotypes. With each subsequent marker, the probability for an admixture related recombination was evaluated via Equation 8. In case of a recombination, a new ancestral source was selected using the π_{sim} admixture proportions and the copying process continued.

We used the Beagle package [5] with default parameters, to phase the training and testing individuals separately. Next, the ancestral background LD model states and parameters were determined through Equations 10 and 11 by examining Beagle's DAG output. To build an efficient background LD model for ALLOY, we selected a subset of ancestry informative markers (AIM), which are genetic variants that carry a population-specific characterizing allele distribution and can be used to efficiently distinguish between genetic segments of different origins. In order to select the set of ancestry informative markers, we used the Shannon Information Content (SIC) criteria [17]. Namely, for a given set of markers and their corresponding allele distribution in the ancestral populations, we measured the mutual information (MI) $I(X_l; Z)$ between ancestry Z and allele X_l at position l. Using the SIC measurement, we followed the marker selection heuristic presented in [22], choosing a constant number of highly informative markers within a window of fixed size. Specifically, in our simulations, we selected the single most informative marker in windows of 0.05 centimorgans. For the YM, YMC, and YMCC data sets, we used SNPs with the highest MI differentiating the YRI and MKK populations; for the CJ, CJC, CJCY admixture scenarios, we selected markers with the highest MI when differentiating the CHB and JPT populations. While ALLOY's background LD model was based on a subset of SNPs, inference was performed on all SNPs, calling the ancestry of the excluded SNPs using a nearest marker approach.

Evaluating ALLOY's Accuracy under Complex and Ancient Admixtures. When performing inference, we modeled the genotyping error rate with $\epsilon = 0.01$, and used $\phi(z) = \frac{z}{10}$ as our smoothing function in Equation 8. We compared the performance of ALLOY to WINPOP [11], a local ancestry inference platform that has been shown to outperform previous state of the art methods such as SABER [21], HAPAA [20] and HAPMIX [14]. We measured the accuracy of ALLOY and WINPOP when inferring local ancestry of simulated admixed individuals under increasingly complex admixtures, and with a varying number of generations since the first admixture event, ranging from recent admixture ($g = 7$) to more ancient admixture ($g = 100$). Accuracy was conservatively measured as the average fraction of SNPs for which the correct ancestry was inferred. As depicted in Figure 3, our results show that ALLOY's accuracy is greater than WINPOP's in nearly all tested scenarios. Our experiments show that applying WINPOP over the full set of markers achieves a higher performance in comparison to analyzing only a subset of ancestry information markers. WINPOP performs SNP selection prior to inference to confirm with their model assumptions, and hence benefits from the larger initial set of markers. We therefore reported WINPOP results corresponding to an analysis applied on the entire HapMap Phase III set of SNPs, rather than the SNP subsets used for training ALLOY.

Exploring Background LD Models. As previously described, the background LD models in ALLOY can capture a wide range of complexities, from simpler models such as those used in AncestryMap [13] and SABER which model zero- and first-order dependencies between markers, respectively, to more complex explicit haplotype models as used by HAPAA and HAPMIX. More importantly,

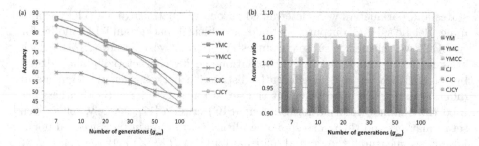

Fig. 3. (a) The performance of ALLOY based on the number of generations since admixture g_{sim} for various admixture configurations with equal ancestral proportions. ALLOY was run with $g = g_{sim}$ and $\pi = \pi_{sim}$, and the accuracy was conservatively measured as the fraction of markers for which the exact ancestry pair was inferred. (b) For the same experiments, ALLOY was compared to WINPOP by measuring the accuracy ratio between them ($\frac{\text{ALLOY's accuracy}}{\text{WINPOP's accuracy}}$). The results clearly demonstrate AL-LOY superior accuracy in the vast majority of tested admixture configurations, with an increase in performance in more than 86% of the tests.

ALLOY is able to capture models of intermediate complexity. We explored the performance of ALLOY using a range of background LD models with varying complexities. Background models of different complexities were generated by applying Beagle on our training data using different values for Beagle's *scale* parameter, which controls the complexity of the generated DAG underlying AL-LOY's model. As *scale* approaches 0, the model approaches the explicit model used in HAPAA, and as the value of *scale* grows, the generated model approaches a zero-order model similar to the one used by AncestryMap. The results, shown in Figure 4, illustrate that the models of intermediate complexity outperform both the more complex as well as the simpler models used by previous methods.

Measuring Robustness to Inaccuracies in Model Parameters. Our method assumes that the admixture parameters, such as the number of generations g and the admixture proportions π are given. When applied on real data, however, the true values for these parameters are unknown. We examined the robustness of ALLOY to inaccuracies in model parameters. Specifically, we measured the impact of misspecified admixture proportion π on the accuracy of inference. To test for robustness, we simulated a YM mixture with $\pi_{sim} = (0.5, 0.5)$ and $g_{sim} = 30$, and evaluated ALLOY's performance varying π between $(0.05, 0.95)$ and $(0.95, 0.05)$ during inference. Our results indicate that ALLOY's performance is robust to inaccuracies in π, yielding the highest accuracy when $\pi = \pi_{sim}$, slightly reducing the accuracy by 0.0029 to its lowest value at the two extremes (i.e., $\pi = (0.95, 0.05)$ and $\pi = (0.05, 0.95)$). We further evaluated ALLOY's performance when g was misspecified. A YM mixture was simulated with $g_{sim} = 20, \pi_{sim} = (0.5, 0.5)$. When $g = g_{sim}$, AL-LOY achieved 73.77% accuracy; for mis-specified values of g between 10 and 40, accuracy ranged from 73.41% to 73.88%, respectively.

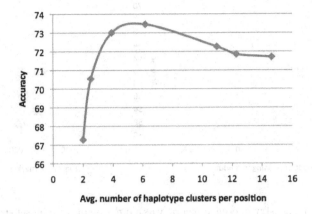

Fig. 4. Ancestry inference accuracy as a function of model complexity, as measured by the average number of haplotype clusters under a certain background LD model. The models range from a simplistic assumption of independence on the left, to more explicit models on the right. The plot illustrates that both over-simplification, corresponding to the LD models used in AncsetryMap and SABER, and over-specification, corresponding to the models leaning towards those used in HAPAA and HAPMIX, yield reduced performance in comparison to a more generalizing local haplotypic model.

Additionally, we explored the sensitivity of ALLOY's performance to different genotyping error rates ϵ. When simulating a YM mixture with $g_{sim} = 20, \pi_{sim} = (0.5, 0.5)$ as above, and performing inference with values of $\epsilon \leq 0.025$, ALLOY's accuracy was at least 73.20%. Assuming a 5% genotyping error rate ($\epsilon = 0.05$), the accuracy decreased by less than 1%.

Evaluating Model Accuracy under Varying Amounts of Training Data.
Currently, the amount of available genotype data is limited in the number of individuals genotyped and the density of SNPs measured. However, the number of genotyped individuals and the SNP density of genotyping technologies are expected to greatly increase in the near future. To evaluate the effect of training set size on ALLOY's performance, we trained our background LD model on sets of individuals with increasing size. Specifically, we derived a model for the YRI and MKK ancestral populations using subsets of the individuals of varying size, and evaluated the inference accuracy. The results, shown in Figure 5a, emphasize the importance of training set size to the improved performance, suggesting that as more samples are collected and genotyped, more accurate background models could be derived, yielding a higher level of accuracy.

We further evaluated the performance of ALLOY with respect to the number of SNPs used during training. We generated subsets of informative SNPs of various sizes by using different window sizes during the AIM selection phase. To evaluate the importance of using AIMs, we also selected random SNP subsets of matching sizes. We simulated individuals from a YM admixed population with $g_{sim} = 30$ generations of admixture, evaluating ALLOY's accuracy when

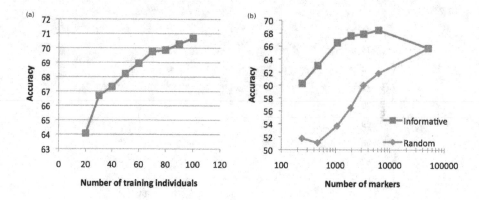

Fig. 5. (a) Local ancestry inference accuracy as a function of training set size. A various number of individuals were used as representatives of the ancestral populations in the computation of the background LD model, demonstrating an increased performance as more samples are used in the training phase. (b) The accuracy of inferring ancestry as a function of the number of markers used. The plot illustrates the significance of using ancestry informative markers in comparison to a randomly chosen set, as for all tested resolutions, the use of the informative set yielded an improved performance. The results also indicate that the addition of non-informative markers reduces performance (demonstrated by the right-most data point) as these are assumed to interfere with the construction of an effective background LD model.

trained using the different SNP subset. Figure 5b shows that ALLOY's accuracy increases as more SNPs are used. The results further demonstrate that ALLOY's performance is significantly higher with informative SNPs compared to random ones. The rightmost point in Figure 5b corresponds to ALLOY's performance when all SNPs are used. These results indicate that using excessive uninformative markers can reduce accuracy in comparison to a model based on informative markers.

4 Discussion

ALLOY represents the LD structure of each population with a highly expressive model that lies between the simpler first-order Markov hidden Markov model in SABER, and the explicit-haplotype model in HAPAA and HAPMIX. The first advantage of this approach is its improved accuracy compared to either extreme, as shown in Figure 4. Additionally, our inference algorithm has higher computational efficiency than explicit-haplotype models. In this work, we derive the population-specific LD structures by generating haplotype clusters through the Beagle package. We translate the produced DAG into prior and transition probabilities that define the parameters of our factorial hidden Markov model. In future work, alternatives to Beagle can be used for modeling LD; for instance, one can develop ancestry-aware methods that produce LD models that emphasize the structural differences between ancestral populations.

ALLOY assumes that the admixture parameters are given. In particular, the number of generations since admixture g, and the relative proportions of the ancestral populations π, are required. We showed through simulations that our method is robust to inaccuracies in the estimation of the admixture proportion. Nonetheless, π can be estimated by direct examination of the sampled individuals' genotype likelihood. Alternatively, given a set of individuals representing a particular admixed population, demographic parameters such as admixture time g and ancestral proportions π, can be derived as a post-processing step. For instance, given the inferred local ancestry of a set of individuals, we can observe the lengths and ancestral proportions of the ancestral segments. Using these observations, maximum likelihood estimates for π and g can be computed. We also note that the flexibility of our FHMM enables different admixture times and proportions to be incorporated separately for the maternal and paternal haplotypes. Hence, pedigrees exhibiting very recent complex admixture at the grandparental level can be explicitly modeled. For example, the parameters of our method can be tuned to accurately infer the ancestry of an admixed individual that has one African-American and one Chinese parent.

In the Methods section we described an inference algorithm with a time complexity that depends on the local ancestral LD structure rather than the number of ancestral haplotypes used when training the background model. Specifically, the algorithm's time complexity is $O(L \cdot C^3)$, where C is an upper bound on the number of states in a single position (i.e. $C = \max_l |A_l|$). In our implementation of ALLOY, we reduced the time complexity by rearranging the calculations corresponding to the transition probabilities in the forward and backward computations, described by Equations 2 and 3, respectively. In particular, transitioning between states corresponding to an admixture recombination event can be collapsed into a single term. For instance, when transitioning between states corresponding to different ancestries in the forward iteration, Equation 7 is reduced to the term $P(R_l) \cdot P(H_l)$. Hence, $\alpha_{l-1}^m(a_l, a'_{l-1})$ can be rewritten as

$$\alpha_{l-1}^m(a_l, a'_{l-1}) = P(R_l) \cdot P(a_l) + P(\bar{R}_l) \cdot \sum_{a_{l-1} \in A_{l-1}^{anc(a_l)}} P_{anc(a_l)}(a_l | a_{l-1}).$$

When such an optimization is applied, the time complexity is reduced to $O(L \cdot C^2 \cdot C_K)$ where C_K is an upper bound on the number of states corresponding to a single population (i.e. $C_K = \max_{l,k} |A_l^k|$). We note that this implementation of ALLOY has a practical running time, completing a single experiment as described in the Results section in approximately one minute.

Our simulations experimented with SNP markers that were found to be polymorphic in 1,184 individuals sampled from 11 populations in the third phase of the HapMap project [2]. However, additional variation exists in these populations beyond the SNPs assayed in this data set. In particular, rare SNPs which have been found to exhibit little sharing among diverged populations [7] and can therefore act as highly informative markers for ancestry inference, are likely to be missing from the panel. Therefore, as additional rare SNPs are discovered and sampled, we expect the accuracy of ALLOY to improve. We further note that the spectrum of human genetic variation ranges beyond SNPs. For instance, copy-number variations (CNV) and other structural variations constitute a large fraction of the total human genomic variation [1]. As with SNPs, rare CNVs are useful for separating ancestries and have been shown to be more abundant than rare SNPs [9]. Our model is not limited to bi-allelic SNPs, and supports the incorporation of markers of higher variability, such as CNVs, by adjusting Equation 4. The construction of the variable-length Markov chain linkage-models, either through Beagle or other methods, can be extended to take such additional genetic variation into account.

ALLOY is a novel method for inferring the local ancestry of admixed individuals, which is an essential task for various applications in human genetics. We have shown that our approach has higher accuracy than the previous state of the art and that its VLMC-based LD model plays a crucial role in its superior performance. Our method is applicable to ancient and complex admixtures, and is capable of separately modeling the maternal and paternal histories. We expect that as the genetic variation of worldwide populations is extensively sampled, ALLOY will be able to better characterize the particular histories of examined individuals. ALLOY is publicly and freely available at *http://alloy.stanford.edu/*.

Acknowledgments. We thank Chuong B. Do for helpful discussions. This material is based upon work supported by the National Science Foundation Graduate Research Fellowship under Grant No. DGE-1147470. This publication was made possible by Grant Number 5RC2HG005570-02 from the NIH. Its contents are solely the responsibility of the authors and do not necessarily represent the official views of the NIH. This material is based upon work supported by the National Science Foundation under Grant No. 0640211. Any opinions, findings, and conclusions or recommendations expressed in this material are those of the authors and do not necessarily reflect the views of the National Science Foundation.

References

[1] Alkan, C., Coe, B.P., Eichler, E.E.: Genome structural variation discovery and genotyping. Nature Reviews. Genetics 12(5), 363–376 (2011)

[2] Altshuler, D.M., Gibbs, R.A., Peltonen, L., Dermitzakis, E., Schaffner, S.F., Yu, F., Bonnen, P.E., De Bakker, P.I.W., Deloukas, P., Gabriel, S.B., et al.: Integrating common and rare genetic variation in diverse human populations. Nature 467(7311), 52–58 (2010)

[3] Baye, T.M., Wilke, R.A.: Mapping genes that predict treatment outcome in admixed populations. The Pharmacogenomics Journal 10(6), 465–477 (2010)

[4] Bercovici, S., Geiger, D.: Inferring ancestries efficiently in admixed populations with linkage disequilibrium. Journal of Computational Biology: A Journal of Computational Molecular Cell Biology 16(8), 1141–1150 (2009)

[5] Browning, S.R., Browning, B.L.: Rapid and accurate haplotype phasing and missing-data inference for whole-genome association studies by use of localized haplotype clustering. The American Journal of Human Genetics 81(5), 1084–1097 (2007)

[6] Ghahramani, Z., Jordan, M.I., Smyth, P.: Factorial hidden markov models. In: Machine Learning. MIT Press (1997)

[7] Gravel, S., Henn, B.M., Gutenkunst, R.N., Indap, A.R., Marth, G.T., Clark, A.G., Yu, F., Gibbs, R.A., Project, T.G., Bustamante, C.D.: Demographic history and rare allele sharing among human populations. Proceedings of the National Academy of Sciences 108(29), 11983–11988 (2011)

[8] Haldane, J.B.S.: The combination of linkage values, and the calculation of distance between the loci of linked factors. J. Genet. 8, 299–309 (1919)

[9] Jakobsson, M., Scholz, S.W., Scheet, P., Gibbs, J.R., VanLiere, J.M., Fung, H.-C., Szpiech, Z.A., Degnan, J.H., Wang, K., Guerreiro, R., et al.: Genotype, haplotype and copy-number variation in worldwide human populations. Nature 451(7181), 998–1003 (2008)

[10] Long, J.C.: The genetic structure of admixed population. Genetics (127), 417–428 (1991)

[11] Pasaniuc, B., Sankararaman, S., Kimmel, G., Halperin, E.: Inference of locus-specific ancestry in closely related populations. Bioinformatics 25, i213–i221 (2009)

[12] Pasaniuc, B., Zaitlen, N., Lettre, G., Chen, G.K., Tandon, A., Kao, W.H.L., Ruczinski, I., Fornage, M., Siscovick, D.S., Zhu, X., Larkin, E., Lange, L.A., Cupples, L.A., Yang, Q., Akylbekova, E.L., Musani, S.K., Divers, J., Mychaleckyj, J., Li, M., Papanicolaou, G.J., Millikan, R.C., Ambrosone, C.B., John, E.M., Bernstein, L., Zheng, W., Hu, J.J., Ziegler, R.G., Nyante, S.J., Bandera, E.V., Ingles, S.A., Press, M.F., Chanock, S.J., Deming, S.L., Rodriguez-Gil, J.L., Palmer, C.D., Buxbaum, S., Ekunwe, L., Hirschhorn, J.N., Henderson, B.E., Myers, S., Haiman, C.A., Reich, D., Patterson, N., Wilson, J.G., Price, A.L.: Enhanced statistical tests for GWAS in admixed populations: assessment using African Americans from CARe and a Breast Cancer Consortium. PLoS Genetics 7(4), e1001371 (2011)

[13] Patterson, N., Hattangadi, N., Lane, B., Lohmueller, K.E., Hafler, D.A., Oksenberg, J.R., Hauser, S.L., Smith, M.W., O'Brien, S.J., Altshuler, D., Daly, M.J., Reich, D.: Methods for high-density admixture mapping of disease genes. American Journal of Human Genetics 74(5), 979–1000 (2004)

[14] Price, A.L., Tandon, A., Patterson, N., Barnes, K.C., Rafaels, N., Ruczinski, I., Beaty, T.H., Mathias, R., Reich, D., Myers, S.: Sensitive detection of chromosomal segments of distinct ancestry in admixed populations. PLoS Genet. 5(6), e1000519 (2009)

[15] Rabiner, L.R.: A tutorial on hidden markov models and selected applications in speech recognition. Proceedings of the IEEE, 257–286 (1989)

[16] Ron, D., Singer, Y., Tishby, N.: On the learnability and usage of acyclic probabilistic finite automata. Journal of Computer and System Sciences, 31–40 (1995)

[17] Rosenberg, N.A., Li, L.M., Ward, R., Pritchard, J.K.: Informativeness of genetic markers for inference of ancestry. The American Journal of Human Genetics (73), 1402–1422 (2003)

[18] Sankararaman, S., Sridhar, S., Kimmel, G., Halperin, E.: Estimating Local Ancestry in Admixed Populations. Journal of Human Genetics, 290–303 (February 2008)

[19] Seldin, M.F., Pasaniuc, B., Price, A.L.: New approaches to disease mapping in admixed populations. Nature Reviews. Genetics 12(8), 523–528 (2011)

[20] Sundquist, A., Fratkin, E., Do, C.B., Batzoglou, S.: Effect of genetic divergence in identifying ancestral origin using HAPAA. Genome Research 18(4), 676–682 (2008)

[21] Tang, H., Coram, M., Wang, P., Zhu, X., Risch, N.: Reconstructing genetic ancestry blocks in admixed individuals. American Journal of Human Genetics 79(1), 1–12 (2006)

[22] Tian, C., Hinds, D.A., Shigeta, R., Kittles, R., Ballinger, D.G., Seldin, M.F.: A genomewide single-nucleotide polymorphism panel with high ancestry information for african american admixture mapping. The American Journal of Human Genetics (79), 640–649 (2006)

[23] Winkler, C.A., Nelson, G.W., Smith, M.W.: Admixture mapping comes of age. Annual Review of Genomics and Human Genetics 11, 65–89 (2010)

Charge Group Partitioning
in Biomolecular Simulation

Stefan Canzar[1,*], Mohammed El-Kebir[1,2,*], René Pool[2,3], Khaled Elbassioni[4],
Alpesh K. Malde[5], Alan E. Mark[5,6], Daan P. Geerke[3],
Leen Stougie[1,7], and Gunnar W. Klau[1]

[1] Centrum Wiskunde & Informatica, Life Sciences Group, Amsterdam, Netherlands
[2] Centre for Integrative Bioinformatics VU, VU University Amsterdam, Netherlands
[3] Division of Molecular Toxicology, VU University Amsterdam, Netherlands
[4] Max-Planck-Institut für Informatik, Saarbrücken, Germany
[5] School of Chemistry and Molecular Biosciences, The University of Queensland,
Brisbane, QLD 4072, Australia
[6] Institute for Molecular Bioscience, The University of Queensland, Brisbane,
QLD 4072, Australia
[7] Department of Operations Research, VU University Amsterdam, Netherlands

Abstract. Molecular simulation techniques are increasingly being used
to study biomolecular systems at an atomic level. Such simulations rely
on empirical force fields to represent the intermolecular interactions.
There are many different force fields available—each based on a different set of assumptions and thus requiring different parametrization
procedures. Recently, efforts have been made to fully automate the assignment of force-field parameters, including atomic partial charges, for
novel molecules. In this work, we focus on a problem arising in the automated parametrization of molecules for use in combination with the
GROMOS family of force fields: namely, the assignment of atoms to charge
groups such that for every charge group the sum of the partial charges
is ideally equal to its formal charge. In addition, charge groups are required to have size at most k. We show $\mathcal{NP}$-hardness and give an exact algorithm capable of solving practical problem instances to provable
optimality in a fraction of a second.

Keywords: charge groups, atomic force fields, GROMOS, biomolecular
simulation, tree-decomposition, dynamic programming.

1 Introduction

In the context of drug development, biomolecular systems such as protein-
peptide [29], protein-ligand [24] and protein-lipid interactions [8] can be studied
with the use of molecular simulations [2, 26] using a force field model that describes the interatomic interactions. Many biomolecular force fields are available,
including AMBER [10], CHARMM [9], OPLS [16] and GROMOS [19, 21, 23]. These

---

* Joint first authorship.

B. Chor (Ed.): RECOMB 2012, LNBI 7262, pp. 29–43, 2012.

force fields have in common that the non-bonded intermolecular interactions are represented in terms of interatomic pair potentials.

Typically, the number of atoms in biomolecular systems are in the range of 10^4 to 10^6. To observe relevant biological phenomena, time scales in the order of nano- to milliseconds need to be simulated. For such large-scale systems, evaluating all atom-atom interactions is practically infeasible. One way of dealing with this is to only consider interactions of atoms whose distance is within a pre-specified cut-off radius. Since not all interactions are considered, an error is introduced. The magnitude of the *error* due to omitting atom-atom interactions is inversely proportional to the distance between the atoms. More problematically, there are *discontinuities* as atoms move in and out of the cut-off radius.

Errors and discontinuities are reduced by combining atoms into *charge groups*, for which individual centers of geometry are determined. If the distance between two centers of geometry lies within the cut-off distance then all interactions between the atoms of the involved charge groups are considered. Ideally, charge groups should be neutral as interactions are then reduced to dipole-dipole interactions that scale inversely proportional to the cubed interatomic distance. Charge groups should not be too large. This is because the effective cut-off distance of an individual atom in a given charge group is given by the cut-off distance minus the distance to the center of geometry of the charge group. If the distance of an atom to the center of geometry becomes large, the effective cutoff becomes small, leading to errors and discontinuities as described above. For the same reason, charge groups should be connected as interatomic bonds indicate spatial proximity.

To simulate a molecule, a force field requires a specific *topology*, which includes the atom types, bonds and angles, the atomic charges and the charge group assignment. Most biomolecular force fields come with a set of topologies for frequently simulated molecules such as amino acids, lipids, nucleotides and cofactors. Novel molecules and compounds, however, require the construction of novel topologies. Such a situation occurs, for instance, when assessing the binding affinity of a novel drug-like compound to a certain protein.

Manually building topologies for novel compounds is a tedious and time-consuming task especially when a large chemical library needs to be screened, for example when determining binding affinities for large sets of potential drug compounds to a newly discovered protein target. Therefore, automated approaches are needed.

Here, we focus on the GROMOS family of force fields, which has been specifically tailored to simulate biochemical processes, including protein-drug binding and peptide folding. A widely used topology generator for the GROMOS force field is PRODRG [22]. However, the charge group assignment by PRODRG for amino acid topologies contained several large charge groups comprising disconnected atoms, which is inconsistent with GROMOS [17]. The Automated Topology Builder (ATB) is a recent method for automated generation of GROMOS topologies [18]. The assignment of atomic charges and charge groups by the ATB proceeds in three consecutive stages. Firstly, partial charges are computed using quantum calculations. Subsequently, the symmetry of the molecule is exploited to ensure that symmetric atoms have identical charges. Finally, the molecule is partitioned into charge groups using a greedy algorithm. The ATB method was experimentally verified for a set of biologically relevant molecules [18]. For some large molecules, such as the cofactor Adenosine-5'-triphosphate (ATP), however, the ATB assigns too large charge groups, which leads to instabilities during simulation as described above.

As existing automated procedures such as PRODRG and the ATB fail in assigning appropriate charge groups, we have investigated the problem in detail. Our contribution is threefold: (1) We introduce the charge group partitioning problem and give a sound mathematical problem definition resulting in charge groups of small size and zero charge. We prove $\mathcal{NP}$-hardness of the problem and identify important special cases, for which we give polynomial time algorithms. (2) Exploiting the properties of molecular structures enables us to present a tree decomposition-based algorithm that solves typical practical problem instances to optimality within fractions of a second. (3) We evaluate the performance of our method by running simulations using the resulting charge group assignments of amino acid side chains, which yield results consistent with experimentally known values. Moreover, for large, highly charged, molecules such as ATP we obtain charge groups which are both suitable for use in simulations as well reasonable from a chemical perspective.

2   Problem Statement and Complexity

In this section we give a formal definition of the problem associated with assigning appropriate charge groups within a molecule. Our aim is to capture the two important aspects of chemical intuition discussed above: (1) the number of atoms in a charge group should not exceed a given integer k and (2) the sum of partial and formal charges of a charge group is ideally equal. Mathematically, the latter condition is equivalent to requiring the sum of differences of formal and partial charges in a charge group to be close to zero. We prove $\mathcal{NP}$-hardness of the problem even if we take into account special characteristics of graphs representing a molecular structure. For the special case $k = 2$ we obtain a polynomial-time algorithm by reducing the problem to a minimum cost perfect matching problem.

A molecular structure can be modeled as a degree-bounded graph $G = (V, E)$, where the nodes correspond to atoms and the edges to chemical bonds. In addition, we consider node weights $\delta : V \to \mathbb{R}$, where $\delta(v)$ corresponds to the difference between formal and partial charge of the atom v. A formal definition of the *charge group partitioning problem* is as follows:

Definition 1 (Charge group partitioning, CGP). *Given a graph $G = (V, E)$, node weights $\delta : V \to \mathbb{R}$, and an integer $2 \leq k \leq |V| - 1$, find a partition $\mathcal{V}$ of V such that for all $V' \in \mathcal{V}$ it holds $|V'| \leq k$, the subgraph $G[V']$ induced by V' is connected, and which has minimal total error*

$$c(\mathcal{V}) := \sum_{V' \in \mathcal{V}} \left| \sum_{v \in V'} \delta(v) \right| .$$

Each subset $V' \in \mathcal{V}$ of the nodes in the partition corresponds to a charge group. The following theorem shows $\mathcal{NP}$-hardness of the problem.

Theorem 1. CGP *is $\mathcal{NP}$-hard, even in the restricted case where G is planar, $k = 4$, the maximum degree of a node in the graph is 4, and the node weights are $\mathcal{O}(1)$.*

Proof. Clearly, the problem belongs to $\mathcal{NP}$. Consider the following problem.

Definition 2 (Planar 3-dimensional matching, PLANAR 3DM). *Given disjoint sets X_1, X_2, X_3 with $|X_1| = |X_2| = |X_3| = m$ and a set of n triples $\mathcal{T} \subset X_1 \times X_2 \times X_3$. The bipartite graph B, with $\mathcal{T}$ as its one color class and $X = X_1 \cup X_2 \cup X_3$ as its other color class and an edge between $T \in \mathcal{T}$ and $x \in X$ if and only if $x \in T$, is planar. Each element of X appears in 2 or 3 triples only. Does there exist a perfect matching in $\mathcal{T}$; i.e. a subset $M \subset \mathcal{T}$ of m triples such that each element of X occurs uniquely in a triple in M?*

This problem has been shown $\mathcal{NP}$-complete by Dyer and Frieze [12]. We reduce it to CGP. Take the bipartite graph B in the definition of PLANAR 3DM with $\mathcal{T}$ and X as color classes. Give each $x \in X$ a weight $\delta(x) = -1$ and each $T \in \mathcal{T}$ a weight $\delta(T) = 3$. For each $T \in \mathcal{T}$ we introduce three extra vertices s_1^T, s_2^T, s_3^T with weights $\delta(s_1^T) = \epsilon$, $\delta(s_2^T) = \epsilon$, $\delta(s_3^T) = -3\epsilon$, for an arbitrary $0 < \epsilon < 1$, and connect them by the path (T, s_1^T, s_2^T, s_3^T), which we call the *tail* of T. Clearly, the resulting graph G remains planar (and bipartite). Since each $x \in X$ is in at most three triples it is easy to see that G has bounded degree 4.

Given a feasible partition to the CGP-instance, we say a group is of type i if it contains exactly i nodes from X, $i \in \{0, 1, 2, 3\}$ and exactly one node from $\mathcal{T}$. Notice that, for $i = 1, 2, 3$, each type i group contributes error $(3 - i)$ by itself, and because it covers a $\mathcal{T}$-node and therefore leaves a tail-path it contributes indirectly an extra error ϵ (the alternative of including one of the tail nodes into the group with the triple node does not decrease the sum of the two errors). A type 0 group consists of a $\mathcal{T}$-node only and therefore will be combined with its tail to yield an error of $3 - \epsilon$. Let y_i denote the number of type i group, $i \in \{0, 1, 2, 3\}$. Let y denote the number of X-vertices that form a group on their own. Then the feasible solution has total error

$$W = y_0(3 - \epsilon) + y_1(2 + \epsilon) + y_2(1 + \epsilon) + y_3\epsilon + y. \tag{1}$$

We show that there exists a matching if and only if G admits a partition with total error

$$W = m\epsilon + (n - m)(3 - \epsilon).$$

Suppose $M \subset \mathcal{T}$ is a matching. For every triple $T_i \in M$ we create a type 3 group consisting of the corresponding vertex T_i in G and the three vertices corresponding to its three elements. Hence $y_3 = m$. By the properties of the matching all X-vertices of G are now covered, and $n - m$ triple-vertices of G remain uncovered. The latter necessarily form $n - m$ type 0 groups: $y_0 = n - m$. Insertion in (1) yields $W = m\epsilon + (n - m)(3 - \epsilon)$.

Now assume that no perfect matching exists. First, note that in any optimal solution to the CGP-instance $y = 0$. Assume $y > 0$ and let $x \in X$ be such a vertex. Then every neighbor of x in $\mathcal{T}$ is contained in a group of type i, with $i \leq 2$. Therefore, adding x to any such group would decrease the cost of the solution by at least $2(1 - \epsilon)$. Furthermore, every group that contains two nodes from $\mathcal{T}$ can be split into two groups without increasing the cost of the solution. Now, since there exists no perfect matching, we need $m + c$ groups of type $1, 2$, or 3, for some $c \geq 1$, to cover all vertices in X. Using equations

$$y_1 + y_2 + y_3 = m + c \tag{2}$$
$$y_1 + 2y_2 + 3y_3 = 3m \tag{3}$$

we get

$$y_3 = m - 2c + y_1 \tag{4}$$
$$y_2 = 3c - 2y_1 \tag{5}$$

and the cost contributed to (1) by type 1, 2, and 3 groups becomes equal to $m\epsilon + c(3 + \epsilon)$. Together with the remaining $n - m - c$ groups of type 0 the total weight becomes

$$m\epsilon + (n - m)(3 - \epsilon) + 2c\epsilon.$$

$\square$

Using the same reduction, but extending the tails to length $k - 1$ paths with ϵ weight on the internal vertices and $-(k - 1)\epsilon$ weight on the leaf, proves the problem to be hard for any $k \geq 4$.

CGP with $k = 2$ can be solved by formulating a minimum cost perfect matching problem. Starting from $G = (V, E)$, we assign a weight to the edges that is equal to the error that the pair of vertices will contribute if chosen as a group of the partition. For each vertex $v \in V$ create a shadow vertex v' with $\delta(v') = 0$. The weight on the edge $\{v, v'\}$ is then $|\delta(v)|$, the error if v is chosen as a single vertex group. Additionally we insert an edge $\{u', v'\}$ of weight 0 if and only if $\{u, v\} \in E$. It is not difficult to see that a minimum cost perfect matching in this graph corresponds to an optimal partition, where an edge in the matching between a vertex and its shadow vertex signifies a single vertex group in the partition.

Intriguingly, this leaves $k = 3$ as the only case for which the complexity is unknown.

3   Dynamic Programming for Bounded Treewidth

While problem CGP is $\mathcal{NP}$-hard in general as shown in the previous section, we can solve it by a dynamic program in polynomial time if the molecule graph is a tree. Starting from the leaves we proceed towards an arbitrarily chosen root node. At a given node i we guess the group V' that contains i in the optimal solution to the subproblem induced by the subtree rooted at i and recurse on the subtrees obtained when removing V'. Due to the size restriction $|V'| \leq k$ we only have to consider a polynomial number of groups.

Although the structural formula of biomolecules is not always a tree, as we will see later, it is usually still tree-like, which has already been exploited in [11]. Formally, this property is captured by the *treewidth* of a graph [20]. The definition is as follows.

Definition 3. *A* tree decomposition (T, X) *of a graph* $G = (V, E)$ *consists of a tree* T *and sets* X_i *for all* $i \in V(T)$, *called* bags, *satisfying the three following properties:*

1. *Every vertex in* G *is associated with at least one node in* T: $\bigcup_{i \in V(T)} X_i = V$.
2. *For every edge* $\{u, v\} \in E$, *there is an* $i \in V(T)$ *such that* $\{u, v\} \subseteq X_i$.
3. *The nodes in* T *associated with any vertex in* G *define a subtree of* T.

The width *of a tree decomposition is* $\max_i |X_i| - 1$. *The* treewidth *of* G *is the minimum width of any tree decomposition of* G.

In this section, we propose a tree-decomposition based dynamic program for problem CGP whose running time grows exponentially with the treewidth of G. Therefore, a tree-decomposition of small width is crucial for the efficiency of our approach. Unfortunately, computing a tree decomposition of minimum width is $\mathcal{NP}$-hard [3]. However, for the class of r-outerplanar graphs an optimal tree-decomposition can be determined in time $\mathcal{O}(r \cdot n)$ [1]. A graph is r-outerplanar if, after removing all vertices on the boundary face, the remaining graph is $(r-1)$-outerplanar. A graph is 1-outerplanar if it is outerplanar, that is, if it admits a crossing-free embedding in the plane such that all vertices are on the same face. Interestingly enough, most molecule graphs of biomolecules are r-outerplanar for some small integer r. For example, Horváth et al. [15] have observed that 94.3% of the molecules in the NCI[1] database are 1-outerplanar. Even more, every r-outerplanar graph has treewidth at most $3r - 1$ [7]. Therefore, not surprisingly, Yamaguchi et al. [28] observed that out of 9,712 chemical compounds in the KEGG LIGAND database [14], all but one had treewidth between 1 and 3, with a single molecule having treewidth 4. In fact, among the molecules considered here the maximal treewidth was 2. As a result, our tree-decomposition based dynamic program found an optimal charge group partitioning in well under one second.

Let (T, X) be a tree decomposition of width ℓ for graph $G = (V, E)$. The high-level idea of the algorithm is as follows, see also Figure 1. For an arbitrarily chosen root i of the tree decomposition we guess the groups that intersect X_i, denoted by the dashed lines in the figure. After removing these groups, G falls apart into connected components, denoted by the filled regions in the figure. By the properties of a tree decomposition, these connected components will correspond one-to-one to the subtrees of the tree decomposition obtained by removing bags that became empty. Recursing on the roots of these new subtrees yields the overall optimal solution.

---

[1] National Cancer Institute (http://cactus.nci.nih.gov/)

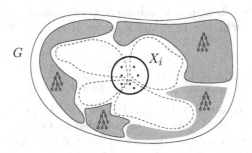

Fig. 1. Illustration of the tree-decomposition based dynamic programming algorithm. A graph G falls apart into connected components (grey regions) by removing the groups (dashed lines) that intersect bag X_i.

Without loss of generality we assume that T has at most $n := |V|$ vertices and depth $\mathcal{O}(\log n)$ [6], with r being the root of T. In the following, we let $V_i = \bigcup_{j \in T_i} X_j$, where T_i denotes the subtree rooted at i, and write $V(T_i)$ for the set of nodes in T_i. We define an extension of a partition of a vertex set $V_1 \subseteq V$ with nodes in $V_2 \setminus V_1$ into connected subgraphs of G of size at most k:

Definition 4. *For vertex sets $V_1 \subseteq V_2 \subseteq V$, set $\mathcal{L}_G(V_1, V_2)$ contains all sets $\mathcal{V} \in 2^{V_2}$ with $V_1 \subseteq \bigcup_{V' \in \mathcal{V}} V'$, all sets in $\mathcal{V}$ being disjoint, and all $V' \in \mathcal{V}$ satisfying: (i) $G[V']$ is connected, (ii) $|V'| \leq k$ and (iii) $V_1 \cap V' \neq \emptyset$.*

Furthermore, by $r(S)$ we denote the root of a subtree S of T, and for any node i in T and any vertex set $A \subseteq V$ we denote by $\mathcal{S}(i, A)$ the set of trees, corresponding to the connected components of $T_i[j \in V(T_i) \mid X_j \setminus A \neq \emptyset]$ whose roots are not a descendant of another subtree in S (i.e. there are no $S_i, S_j \in S$ for which $V_{r(S_i)} \subseteq V_{r(S_j)}$). With a slight abuse of notation, for sets $A \subseteq V$ and $\mathcal{V} \subseteq 2^V$ we will write $A \cup \mathcal{V}$ instead of $\bigcup_{V' \in \mathcal{V}} V' \cup A$, when the meaning is clear from the context. Then for any node i of T and any subset $A \subseteq V$ the cost of an optimal solution to CGP on graph $G[V_i \setminus A]$, denoted by $cgp(i, A)$, can be described by the recurrence

$$cgp(i, A) = \min_{\mathcal{V} \in \mathcal{L}_G(X_i \setminus A, V_i \setminus A)} \left\{ c(\mathcal{V}) + \sum_{S \in \mathcal{S}(i, A \cup \mathcal{V})} cgp(r(S), A \cup \mathcal{V}) \right\}, \quad (6)$$

which also holds in the base case where $\mathcal{S}(i, A \cup \mathcal{V}) = \emptyset$, in particular when i is a leaf of T. The optimal partition has cost $cgp(r, \emptyset)$. We can solve the recurrence relation (6) using dynamic programming.

Theorem 2. *The cost of an optimal solution to CGP on a graph of treewidth ℓ and maximum degree d can be computed in time $n \cdot \mathcal{O}(e^{2k} \ell^4 d^{4k-2} \cdot \log n)^\ell$.*

Proof. Let (T, X) be a tree decomposition of G of width k and depth $\mathcal{O}(\log n)$. Consider an arbitrary node i in T and a subset $A \subseteq V$, for which $X_i \setminus A \neq \emptyset$. We first observe that

$$|\mathcal{L}_G(X_i \setminus A, V_i \setminus A)| \leq \left(\frac{e^k d^{2k-1}(\ell+1)^2}{(d-1)k}\right)^{\ell+1}. \tag{7}$$

Indeed, for each partition $\mathcal{Y} = \{Y_1, \ldots, Y_h\}$ of $X_i \setminus A$, the number of possible extensions in $\mathcal{L}_G(X_i \setminus A, V_i \setminus A)$ can be bounded as follows. For $j = 1, \ldots, h$, let B_j be the set of vertices at distance at most $k - 1$ from Y_j in the graph $G_j = G[V_i \setminus (A \cup X_i) \cup Y_j]$ (this set can be found by contracting Y_j to a single vertex y_j and performing BFS in G_j starting from y_j). Each possible extension is then given by a family of pairwise-disjoint sets $Z_1, \ldots, Z_h$, where $Z_j \subseteq B_j$, $G[Z_j \cup Y_j]$ is connected and $|Y_j \cup Z_j| \leq k$. Since the degree of each vertex is at most d, it follows that $|B_j| \leq |Y_j| d^{k-1}$. Consequently, the total number of choices of sets Z_j is at most $(\ell + 1)e^k d^{2k-1}/(k(d-1))$ (and all these choices can be enumerated in time $\mathcal{O}(d^{2k} k^2 (\ell+1)^2)$ and space $\mathcal{O}(d^{2k}(\ell+1)^2)$; see [25]). Since $h \leq \ell + 1$, the overall number of choices we consider is bounded by (7).

Since every $\mathcal{V}$ considered in (6) intersects $X_i \setminus A$ (requirement 3), and due to the properties of a tree decomposition and the connectivity of all parts $V' \in \mathcal{V}$ (in G), the induced subgraph $T_i[j \in V(T_i) \mid X_j \cap V' \neq \emptyset]$, for all $V' \in \mathcal{V}$, is a subtree of T_i rooted at i. Keeping this crucial observation in mind, let us focus our attention on a particular node i in T, and bound the number of sets A that we need to consider on the left hand side of equation (6). To this end, it is convenient to consider the computation tree $\mathbf{T}$ for (6) (that is, the recursion tree obtained when solving (6)) in a *top-down* fashion. We can label each node in this tree by (j, A), where j is a node in T and A is a subset of V. The root of $\mathbf{T}$ is $(r, \emptyset)$ and the children of node (j, A) are labeled by the elements of the set $\{(r(S), A \cup \mathcal{V}) \ : \ S \in \mathcal{S}(i, \mathcal{V}), \mathcal{V} \in \mathcal{L}_G(X_i \setminus A, V_i \setminus A)\}$.

Consider node (i, A) in $\mathbf{T}$, and let $(j_1, A_1), \ldots, (j_h, A_h)$ be its ancestors. It is clear that every vertex $v \in A$ belongs to *exactly one* connected component (group) V' that originated at some ancestor (j_r, A_r), i.e, $v \in V' \in \mathcal{V} \in \mathcal{L}_G(X_{j_r} \setminus A_r, V_{j_r} \setminus A_r)$; we say in this case that ancestor (j_r, A_r) contributes to (i, A). Since $X_i \setminus A \neq \emptyset$ (by our assumption that (i, A) appears in the computation tree), it follows by our observation above that the number of ancestors that contribute to (i, A) is at most ℓ (since each such ancestor contributes at least one component that has a non-empty intersection with X_i). In other words, A can be partitioned into at most ℓ parts, such that each part belongs to a connected component that originated at some ancestor of (i, A), and hence, $|A| \leq k\ell$. The number of choices for the contributing ancestors is at most $\mathrm{depth}(T)^\ell$. Using an argument similar to the one used to derive (7), we can conclude that for each vertex v in one of the chosen ancestors, the number of connected components originating at v is at most $e^k d^{2k-1}/(k(d-1))$ and thus we obtain $\left(e^k d^{2k-1}\ell \cdot \mathrm{depth}(\mathbf{T})/(k(d-1))\right)^\ell$ for the total number of choices for A. For each such choice we have to evaluate a number of sets $\mathcal{V}$ bounded by (7), whose properties 1-3 can be verified in time $\mathcal{O}(n)$. Determining the roots of subtrees in $\mathcal{S}(i, A \cup \mathcal{V})$ takes time $\mathcal{O}(n\ell)$.   □

Additionally storing, along with each entry $cgp(i, A)$, the partition $\mathcal{V} \in \mathcal{L}_G(X_i \setminus A, V_i \setminus A)$ minimizing the right hand side in (6), allows us to finally reconstruct a charge group partition that gives the optimal cost.

4   Experimental Evaluation

We implemented the dynamic programming method for bounded treewidth in C++ using the LEMON graph library (http://lemon.cs.elte.hu). We used libtw (http://www.treewidth.com/) to obtain bounded treewidth decompositions of the input molecules. In our implementation we solve the dynamic programming recurrence (6) in a top-down fashion by employing memoization.

4.1   Hydration Free Energy of Amino Acid Side Chains

We tested the quality of charge group assignments by comparing the calculated free energies of solvation in water of a set of 14 charge-neutral amino acid side chain analogs to experimental values, which are denoted by $\Delta G_{\mathrm{hyd,exp}}$ [13, 19]. For each analog, we used the GROMOS 53A6 covalent and van der Waals parameters [19] and partial atomic charges symmetrized by the ATB [18]. For comparison, we also include the manually parametrized solution that the GROMOS 53A6 force field provides [19]. The topologies are derived from the amino acid structures by truncating at the C_α–C_β bond and replacing the C_α atom by a hydrogen atom. For simplicity, we refer to these analogs by their parent amino acid.

Using the GROMACS 4.5.1 package [4], we computed the free energy of hydration $\Delta G_{\mathrm{hyd,calc}}$ using the thermodynamic integration method [5]. A series of simulations were performed at a constant pressure of $p = 1$ bar and a constant temperature $T = 298.15$ K. The free energy was calculated for the process $A \to B$ which involved switching off all non-bonded interactions of the solute in water and in the gas phase. The hydration free energy is calculated as $\Delta G_{\mathrm{hyd,calc}} = \Delta G_{\mathrm{AB,solution}} - \Delta G_{\mathrm{AB,gas}}$ [27]. The simulations were performed in cubic periodic boxes of length $L = 3$ nm. Depending on the analog, the solvated system contained ~ 900 SPC [27] water molecules.

As described in the introduction, neutral charge groups lead to more accurate simulation results. In our problem definition we aim to identify a charge group assignment where the constituent charge groups have small residual error, which is the absolute difference between the sum of the formal charges and the sum of the partial charges of the atoms in the charge group. To ensure neutral charge groups where possible, we adjust the partial charges slightly by redistributing the residual error of every charge group over its atoms.

The results are presented in Table 1 and Figure 2. The GROMOS 53A6 simulation results (ffG53A6 in Table 1) for the studied analogs show good agreement with experiment, which is not surprising as the force field has been parametrized to reproduce the hydration free energy [19]. Using the ATB charge group assignment solution (ATB in Table 1) leads to slightly larger deviations from

Table 1. Comparison of hydration free energies ΔG_{hyd} of amino acid (AA) analogs. All free energy values are given in kJ/mol. When two values separated by a semicolon are given, two experimental values were found. The absolute free energy differences between simulation outcomes and the experimental values are given in parentheses. The average values of these differences are given in the bottom line. "ffG53A6" denotes results using the default GROMOS force field parameters for the analog, "ATB" denotes those using the ATB charge group assignment, "$k = 5$" denotes those using our method.

AA analog	$\Delta G_{hyd,exp}$	$\Delta G_{hyd,calc}$		
		ffG53A6	ATB	$k = 5$
Asn	-40.6	-42.7 (2.1)	-40.5 (0.1)	-47.0 (6.4)
Asp	-28.0	-30.1 (2.1)	-29.1 (1.1)	-28.6 (0.6)
Cys	-5.2	-4.9 (0.3)	-7.0 (1.8)	-7.1 (1.9)
Gln	-39.4	-40.4 (1.0)	-35.9 (3.5)	-35.9 (3.5)
Glu	-27.0	-27.0 (0.0)	-28.2 (1.2)	-32.1 (5.1)
His	-42.9	-44.8 (1.9)	-43.7 (0.8)	-40.9 (2.0)
Ile	8.7; 8.8	9.1 (0.3)	6.3 (2.5)	6.7 (2.1)
Leu	9.4; 9.7	10.8 (1.2)	7.4 (2.2)	7.1 (2.5)
Lys	-18.3	-18.1 (0.2)	-7.2 (11.1)	-7.2 (11.1)
Met	-6.2	-7.4 (1.2)	2.5 (8.7)	2.6 (8.8)
Phe	-3.1	-1.3 (1.8)	1.8 (4.9)	0.6 (3.7)
Trp	-24.7	-25.9 (1.2)	-20.9 (3.8)	-19.7 (5.0)
Tyr	-26.6	-26.9 (0.3)	-30.1 (3.5)	-39.5 (12.9)
Val	8.2	8.5 (0.3)	8.0 (0.2)	8.0 (0.2)
average		(1.1)	(3.2)	(4.7)

experiment, but the average deviation is also within the experimental error of approximately 5 kJ/mol [18]. Although the current method leads to values close to those obtained experimentally, they deviate slightly more from experiment than the ATB values.

4.2   Adenosine Tri-Phosphate (ATP)

Although showing good performance on the amino acid side chains, the ATB method may lead to unacceptably large charge groups, in particular for large highly-charged molecules. An example is the cofactor Adenosine-5'-triphosphate (ATP), for which the ATB combined all phosphate groups and part of the ribose and nucleotide ring systems into a single charge group, see Figure 3(c). In Figure 3(b), the GROMOS 53A6 charge group assignment is given. For comparison, our solution is presented in Figure 3(a), and shows that the phosphate groups have been sorted in separate charge groups, in agreement with the 53A6 assignment and in line with chemical intuition where one expects functional group such as phosphate, amino, and hydroxyl to form separate charge groups.

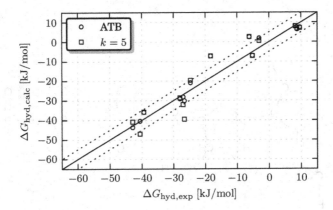

Fig. 2. Calculated ΔG_{hyd} values versus experimental ones, showing the effect of the charge group assignment on the simulated hydration free energy. The labels in the legend refer to Table 1. The solid line represents perfect agreement with experiment, dotted lines indicate the ± 5 kJ/mol approximate experimental error.

5 Discussion

In this work we have formally introduced the charge group partitioning problem which arises in the development of atomic force fields, and more generally, in the identification of functional groups in molecules. The problem is to assign atoms to charge groups of size at most k and such that for every charge group the sum of its partial charges is close to the sum of its formal charges. We showed $\mathcal{NP}$-hardness for $k \geq 4$ and proposed and implemented an exact algorithm capable of solving practical problem instances to provable optimality. With this combination of rigorous definition and exact solution approach, we have made a first step towards formalizing and quantifying some of the aspects that make up "chemical intuition".

Algorithmically, we showed that the case $k = 2$ is solvable in polynomial time. In addition, we have presented a polynomial-time algorithm for bounded charge group size in cases where the molecular graph is a tree. Based on the observation that molecular graphs have bounded treewidth in practice and exploiting further properties such as outerplanarity and bounded degree, we developed a practical dynamic programming algorithm, which is based on a tree-decomposition of the graph corresponding to the chemical structure of interest. An interesting open question is to settle the complexity status for the case $k = 3$.

Our experiments have shown that taking into account charge group size and neutrality already gives good results, especially for large highly-charged molecules such as ATP, where other methods fail to produce meaningful solutions. Still, the greedy partitioning algorithm built into the ATB performs better on the set of smaller amino acid side chain molecules, which is due to the fact that this method exploits additional chemical knowledge. It is thus able, for instance, to deal with a symmetric molecule such as the Tyrosine side chain,

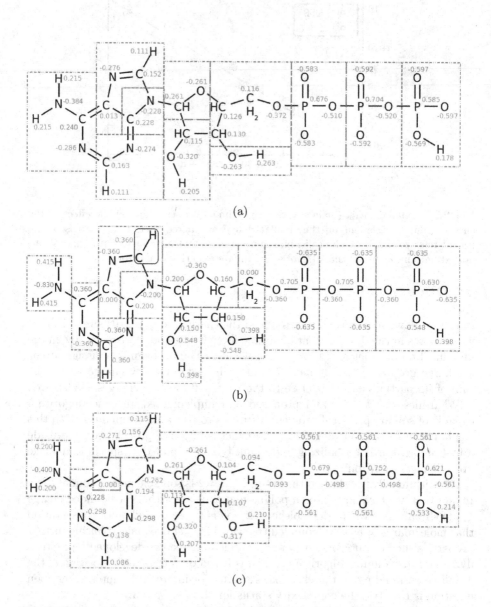

Fig. 3. Charge group assignments for Adenosine Tri-Phosphate (ATP) at pH 5.0. The total molecular charge is -3. The partial charges are shown in grey. (a) Our optimal assignment according to Def. 1 obtained with $k = 5$, (b) GROMOS 53A6 assignment, and (c) assignment by the ATB. Note that the C–H segments indicated by the rounded boxes are considered as single atom types in the GROMOS assignment, whereas they comprise two atoms in the other assignments.

where the charge group assignment of our new method resulted in a large deviation because we do not consider symmetry in our problem definition. As our approach is general and flexible and thus able to accommodate further constraints that are based on chemical knowledge or intuition, we plan to integrate more of these aspects into our model as part of future work. For example, we will investigate the effect of bounding the error per charge group. In addition to symmetry in terms of atom types and bonds, we plan to integrate constraints that take spatial geometry into account. We would like to stress that only through a proper problem definition together with a method capable of obtaining provably optimal solutions, one is able to make progress in answering the question how a good charge group partition should look like.

Acknowledgments. We thank SARA Computing and Networking Services (www.sara.nl) for their support in using the Lisa Compute Cluster. In addition, we are grateful to the referees for helpful comments. The research leading to these results has received support from the Innovative Medicines Initiative Joint Undertaking under grant agreement no. 115002 (eTOX), resources of which are composed of financial contribution from the European Union's Seventh Framework Programme (FP7/20072013) and EFPIA companies' in kind contribution.

References

1. Alber, J., Dorn, F., Niedermeier, R.: Experimental evaluation of a tree decomposition-based algorithm for vertex cover on planar graphs. Discrete Applied Mathematics 145(2), 219–231 (2005)
2. Allen, M.P., Tildesley, D.J.: Computer Simulation of Liquids. Oxford University Press, Oxford (1987)
3. Arnborg, S., Corneil, D.G., Proskurowski, A.: Complexity of finding embeddings in a k-tree. SIAM J. Algebraic Discrete Methods 8(2), 227–284 (1987)
4. Berendsen, H.J.C., van der Spoel, D., Van Drunen, R.: GROMACS: a message-passing parallel molecular dynamics implementation. Com. Phys. Comm. 91(1-3), 43–56 (1995)
5. Beveridge, D.L., DiCapua, F.M.: Free energy via molecular simulation: applications to chemical and biomolecular systems. Annu. Rev. Biophys. Biophys. Chem. 18(1), 431–492 (1989)
6. Bodlaender, H.L.: NC-Algorithms for Graphs with Small Treewidth. In: van Leeuwen, J. (ed.) WG 1988. LNCS, vol. 344, pp. 1–10. Springer, Heidelberg (1989)
7. Bodlaender, H.L.: A partial k-arboretum of graphs with bounded treewidth. Theor. Comput. Sci. 209(1-2), 1–45 (1998)
8. Boggara, M.B.B., Faraone, A., Krishnamoorti, R.: Effect of pH and ibuprofen on the phospholipid bilayer bending modulus. J. Phys. Chem. B 114(24), 8061–8066 (2010)
9. Brooks, B.R., Brooks III, C.L., Mackerell Jr., A.D., Nilsson, L., Petrella, R.J., Roux, B., Won, Y., Archontis, G., Bartels, C., Boresch, S., Caflisch, A., Caves, L., Cui, Q., Dinner, E.: CHARMM: The Biomolecular Simulation Program. J. Comput. Chem. 30(10, sp. iss. si), 1545–1614 (2009)

10. Cornell, W.D., Cieplak, P., Bayly, C.I., Gould, I.R., Merz, K.M., Ferguson, D.M., Spellmeyer, D.C., Fox, T., Caldwell, J.W., Kollman, P.A.: A second generation force field for the simulation of proteins, nucleic acids, and organic molecules. J. Am. Chem. Soc. 117, 5179–5197 (1995)
11. Dehof, A.K., Rurainski, A., Bui, Q.B.A., Böcker, S., Lenhof, H.-P., Hildebrandt, A.: Automated bond order assignment as an optimization problem. Bioinformatics 27(5), 619–625 (2011)
12. Dyer, M., Frieze, A.: On the complexity of partitioning graphs into connected subgraphs. Discrete Applied Mathematics 10(2), 139–153 (1985)
13. Gerber, P.R.: Charge distribution from a simple molecular orbital type calculation and non-bonding interaction terms in the force field mab. J. Comput.-Aided Mol. Des. 12(1), 37–51 (1998)
14. Goto, S., Okuno, Y., Hattori, M., Nishioka, T., Kanehisa, M.: LIGAND: database of chemical compounds and reactions in biological pathways. Nucleic Acids Res. 30(1), 402–404 (2002)
15. Horváth, T., Ramon, J., Wrobel, S.: Frequent subgraph mining in outerplanar graphs. Data Min. Knowl. Discov. 21(3), 472–508 (2010)
16. Jorgensen, W.L., Maxwell, D.S., Tirado-Rives, J.: Development and testing of the OPLS all-atom force field on conformational energetics and properties of organic liquids. J. Am. Chem. Soc. 118(45), 11225–11236 (1996)
17. Lemkul, J.A., Allen, W.J., Bevan, D.R.: Practical considerations for building GROMOS-compatible small-molecule topologies. J. Chem. Inf. Model. 50(12), 2221–2235 (2010)
18. Malde, A.K., Zuo, L., Breeze, M., Stroet, M., Poger, D., Nair, P.C., Oostenbrink, C., Mark, A.E.: An automated force field topology builder (ATB) and repository: version 1.0. J. Chem. Theory Comput. 7(12), 4026–4037 (2011)
19. Oostenbrink, C., Villa, A., Mark, A.E., van Gunsteren, W.F.: A biomolecular force field based on the free enthalpy of hydration and solvation: The GROMOS force-field parameter sets 53A5 and 53A6. J. Comp. Chem. 25(13), 1656–1676 (2004)
20. Robertson, N., Seymour, P.D.: Graph minors. II. Algorithmic aspects of tree-width. J. Algorithms 7(3), 309–322 (1986)
21. Schmid, N., Eichenberger, A., Choutko, A., Riniker, S., Winger, M., Mark, A.E., van Gunsteren, W.F.: Definition and testing of the GROMOS force-field versions 54A7 and 54B7. Eur. Biophys. J. 40, 843–856 (2011)
22. Schüttelkopf, A.W., van Aalten, D.M.: PRODRG: a tool for high-throughput crystallography of protein-ligand complexes. Acta Crystallogr. 60, 1355–1363 (2004)
23. Scott, W.R.P., Hunenberger, P.H., Tironi, I.G., Mark, A.E., Billeter, S.R., Fennen, J., Torda, A.E., Huber, T., Kruger, P., van Gunsteren, W.F.: The GROMOS biomolecular simulation program package. J. Phys. Chem. A 103(19), 3596–3607 (1999)
24. Sharma, M., Khanna, S., Bulusu, G., Mitra, A.: Comparative modeling of thioredoxin glutathione reductase from Schistosoma mansoni: a multifunctional target for antischistosomal therapy. J. Mol. Graphics Model. 27(6), 665–675 (2009)
25. Uehara, R., Factor, K.T.: The number of connected components in graphs and its applications
26. van Gunsteren, W.F., Bakowies, D., Baron, R., Chandrasekhar, I., Christen, M., Daura, X., Gee, P., Geerke, D.P., Glttli, A., Hünenberger, P.H., Kastenholz, M.A., Oostenbrink, C., Schenk, M., Trzesniak, D., van der Vegt, N.F.A., Yu, H.B.: Biomolecular modeling: Goals, problems, perspectives. Angew. Chem. Int. Ed. 45(25), 4064–4092 (2006)

27. Villa, A., Mark, A.E.: Calculation of the free energy of solvation for neutral analogs of amino acid side chains. J. Comp. Chem. 23(5), 548–553 (2002)
28. Yamaguchi, A., Aoki, K.F., Mamitsuka, H.: Graph complexity of chemical compounds in biological pathways. Genome Inform. 14, 376–377 (2003)
29. Yang, C., Zhu, X., Li, J., Shi, R.: Exploration of the mechanism for LPFFD inhibiting the formation of beta-sheet conformation of A beta(1-42) in water. J. Mol. Model. 16(4), 813–821 (2010)

Increased Methylation Variation in Epigenetic Domains across Cancer Types

Hector Corrada Bravo

University of Maryland, USA

Tumor heterogeneity is a major barrier to effective cancer diagnosis and treatment. We recently identified cancer-specific differentially DNA-methylated regions (cDMRs) in colon cancer, which also distinguish normal tissue types from each other, suggesting that these cDMRs might be generalized across cancer types. Here we show stochastic methylation variation of the same cDMRs, distinguishing cancer from normal tissue, in colon, lung, breast, thyroid and Wilms' tumors, with intermediate variation in adenomas. Whole-genome bisulfite sequencing shows these variable cDMRs are related to loss of sharply delimited methylation boundaries at CpG islands. Furthermore, we find hypomethylation of discrete blocks encompassing half the genome, with extreme gene expression variability. Genes associated with the cDMRs and large blocks are involved in mitosis and matrix remodeling, respectively. We suggest a model for cancer involving loss of epigenetic stability of well-defined genomic domains that underlies increased methylation variability in cancer that may contribute to tumor heterogeneity.

http://www.nature.com/ng/journal/v43/n8/full/ng.865.html

B. Chor (Ed.): RECOMB 2012, LNBI 7262, p. 44, 2012.
© Springer-Verlag Berlin Heidelberg 2012

Estimating the Accuracy of Multiple Alignments and its Use in Parameter Advising*

Dan F. DeBlasio[1], Travis J. Wheeler[2], and John D. Kececioglu[1],**

[1] Department of Computer Science, The University of Arizona, USA
[2] Janelia Farm Research Campus, Howard Hughes Medical Institute, USA
kece@cs.arizona.edu

Abstract. We develop a novel and general approach to estimating the accuracy of protein multiple sequence alignments without knowledge of a reference alignment, and use our approach to address a new problem that we call *parameter advising*. For protein alignments, we consider twelve independent features that contribute to a quality alignment. An accuracy estimator is learned that is a polynomial function of these features; its coefficients are determined by minimizing its error with respect to true accuracy using mathematical optimization. We evaluate this approach by applying it to the task of parameter advising: the problem of choosing alignment scoring parameters from a collection of parameter values to maximize the accuracy of a computed alignment. Our estimator, which we call Facet (for "feature-based accuracy estimator"), yields a parameter advisor that on the hardest benchmarks provides more than a 20% improvement in accuracy over the best default parameter choice, and outperforms the best prior approaches to selecting good alignments for parameter advising.

1 Introduction

Estimating the accuracy of a computed multiple sequence alignment without knowing the correct alignment is an important problem. A good accuracy estimator has broad utility, from building a meta-aligner that selects the best output of a collection of aligners, to boosting the accuracy of a single aligner by choosing the best values for alignment parameters. The accuracy of a computed alignment is typically determined with respect to a reference alignment, by measuring the fraction of substitutions in the core columns of the reference alignment that are present in the computed alignment. We estimate accuracy without knowing the reference by learning a function that combines several easily-computable features of an alignment into a single value.

In the literature, several approaches have been presented for assessing the quality of a computed alignment without knowing a reference alignment for its sequences. These approaches follow two general strategies for estimating the accuracy of a computed alignment in recovering a correct alignment (where correctness is either with respect to the unknown structural alignment, as in our

---

* Research supported by US NSF Grant IIS-1050293 and DGE-0654435.
** Corresponding author.

B. Chor (Ed.): RECOMB 2012, LNBI 7262, pp. 45–59, 2012.

present study, or the unknown evolutionary alignment, as in simulation studies). One strategy is to develop a new *scoring function* on alignments that ideally is correlated with accuracy. These scoring functions combine local features of an alignment into a score, where the features typically include a measure of the conservation of amino acids in alignment columns. Scoring-function-based approaches include AL2CO by Pei and Grishin [9], NorMD by Thompson et al. [11], and PredSP by Ahola et al. [1]. The other general strategy is to (a) examine a collection of alternate alignments of the same sequences, where the collection can be generated by changing the method used for computing the alignment or changing the input to a method, and (b) measure the *support* for the computed alignment among the collection [8,7,10,6]. In this strategy the support for the computed alignment, which essentially measures the stability of the alignment to changes in the method or input, serves as a proxy for accuracy. Support-based approaches include MOS by Lassmann and Sonnhammer [8], HoT by Landan and Graur [7], GUIDANCE by Penn et al. [10], and PSAR by Kim and Ma [6]. In our experiments, among scoring-function-based approaches, we compare just to NorMD and PredSP, since AL2CO is known [8] to be dominated by NorMD. Among support-based approaches, we compare just to MOS and HoT, as GUIDANCE requires alignments of at least four sequences (ruling out the many three-sequence protein reference aligments in suites such as BENCH [3]), while PSAR is not yet implemented for protein alignments.

The new approach we develop for accuracy estimation significantly improves on prior approaches, as we demonstrate through its performance on parameter advising. Compared to prior scoring-function-based approaches, we (a) introduce several *novel feature functions* that measure non-local properties of an alignment that show stronger correlation with accuracy, (b) consider *larger classes of estimators* beyond linear combinations of features, and (c) develop *new regression approaches* for learning a monotonic estimator from examples. Compared to support-based approaches, our estimator does not degrade on difficult alignment instances, where for parameter advising, good accuracy estimation is most needed. As shown in our advising experiments, support-based approaches lose the ability to detect accurate alignments of hard-to-align sequences, since most alternate alignments are poor and lend little support to the best alignment.

2   Accuracy Estimators

Without knowing a reference alignment that establishes the ground truth against which the true accuracy of an alignment is measured, we are left with only being able to estimate the accuracy of an alignment. Our approach to obtaining an estimator for alignment accuracy is to (a) identify multiple *features* of an alignment that tend to be correlated with accuracy, and (b) combine these features into a single accuracy estimate. Each feature, as well as the final accuracy estimator, is a real-valued function of an alignment.

The simplest estimator is a linear combination of feature functions, where features are weighted by coefficients. These coefficients can be learned by training the estimator on example alignments whose true accuracy is known. This training process will result in a fixed coefficient or weight for each feature. Alignment accuracy is usually represented by a value in the range $[0, 1]$, with 1 corresponding to perfect accuracy. Consequently the value of the estimator on an alignment should be bounded, no matter how long the alignment or how many sequences it aligns. For boundedness to hold when using fixed feature weights, the feature functions themselves must also be bounded. Hence we assume that the feature functions also have the range $[0, 1]$. We can then guarantee that the estimator has range $[0, 1]$ by ensuring that coefficients found by the training process yield a convex combination of features.

In general we consider estimators that are polynomial functions of alignment features. More precisely, suppose the features that we consider for alignments A are measured by the k feature functions $f_i(A)$ for $1 \leq i \leq k$. Then our accuracy estimator $E(A)$ is a polynomial in the k variables $f_i(A)$. For example, for a degree-2 polynomial, $E(A) := a_0 + \sum_{1 \leq i \leq k} a_i f_i(A) + \sum_{1 \leq i, j \leq k} a_{ij} f_i(A) f_j(A)$. Learning an estimator from example alignments, as discussed in Section 3, corresponds to determining the coefficients for its terms. We can efficiently learn optimal values for the coefficients, that minimize the error between the estimate $E(A)$ and the actual accuracy of alignment A on a set of training examples, even for estimators that are polynomials of *arbitrary* degree d. The key is that such an estimator can always be reduced to the linear case by a change of feature functions. For *each* term in the degree-d estimator, where the term is specified by the powers p_i of the f_i, define a *new* feature function $g_j(A) := \prod_{1 \leq i \leq k} \left(f_i(A) \right)^{p_i}$. Then the original degree-d estimator is equivalent to the linear estimator $E(A) = c_0 + \sum_{1 \leq j < t} c_j g_j(A)$, where t is the number of terms in the original polynomial. For a degree-d estimator with k original feature functions, the number of coefficients t in the linearized estimator is at least $\mathcal{P}(d, k)$, the number of integer partitions of d with k parts. This number of coefficients grows very fast with d, so overfitting quickly becomes an issue. (Even a cubic estimator on 10 features already has 286 coefficients.) In our experiments, we focus on *linear* and *quadratic* estimators.

3   Learning the Estimator from Examples

To learn an accuracy estimator, we collect a training set of alignments whose true accuracy is known, and find coefficients for the estimator that give the best fit to true accuracy. We form the training set for our experiments by (1) collecting reference alignments from standard suites of benchmark protein multiple alignments; (2) for each such reference alignment, generating alternate alignments of its sequences by calling a multiple sequence aligner with differing values for the parameters of its alignment scoring function, in particular by varying its gap penalties, which can yield markedly different alignments; and (3) labeling each such alternate alignment by its accuracy with respect to the reference alignment for its sequences. We are careful to use suites of benchmarks for which the reference alignments are obtained by structural alignment of the proteins using their known three-dimensional structures. These alternate alignments together with their labeled accuracies form the *examples* in our training set.

Fitting to Accuracy Values. A natural fitting criterion is to minimize the error on the example alignments between the estimator and the true accuracy value. For alignment A in our training set $\mathcal{S}$, let $E_c(A)$ be its estimated accuracy where vector $c = (c_0, \ldots, c_{t-1})$ specifies the values for the coefficients of the estimator polynomial, and let $F(A)$ be the *true accuracy* of example A.

Formally, minimizing the weighted error between estimated accuracy and true accuracy yields estimator $E^* := E_{c^*}$ with coefficient vector c^* that minimizes $\sum_{A \in \mathcal{S}} w_A \left| E_c(A) - F(A) \right|^p$, where power p controls the degree to which large accuracy errors are penalized. Weights w_A correct for sampling bias among the examples, as explained below.

When $p = 2$, this corresponds to minimizing the L_2 norm between the estimator and the true accuracies. The absolute value in the objective function may be removed, and the formulation becomes a *quadratic programming problem* in variables c, which can be efficiently solved. (Note that E_c is linear in c.) If the feature functions all have range $[0, 1]$, we can ensure that the resulting estimator E^* also has range $[0, 1]$ by adding to the quadratic program the linear inequalities $c \geq 0$ and $\sum_{0 \leq i < t} c_i \leq 1$.

When $p = 1$, the formulation corresponds to minimizing the L_1 norm. This is less sensitive to outliers than the L_2 norm, an advantage when the underlying features are noisy. Minimizing the L_1 norm can be reduced to a *linear programming problem* as follows. In addition to variables c, we have a second vector of variables e with an entry e_A for each example $A \in \mathcal{S}$ to capture the absolute value in the L_1 norm, along with inequalities $e_A \geq E_c(A) - F(A)$ and $e_A \geq F(A) - E_c(A)$, which are linear in variables c and e. We then minimize the linear objective function $\sum_{A \in \mathcal{S}} w_A e_A$. For n examples, the linear program has $n + t$ variables and $O(n)$ inequalities, which is solvable even for very large numbers of examples. We can also add inequalities that ensure E^* has range $[0, 1]$.

The weights w_A on examples aid in finding an estimator that is good across all accuracies. In the suites of protein alignment benchmarks that are commonly available, a predominance of the benchmarks consist of sequences that are easily alignable, meaning that standard aligners find high-accuracy alignments for these benchmarks. (This is mainly a consequence of the fact that proteins for which reliable structural reference alignments are available tend to be closely related, and hence easier to align; it does not mean that typical biological inputs are easy!) In this situation, when training set $\mathcal{S}$ is generated as described earlier, most examples have high accuracy, with relatively few at moderate to low accuracies. Without weights on examples, the resulting estimator E^* is strongly biased towards optimizing the fit for high accuracy alignments, at the expense of a poor fit at lower accuracies. To prevent this, we bin the examples in $\mathcal{S}$ by their true accuracy, where $\mathcal{B}(A) \subseteq \mathcal{S}$ is the set of alignments falling in the bin for example A, and then weight the error term for A by $w_A := 1/|\mathcal{B}(A)|$. (In our experiments, we form 10 bins equally spaced at 10% increments in accuracy.) In the objective function this weights *bins* uniformly (rather than weighting *examples* uniformly) and weights the error equally across the full range of accuracies.

Fitting to Accuracy Differences. Many applications of an accuracy estimator E will use it to choose from a set of alignments the one that is estimated to be most accurate. (This occurs, for instance, in parameter advising as discussed in Section 5.) In such applications, the estimator is effectively ranking alignments, and all that is needed is for the estimator to be *monotonic* in true accuracy. Accordingly, rather than trying to fit the estimator to match accuracy *values*, we can instead fit it so that *differences* in accuracy are reflected by at least as large differences in the estimator. This fitting to differences is less constraining than fitting to values, and hence might be better achieved.

More precisely, suppose we have selected a set $\mathcal{P} \subseteq \mathcal{S}^2$ of ordered pairs of example alignments, where every pair $(A, B) \in \mathcal{P}$ satsifies $F(A) < F(B)$. Set $\mathcal{P}$ holds pairs of examples on which accuracy F increases for which we desire similar behavior from our estimator E. (Later we discuss how we select a small set $\mathcal{P}$ of important pairs.) If estimator E increases at least as much as accuracy F on a pair in $\mathcal{P}$, this is a success, and if it increases less than F, we consider the amount it falls short an error, which we try to minimize. Notice this tries to match large accuracy increases, and penalizes less for not matching small increases.

We formulate fitting to differences as finding the optimal estimator $E^* := E_{c^*}$ given by coefficients c^* that minimize $\sum_{(A,B) \in \mathcal{P}} w_{AB}\, e_{AB}^p$, where e_{AB} is defined to be $\max\{\big(F(B) - F(A)\big) - \big(E_c(B) - E_c(A)\big),\, 0\}$, and w_{AB} weights the error term for a pair. When power p is 1 or 2, we can reduce this optimization problem to a linear or quadratic program as follows. We introduce a vector of variables e with an entry e_{AB} for each pair $(A, B) \in \mathcal{P}$, along with inequalities $e_{AB} \geq 0$ and $e_{AB} \geq \big(F(B) - F(A)\big) - \big(E_c(B) - E_c(A)\big)$, which are linear in variables c and e. We then minimize the objective function $\sum_{(A,B) \in \mathcal{P}} w_{AB}\, e_{AB}^p$, which is linear or quadratic in the variables for $p = 1$ or 2.

For a set $\mathcal{P}$ of m pairs, these programs have $m + t$ variables and m inequalities, where $m = O(n^2)$ in terms of the number of examples n. For the programs to be manageable for large n, set $\mathcal{P}$ must be quite sparse.

We select a sparse set $\mathcal{P}$ of pairs as follows. Recall that the training set $\mathcal{S}$ of examples consists of alternate alignments of the sequences in benchmark reference alignments, where the alternates are generated by aligning the benchmark under a constant number of different parameter choices. A monotonic estimator, especially one that is used for parameter advising, should properly rank the set of alternate alignments for each benchmark. Consequently for each benchmark, we select all pairs of alternates whose difference in accuracy is at least ϵ, where ϵ is a tunable threshold. For the estimator to generalize outside the training set, it helps to also properly rank alignments between benchmarks. To include some pairs between benchmarks, we choose the minimum, maximum, and median accuracy alignments for each benchmark, and form one list L of all these chosen alignments, ordered by increasing accuracy. Then for each alignment A in L, we scan L to the right to select the first k pairs (A, B) for which $F(B) \geq F(A) + i\,\delta$ where $i = 1, \ldots, k$, and for which B is from a different benchmark than A. While the constants $\epsilon \geq 0$, $\delta \geq 0$, and $k \geq 1$ control the specific pairs that this procedure selects for $\mathcal{P}$, it always selects $O(n)$ pairs on the n examples.

Weighting Pairs for Difference Fitting. When fitting to accuracy differences we again weight the error terms, which are now associated with pairs, to correct for sampling bias within $\mathcal{P}$. We want the weighted pairs to treat the entire accuracy range equally, so the fitted estimator performs well at all accuracies. As when fitting to accuracy values, we partition the example alignments in $\mathcal{S}$ into bins $\mathcal{B}_1, \ldots, \mathcal{B}_k$ according to their true accuracy. To model equal weighting of accuracy bins by pairs, we consider a pair $(A, B) \in \mathcal{P}$ to have half its weight w_{AB} on the bin containing A, and half on the bin containing B. (So in this model, a pair (A, B) with both ends A, B in the same bin $\mathcal{B}$, places all its weight w_{AB} on $\mathcal{B}$.) Then we want to find weights $w_{AB} > 0$ that, for all bins $\mathcal{B}$, satisfy $\sum_{(A,B) \in \mathcal{P}: A \in \mathcal{B}} \frac{1}{2} w_{AB} + \sum_{(A,B) \in \mathcal{P}: B \in \mathcal{B}} \frac{1}{2} w_{AB} = 1$. In other words, the pairs should weight bins uniformly. We say a collection of weights w_{AB} are *balanced* if they satisfy the above property. While balanced weights do not always exist in general, we can identify an easily-satisfied condition that guarantees they do exist, and in this case find balanced weights by the following graph algorithm.

Construct an undirected graph G whose vertices are the bins $\mathcal{B}_i$ and whose edges (i, j) go between bins $\mathcal{B}_i, \mathcal{B}_j$ that have an alignment pair (A, B) in $\mathcal{P}$ with $A \in \mathcal{B}_i$ and $B \in \mathcal{B}_j$. (Notice G has self-loops when pairs have both alignments in the same bin.) Our algorithm first computes weights ω_{ij} on the edges (i, j) in G, and then assigns weights to pairs (A, B) by setting $w_{AB} := 2\omega_{ij}/c_{ij}$, where bins $\mathcal{B}_i, \mathcal{B}_j$ contain alignments A, B, and c_{ij} counts the number of pairs in $\mathcal{P}$ between bins $\mathcal{B}_i$ and $\mathcal{B}_j$. (The factor of 2 is due to a pair only contributing weight $\frac{1}{2} w_{AB}$ to a bin.) A consequence is that all pairs (A, B) that go between the same bins get the same weight w_{AB}.

During the algorithm, an edge (i, j) in G is said to be *labeled* if its weight ω_{ij} has been determined; otherwise it is *unlabeled*. We call the *degree* of a vertex i the total number of endpoints of edges in G that touch i, where a self-loop contributes two endpoints to the degree. Initially all edges of G are unlabeled. The algorithm sorts the vertices of G in order of nonincreasing degree, and then processes the vertices from highest degree on down. In the general step, the algorithm processes vertex i as follows. It accumulates w, the sum of the weights ω_{ij} of all *labeled* edges that touch i; counts u, the number of *unlabeled* edges touching i that are not a self-loop; and determines d, the degree of i. To the unlabeled edges (i, j) touching i, the algorithm assigns weight $\omega_{ij} := 1/d$ if the edge is not a self-loop, and weight $\omega_{ii} := \frac{1}{2}(1 - w - \frac{u}{d})$ otherwise.

This algorithm assigns *balanced weights* if in graph G, every bin has a self-loop, as stated in the following theorem. The proof is omitted due to page limits.

Theorem 1 (Finding Balanced Weights). *Suppose every bin $\mathcal{B}$ has some pair (A, B) in $\mathcal{P}$ with both alignments A, B in $\mathcal{B}$. Then the above graph algorithm finds balanced weights, and runs in $O(k + m)$ time for k bins and m pairs in $\mathcal{P}$.*

Notice that we can ensure the condition in Theorem 1 holds if every bin has at least two example alignments: simply add a pair (A, B) to $\mathcal{P}$ where both alignments are in the bin, if the procedure for selecting a sparse $\mathcal{P}$ did not already. When the training set $\mathcal{S}$ of example alignments is sufficiently large compared to the number of bins (which is within our control), every bin is likely to have at least two examples. So Theorem 1 essentially guarantees that in practice we can fit our estimator using balanced weights.

4   Estimator Features

The quality of the estimator that results from our approach ultimately rests on the quality of the features that we consider. We consider a dozen features of an alignment, the majority of which are novel. All are efficiently computable, so the resulting estimator is fast to evaluate. The strongest feature functions make use of predicted *secondary structure* (which is not surprising, given that protein reference alignments are structural alignments). Another aspect of the best alignment features is that they tend to use *nonlocal information*. This is in contrast to standard ways of scoring sequence alignments, such as with amino acid substitution scores or gap open and extension penalties, which are often a function of a single alignment column or two adjacent columns. While a good accuracy estimator would make an ideal scoring function for *constructing* a sequence alignment, computing an optimal alignment under such a nonlocal scoring function seems prohibitive. Nevertheless, given that our estimator can be efficiently evaluated on any constructed alignment, it is well suited for *selecting* a sequence alignment from among several alternate alignments, as we discuss in Section 5 in the context of parameter advising.

The following are the feature functions we consider for our estimator.

Secondary Structure Blockiness. The reference alignments in the most reliable suites of protein alignment benchmarks are computed by structural alignment of the known three-dimensional structures of the proteins. The so-called *core blocks* of these reference alignments, which are the columns in the reference to which an alternate alignment is compared when measuring its true accuracy, are typically defined as the regions of the structural alignment in which the residues of the different proteins are all within a small distance threshold of each other in the superimposed structures. These regions of structural agreement are usually in the embedded core of the folded proteins, and the secondary structure of the core usually consists of α-helices and β-strands. (There are three basic types of *secondary structure* that a residue can have: α-helix, β-strand, and *coil*, which stands for "other.") As a consequence, in the reference sequence alignment, the sequences in a core block often share the same secondary structure, and the type of this structure is usually α-helix or β-strand.

We measure the degree to which a multiple alignment displays this pattern of structure by a feature we call *secondary structure blockiness*. Suppose that for the protein sequences in a multiple alignment we have predicted the secondary structure of each protein, using a standard prediction tool such as PSIPRED [4]. Then in multiple sequence alignment A and for given integers $k, \ell > 1$, define a secondary structure *block* $\mathcal{B}$ to be (i) a contiguous interval of at least ℓ columns of A, together with (ii) a subset of at least k sequences in A, such that on all columns in this interval, in all sequences in this subset, all the entries in these columns for these sequences have the same predicted secondary structure type, and this shared type is all α-helix or all β-strand. We call $\mathcal{B}$ an α-block or a β-block according to the common type of its entries. (Parameter ℓ, which controls the minimum width of a block, relates to the minimum length of α-helices and β-strands.) A *packing* for alignment A is a set $\mathcal{P} = \{\mathcal{B}_1, \ldots, \mathcal{B}_b\}$ of secondary structure blocks of A, such that the column intervals of the $\mathcal{B}_i \in \mathcal{P}$ are all disjoint. (In other words, in a packing, each column of A is in at most one block. The sequence subsets for the blocks can differ arbitrarily.) The *value* of a block is the total number of residue pairs (or equivalently, substitutions) in its columns; the value of a packing is the sum of the values of its blocks. Finally, the *blockiness* of an alignment A is the maximum value of any packing for A, divided by the total number of residue pairs in the columns of A. In other words, blockiness measures the percentage of substitutions that are in α- or β-blocks.

While at first glance measuring blockiness might seem hard (as most combinatorial packing problems are in the worst-case), it can actually be computed in *linear time*, as the following theorem states. The key insight is that evaluating blockiness can be reduced to solving a longest path problem on a directed acyclic graph of linear size. The proof is omitted due to page limits.

Theorem 2 (Evaluating Blockiness). *Given a multiple alignment A of m protein sequences and n columns, where the sequences are annotated with predicted secondary structure, the blockiness of A can be computed in $O(mn)$ time.*

Other Features. The remaining features are simpler than blockiness.

Secondary Structure Agreement. The secondary structure prediction tool PSIPRED [4] outputs confidence values at each residue that are intended to reflect the probability that the residue has each of the three secondary structure types. Denote these three confidences for a residue i, normalized so they add up to 1, by $p_\alpha(i)$, $p_\beta(i)$, and $p_\gamma(i)$. Then we can estimate the probability that two residues i, j in a column have the same secondary structure type that is not coil by $P(i, j) := p_\alpha(i)\, p_\alpha(j) + p_\beta(i)\, p_\beta(j)$. To measure how strongly the secondary structure locally agrees around two residue positions, we compute a weighted average of P in a window of width $2\ell + 1$ centered at the positions: $Q(i, j) := \sum_{-\ell \leq k \leq \ell} w_k P(i+k, j+k)$, where the weights w_k form a discrete distribution that peaks at $k = 0$ (centered on i and j) and is symmetric. The value of the secondary structure agreement feature is then the average of $Q(i, j)$ over all residue pairs i, j in all columns.

Gap Features. A gap in a pairwise alignment is a maximal run of either insertions or deletions. We consider the following four features on gaps. In *Gap Coil Density*, for every pair of sequences, we measure the fraction of residues involved in gaps in the pairwise alignment induced by the sequence pair that are predicted by PSIPRED to be of secondary structure type *coil*, and average this measure over all induced pairwise alignments. *Gap Extension Density* counts the number of null characters in the alignment (the dashes that denote gaps), normalized by the total number of alignment entries. *Gap Open Density* counts the number of *runs* of null characters in the rows of the alignment, normalized by the total length of all such runs. For *Gap Compatibility*, as in cladistics, we encode the gapping pattern in the columns of an alignment by a binary state: residue, or null character. For an alignment in this encoding we then collapse together adjacent columns that have the same gapping pattern. We evaluate the reduced set of columns for compatibility by checking whether a perfect phylogeny exists on them, using the so-called four-gametes test on pairs of columns. The feature measures the fraction of pairs of reduced columns that pass the test.

Conservation Features. We consider the following six features that measure conservation within alignment columns. In *Substitution Compatibility*, similar to Gap Compatibility, we encode the substitution pattern in the columns of an alignment by a binary state: using a reduced amino acid alphabet of equivalency classes, residues in the most prevalent equivalency class in the column are mapped to 1 and all others to 0. The feature measures the fraction of encoded column pairs that pass the four-gametes test. We considered the standard reduced alphabets with 6, 10, 15, and 20 equivalency classes, and used the 10-class alphabet, which gave the strongest correlation with accuracy. *Amino Acid Identity* is usually called percent-identity. In each induced pairwise alignment, we measure the fraction of substitutions in which the residues have the same amino acid equivalency class. The feature averages this over all induced pairwise alignments. *Secondary Structure Identity* is like amino acid identity, except that instead of the protein's amino acid sequence, we use the secondary structure sequence predicted for the protein by PSIPRED [4] (a string over a 3-letter alphabet). *Average Substitution Score* computes the average score of all substitutions in the alignment using a BLOSUM62 substitution-scoring matrix that has been shifted and scaled so the amino acid similarity scores are in the range $[0, 1]$. *Core Column Density* predicts core columns as those that do not contain null characters and whose fraction of pairs of residues that have the same amino acid equivalency class is above a threshold. The feature then normalizes the count of predicted core columns by the total number of columns in the alignment. *Information Content* measures the average entropy of the alignment, by summing over the columns the log of the ratio of the abundance of a specific amino acid in the column over the background distribution for that amino acid, normalized by the number of columns in the alignment.

5   Application to Parameter Advising

In characterizing six stages in constructing a multiple sequence alignment, Wheeler and Kececioglu [12] gave as the first stage choosing the parameter values for the alignment scoring function. While many alignment tools allow the user to specify scoring function parameter values, such as affine gap penalties or substitution scoring matrices, typically only the default parameter values that the aligner provides are used. This default parameter choice is often tuned to optimize the average accuracy of the aligner over a collection of alignment benchmarks. While the default parameter values might be the single choice that works best on average on the benchmarks, for specific input sequences there may be a different choice on which the aligner outputs a much more accurate alignment.

This leads to the task of *parameter advising*: given particular sequences to align, and a set of possible parameter choices, recommend a parameter choice to the aligner that yields the most accurate alignment of those sequences. Parameter advising has three components: the set S of input sequences, the set P of parameter choices, and the aligner $\mathcal{A}$. (Here a *parameter choice* $p \in P$ is a vector $p = (p_1, \ldots, p_k)$ that specifies values for *all* free parameters in the alignment scoring function.) Given sequences S and parameter choice $p \in P$, we denote the alignment output by the aligner as $\mathcal{A}_p(S)$.

Wheeler and Kececioglu [12] call a procedure that takes the set of input sequences S and the set of parameter choices P, and outputs a parameter recommendation $p \in P$, an *advisor*. A perfect advisor, that always recommends the choice $p^* \in P$ that yields the highest accuracy alignment $\mathcal{A}_{p^*}(S)$, is called an *oracle*. In practice, constructing an oracle is impossible, since for any real set S of sequences that we want to align, a reference alignment for S is unknown (as otherwise we would not need to align them), so the true accuracy of any alignment of S cannot be determined. The concept of an oracle is useful, however, for measuring how well an actual advisor performs.

A natural approach for constructing a parameter advisor is to use an accuracy estimator E as a proxy for true accuracy, and recommend the parameter choice $\widetilde{p} := \operatorname{argmax}_{p \in P} E(\mathcal{A}_p(S))$. In its simplest realization, such an advisor will run the aligner $\mathcal{A}$ repeatedly on input S, once for each possible parameter choice $p \in P$, to select the output that has best estimated accuracy. Of course, to yield a quality advisor, this requires two ingredients: a good estimator E, and a good set P of parameter choices. The preceding sections have addressed how to design estimator E; we now turn to how to find a good set P.

Finding an Optimal Parameter Set. Since the above advisor computes as many alignments as there are parameter choices in P, set P must be small for such an advisor to be practical. Given a bound on $|P|$, we would like to find a set P that, say, maximizes the true accuracy of the aligner $\mathcal{A}$ using the advisor's recommendation from P, averaged over a training collection of benchmarks S. Finding such an optimal set P is prohibitive, however, because the behavior of the advisor is an indirect function of the entire set P (through the estimator E), which effectively forces us to enumerate all possible sets P to find an optimal one.

Instead, we can directly find a set P that maximizes the true accuracy of aligner $\mathcal{A}$ using an *oracle* on P. The true accuracy of aligner $\mathcal{A}$ on a given input S, using an oracle on set P, is simply $\max_{p \in P} F(\mathcal{A}_p(S))$, where again $F(A)$ is the true accuracy of alignment A. We can then find a set P^* that maximizes the average of the above quantity over all inputs S in a collection of benchmarks. We use this criterion, which seeks to maximize the performance of an *oracle* on the parameter set, for our definition of an optimal parameter set P.

We formulate this model as follows. Let U be the universe of parameter choices from which we want to find a parameter set $P \subseteq U$. For a parameter choice $p \in U$ and sequences S, denote the true accuracy of the aligner using p on S as $a_{p,S} := F(\mathcal{A}_p(S))$. Given a bound k on the size of P, we want to find the optimal set $P^* := \mathrm{argmax}_{P \subseteq U} \sum_S \mathrm{argmax}_{p \in P} \, a_{p,S}$, where the summation is over all benchmarks S in a training set. (This is equivalent to maximizing the average true accuracy of the aligner over the training set using the oracle on P.) We call this problem, *Optimal Parameter Set*.

As might be expected, Optimal Parameter Set is NP-complete: it is equivalent to a classic problem from the field of operations research called *Facility Location*. Both problems can be tackled, however, by expressing them as an *integer linear program*. Page limits prevent us from giving the formulation of the integer linear program, but for a universe U of t parameter choices and a training set of n benchmarks, it has $O(tn)$ variables and $O(tn)$ constraints. For a large universe and a large training set, the resulting integer program can get very big.

In our computational experiments, we tackle the integer linear program as follows. First we solve the relaxed *linear program*, where the integer variables are allowed to take on *real values*. We then examine the optimal solution to this linear program, and for all the parameter choices p not chosen in the solution to the relaxed linear program, we *throw out* p from U, forming a new, reduced universe of parameter choices $\widetilde{U}$. Finally we solve the full integer linear program on the reduced universe $\widetilde{U}$. This approach, which we call solving the *reduced integer program*, is essentially a very accurate rounding heuristic for tackling the integer linear program. In our experiments, it is remarkably fast, allowing us to find high-quality solutions for very large instances with a universe of over 500 parameter choices and a training set of over 800 benchmarks.

6   Experimental Results

We evaluate our approach for deriving an accuracy estimator, and the quality of the resulting parameter advisor, through experiments on a collection of benchmark protein multiple sequence alignments. In these experiments, we compare parameter advisors that use our estimator and four estimators from the literature: NorMD [11], MOS [8], HoT [7], and PredSP [1]. (In terms of our earlier categorization of estimators, NorMD and PredSP are scoring-function-based, while MOS and HoT are support-based.) We refer to our estimator by the acronym Facet (short for "**F**eature-based **ac**curacy **est**imator").

In our experiments, for the collection of alignment benchmarks we used the BENCH suite of Edgar [3], which consists of 759 benchmarks, supplemented by a selection of 102 benchmarks from the PALI suite of Balaji et al. [2]. Both BENCH and PALI consist of protein multiple sequence alignments mainly induced by structural alignment of the known three-dimensional structures of the proteins. The entire benchmark collection consists of 861 reference alignments.

For each reference alignment in this benchmark collection, we generated alternate multiple alignments of the sequences in the reference using the multiple alignment tool `Opal` [12,13] with varying parameter choices. Each parameter choice is a vector of four integer values for scoring an alignment using affine gap penalties: a gap-open and gap-extension penalty for internal gaps, and the same for external gaps. These were selected from a universe U of 556 parameter choices [12]. (Set U was formed by finding a single optimal parameter choice through inverse parametric alignment [5], rounding this real-valued parameter choice to integers, and enumerating the Cartesian product of integer choices in the neighborhood of this central choice.) `Opal` also provides a default parameter choice from U that yields the highest average accuracy across the benchmarks.

For the experiments, we measure the *difficulty* of a benchmark S by the true accuracy of the alignment computed by `Opal` on sequences S using its default parameter choice, where the computed alignment is compared to the benchmark's reference alignment on its core columns. Using this measure, we binned the 861 benchmarks by difficulty, where we divided up the full range $[0, 1]$ of accuracies into 10 bins with difficulties $[(i-1)/10, i/10]$ for $i = 1, \ldots, 10$. As is common in benchmark suites, easy benchmarks are highly over-represented compared to hard benchmarks; the number of benchmarks falling in bins $[0, .1]$ through $[.9, 1]$ are respectively: $13, 9, 24, 40, 30, 44, 66, 72, 129, 434$. To correct for this bias in oversampling of easy benchmarks, our approaches for learning an estimator nonuniformly weight the training examples, as described earlier.

To generate training and test sets for our parameter advising experiments on a given set $P \subseteq U$ of parameter choices, we used *three-fold cross validation*. For each bin, we evenly and randomly partitioned the benchmarks in the bin into three groups; we then formed three different splits of the entire set of benchmarks into a training class and a test class, where each split put one group in a bin into the test class and the other two groups in the bin into the training class; finally, for each split we generated a test set and a training set of example alignments by generating $|P|$ alignments from each benchmark S in a training or test class by running `Opal` on S using each parameter choice in P. An estimator learned on the examples in the training set was evaluated on the examples in the associated test set. The results that we report are averages over three runs, where each run is on one of these training and test set pairs. For a set P of 10 parameters, each run has over $5,700$ training examples and $2,800$ test examples.

Selecting Features. Of the features listed in Section 4, not all are equally informative, and some can weaken an estimator. We selected subsets of features for use in accuracy estimators as follows. For each of the twelve features, we first learned an estimator that used that feature alone, and ranked all twelve of these single-feature estimators according to the average accuracy across the difficulty bins of the resulting parameter advisor using that estimator. Then a greedy procedure was used to find a subset of the features, considering them in this order, whose fitted estimator gave an advisor with the highest average accuracy. This procedure was separately run to find good feature subsets for the optimal parameter sets on 5, 10, and 15 parameters. To find a good *overall* feature set that works well for differing numbers of parameters, we examined all subsets of features considered during runs of the greedy procedure, and chose the subset that had the highest accuracy averaged across the 5-, 10-, and 15-parameter sets.

The best overall feature set found by this process is a 5-feature subset consisting of the following features: secondary structure blockiness (BL), secondary structure agreement (SA), secondary structure identity (SI), substitution compatibility (SP), and average substitution score (AS). The corresponding fitted estimator is $(.174) SA + (.172) BL + (.168) SI + (.167) SP + (.152) AS + (.167)$. This 5-feature estimator, when advising using the optimal 10-parameter set, has average accuracy 57.9%. By comparison, the 12-feature estimator has corresponding average accuracy 56.3%. Clearly feature selection helps.

To give a sense of the impact of alternate fitting approaches and higher-degree polynomials, for this 5-feature set, the average accuracy on the 10-parameter set of the linear estimator with difference fitting is 57.9%, and with value fitting is 55.9%; the quadratic estimator with difference fitting is also 57.9%, and with value fitting is 55.8%. In general, we experience only marginal improvement with a quadratic estimator over a linear estimator, even using regularization, though difference fitting does provide a significant improvement over value fitting.

Comparing Estimators to True Accuracy. To examine the fit of an estimator to true accuracy, the scatter plots in Fig. 1 show the value of the Facet, MOS, and PredSP estimators on all example alignments in the 15-parameter test set.

The ideal estimator would be monotonic increasing in true accuracy. Comparing the scatter plots, MOS has the highest slope and spread, PredSP has the lowest slope and spread, while Facet has intermediate slope and spread. This better compromise between slope and spread may be what leads to improved performance for Facet.

Performance on Parameter Advising. To evaluate these methods for parameter advising, we found parameter sets P using our approach that solves the reduced integer linear program, for $|P| = 1, \ldots, 15$. We show results for the resulting parameter advisors that use Facet, MOS, PredSP, HoT, and NorMD. Each advisor is used within Opal to compute an alignment. Advisors are compared by the resulting true accuracy of Opal, first averaged within each bin and then averaged across all bins. We also do the same comparison looking at the rank of the alignment chosen by the advisor, where the alignments generated by the $|P|$ parameter choices are ordered by their true accuracy.

Figure 2 shows the performance of the four best resulting advising methods on the optimal set P of 10 parameters, with respect to accuracy of the alternate alignment chosen by the advisor, averaged over the benchmarks in each difficulty bin. The oracle curve shows what would be achieved by a perfect advisor. The figure also shows the expected performance of a purely *random advisor*, which simply picks a parameter from P uniformly at random. In the figure, note that the advisor that uses Facet is the only one that always performs at or above random in accuracy. The Facet advisor is also strictly better than every other advisor in accuracy on all but one bin, at difficulty $[.6, .7]$. As the figure shows, Facet gives the most help in the hardest bins.

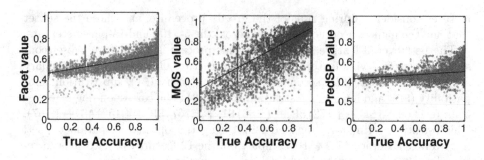

Fig. 1. Correlation of estimators with accuracy. Each scatter plot shows the value of an estimator versus true accuracy for all alignments in the 15-parameter test set.

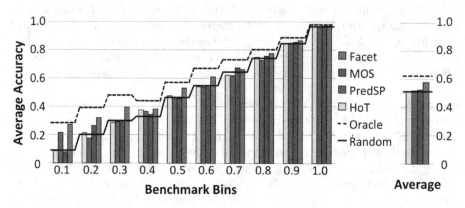

Fig. 2. Advising accuracy across benchmark bins. Each bar shows an estimator's performance in advising, averaged over benchmarks in the bin, for the 10-parameter set.

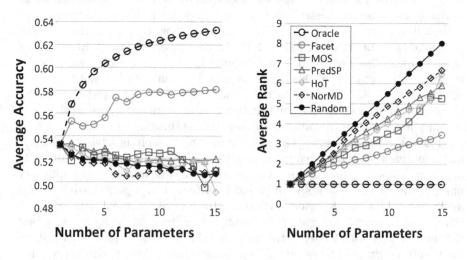

Fig. 3. Advising accuracy and rank for varying numbers of parameters. At each number of parameters, the curves show accuracy or rank averaged over all benchmark bins.

Figure 3 shows the advising performance of the estimators, uniformly averaged across all bins, as a function of the number of parameter choices $k = |P|$ for $k = 1, \ldots, 15$. Notice that with respect to accuracy, the performance of the MOS, PredSP, HoT, and NorMD advisors tends to decline as the number of parameter choices increases (and for some is not far from random). The Facet advisor generally improves with more parameters, though beyond 10 parameters the improvement is negligible. We remark that the performance of the oracle, if its curve is continued indefinitely to the right, reaches a limiting accuracy of 75.3%.

7   Conclusion

We have presented an efficiently-computable estimator for the accuracy of protein multiple alignments. The estimator is a polynomial function of alignment features whose coefficients are learned from example alignments using linear and quadratic programming. We evaluated our estimator in the context of parameter advising, and show it consistently outperforms other approaches to estimating accuracy when used for advising. Compared to using the best default parameter choice, the resulting parameter advisor using a set of 10 parameters provides a 20% improvement in multiple alignment accuracy on the hardest benchmarks.

References

1. Ahola, V., Aittokallio, T., Vihinen, M., Uusipaikka, E.: Model-based prediction of sequence alignment quality. Bioinformatics 24(19), 2165–2171 (2008)
2. Balaji, S., Sujatha, S., Kumar, S.S.C., Srinivasan, N.: PALI: a database of alignments and phylogeny of homologous protein structures. NAR 29(1), 61–65 (2001)
3. Edgar, R.C.: http://www.drive5.com/bench (2009)
4. Jones, D.T.: Protein secondary structure prediction based on position-specific scoring matrices. Journal of Molecular Biology 292, 195–202 (1999)
5. Kim, E., Kececioglu, J.: Learning scoring schemes for sequence alignment from partial examples. IEEE/ACM Trans. Comp. Biol. Bioinf. 5(4), 546–556 (2008)
6. Kim, J., Ma, J.: PSAR: measuring multiple sequence alignment reliability by probabilistic sampling. Nucleic Acids Research 39(15), 6359–6368 (2011)
7. Landan, G., Graur, D.: Heads or tails: a simple reliability check for multiple sequence alignments. Molecular Biology and Evolution 24(6), 1380–1383 (2007)
8. Lassmann, T., Sonnhammer, E.L.L.: Automatic assessment of alignment quality. Nucleic Acids Research 33(22), 7120–7128 (2005)
9. Pei, J., Grishin, N.V.: AL2CO: calculation of positional conservation in a protein sequence alignment. Bioinformatics 17(8), 700–712 (2001)
10. Penn, O., Privman, E., Landan, G., Graur, D., Pupko, T.: An alignment confidence score capturing robustness to guide tree uncertainty. MBE 27(8), 1759–1767 (2010)
11. Thompson, J.D., Plewniak, F., Ripp, R., Thierry, J.C., Poch, O.: Towards a reliable objective function for multiple sequence alignments. JMB 314, 937–951 (2001)
12. Wheeler, T.J., Kececioglu, J.D.: Multiple alignment by aligning alignments. Bioinformatics 23, i559–i568 (2007); Proceedings of the 15th ISMB
13. Wheeler, T.J., Kececioglu, J.D.: Opal: software for aligning multiple biological sequences. Version 2.1.0 (January 2012), http://opal.cs.arizona.edu

Quantifying the Dynamics of Coupled Networks of Switches and Oscillators

Matthew R. Francis[1] and Elana J. Fertig[2,*]

[1] Randolph-Macon College, Ashland, VA, USA
[2] Sidney Kimmel Comprehensive Cancer Center, School of Medicine,
Johns Hopkins University, Baltimore, MD, USA
ejfertig@jhmi.edu

Introduction: The dynamics in systems ranging from intercellular gene regulation to organogenesis are driven by complex interactions (represented as edges) in subcomponents (represented as nodes) in networks. For example, models of coupled switches have been applied to model systems such as neuronal synapses and gene regulatory networks. Similarly, models of coupled oscillators along networks have been used to model synchronization of oscillators which has been observed in synthetic oscillatory fluorescent bacteria, yeast gene regulatory networks, and human cell fate decisions. Moreover, several studies have inferred that biochemical systems contain "network motifs" with both oscillatory and switch-like dynamics. The dynamics of these motifs have been used to model yeast cell cycle regulation and have been further confirmed in synthetic, designed biochemical circuits. Because these heterogeneous network motifs are all identified as components within a single biochemical network, their interactions must drive the global dynamics of the network. Here, we formulate a theory for the network-level dynamics that result from coupling oscillatory and switch-like components have not been studied comprehensively previously.

Theory: Our model of network-wide dynamics from coupled network motifs is based upon introducing cross-coupling between the Kuramoto and Glass models. In Glass networks, active switches will tend to decay unless there are other switches in the network with enduring active states. Similarly, the Kuaramoto model nudges an oscillator's frequency from its natural value to the match the frequency of all oscillators to which it is coupled. In our coupled model, oscillator-switch interactions can feed energy into the switches, nudging switches to turn on when oscillators are in a region of phase space. Analogous to a light switch controlling a ceiling fan, switch-oscillator interactions will induce oscillators to oscillate a their natural frequency when switches are on and freeze when switches are off. By expanding the standard Glass and Kuramoto modeling frameworks, the dynamics of our model can be analyzed with standardized mean-field techniques developed in previously in the nonlinear dynamics literature.

Simulation Results in All-to-All Networks: We observe that the coupled system has four qualitative states in all-to-all networks: (1) oscillators are synchronized and switches are on; (2) switches turn off causing oscillators to freeze; (3) oscillators synchronize and induce concurrent oscillations in switches; and (4)

---

* Corresponding author.

B. Chor (Ed.): RECOMB 2012, LNBI 7262, pp. 60–61, 2012.

oscillators induce transitory oscillations in switches. Each of these states occur stochastically for a given values of parameters controlling the coupling strengths and timing of the oscillators depending on the initial distribution of their states. Parameter values for which switch synchronization would occur in the uncoupled system always induce switch synchronization in the coupled system. However, the probability of obtaining synchronized oscillators or frozen oscillators depends on the relative timing of switch decay and oscillator freezing. System size also affects the distribution of qualitative dynamics. For all parameter values, larger network sizes tend to reduce the probability of observing global, synchronized oscillations in our system.

Results from Modeling Qualitative Dynamics of the Yeast Cell Cycle Derived from Network Motifs: Previously, [1] observed that the dynamics of the fission yeast *Schizosaccharomyces pombe* mitotic cell cycle can be divided into three sequentially interacting modules, triggered by a signal based upon cell size: (1) G1/S transitions with a toggle-switch, (2) S/G2 transitions with a toggle-switch, and (3) G2/M transitions with an oscillator. Although the specific timing differs from [1], we observe similar qualitative dynamics to that observed in [1] when applying our heterogeneous model to evolve the state of these cell cycle stages. We note that the response in this system is consistent with the transitory oscillations observed in all-to-all networks. If we introduce feedback of the G2/M transitions into G1/S, we observe sustained reactivation of the cell cycle regardless of the external signal. These dynamics are analogous to the synchronized dynamics observed in all-to-all networks and consistent with cell growth arising from re-wiring biochemical reactions in cancer cells.

Discussion: Our model of coupled switches and oscillators in all-to-all networks demonstrates that networks with components having heterogeneous dynamics can exhibit synchronization similar to that observed in homogeneous systems. As a result, our model provides a natural mechanism for the coordination of complex machinery such as the initiation of cell-cycle dynamics. However, rewiring the network to introduce feedback will cause the modeled cell cycle machinery to engage continually without regard to the external growth signals, consistent with the malignant rewring in cancer cells. Moroever, in some cases oscillators develop switch-like dynamics and switches oscillatory dynamics. Although these changes in motif dynamics may be only transitory, we hypothesize that dynamics predicted by the structure of subgraphs may be insufficient to describe *in vivo* dynamics if considered in isolation. As a result, we hypothesize that our model of coupled switches and oscillators will provide a natural framework to model the dynamics of gene regulatory networks, neuronal networks, and cardiac systems.

Acknowledgments. This work was sponsored by NCI (CA141053). We thank LV Danilova, B Fertig, AV Favorov, MF Ochs, and E Webster for advice. The full paper is published in MR Francis and EJ Fertig (2012) PLoS One **7**:e29497.

Reference

1. Tyson, J.J., Chen, K., Novak, B.: Nat. Rev. Mol. Cell. Biol. 2, 908 (2001)

lobSTR: A Short Tandem Repeat Profiler
for Personal Genomes

Melissa Gymrek[1,2], David Golan[2], Saharon Rosset[3], and Yaniv Erlich[2,*]

[1] Harvard-MIT Division of Health Sciences and Technology, Cambridge, MA, USA
[2] Whitehead Institute for Biomedical Research, Cambridge, MA 02142, USA
[3] Department of Statistics and Operations Research,
Tel Aviv University, Tel Aviv 69978, Israel

Motivation. Short tandem repeats (STRs), also known as microsatellites, are a class of genetic variations consisting of repetitive elements of 2 to 6 nucleotides that comprise hundreds of thousands of loci in the human genome. The repetitive structure of these loci makes them prone to replication slippage events [5] that can reach a rate of 1/500 mutations per locus per generation [8], 200,000 fold higher than the rate of *de novo* single nucleotide polymorphims (SNPs) [1].

Given their high mutation rate and large allele space, STRs represent a significant source of genetic variation and have been used in a plethora of applications in human genetics including forensics [3], anthropological applications [7], and tracing cancer cell lineages [2]. Additionally, STR expansions are implicated in the etiology of a variety of genetic disorders, such as Huntingon Disease [6] and Fragile-X Syndrome [5].

Despite the plurality of applications, STR variations are not routinely analyzed in whole genome sequencing studies, mainly due to a lack of adequate tools [4]. Here, we present lobSTR, a novel method for profiling STRs in personal genomes.

lobSTR is available at http://jura.wi.mit.edu/erlich/lobSTR/.

lobSTR overcomes STR Profiling Challenges. STRs pose a remarkable challenge to mainstream high throughput sequencing analysis pipelines. First, non-reference STR alleles present as a gapped alignment problem, and mainstream aligners generally exhibit a trade-off between run time and tolerance to insertions/deletions. Second, only reads that fully encompass an STR can be used to determine the exact STR allele present. Third, PCR amplification of an STR locus during library construction can generate stutter noise, confounding the values of true alleles present.

lobSTR addresses these issues in a three step approach. First, during the *sensing step* lobSTR rapidly scans genomic libraries using an *ab initio* procedure to flag reads containing periodic low-complexity regions. It then characterizes their STR sequence and extracts the non-repetitive flanking regions. In the *alignment step*, lobSTR anchors the flanking regions to the genome to reveal the STR position and length. This step avoids a gapped alignment by only aligning flanking regions and is virtually indifferent to the magnitude of STR variation.

---

* Corresponding author.

B. Chor (Ed.): RECOMB 2012, LNBI 7262, pp. 62–63, 2012.

Because flanking regions on both sides of the STR are required for alignment, only fully-spanning, and thus informative, reads are returned. In the *allelotyping step* lobSTR models the stutter noise in order to enhance the signal of the true allelic configuration.

lobSTR Profiling Results. We compared lobSTR to mainstream alignment tools (BWA, Bowtie, Novoalign and BLAT). The speed and reliability of lobSTR exceed the performance of current algorithms for STR profiling. We validated lobSTR's accuracy by measuring deviations from Mendelian inheritance of STR alleles in whole-genome sequencing of a trio and by STR allelotyping using traditional molecular techniques. Encouraged by the speed and accuracy of lobSTR, we used our algorithm to conduct a comprehensive survey of STR variations in a deeply sequenced personal genome. We traced the mutation dynamics of more than 150,000 STR loci and confirmed previous biological hypotheses that were observed in locus-centric STR studies.

As of today, sequence analysis algorithms can detect almost any type of genetic variations, from SNPs and indels, to CNVs and chromosomal translocations. lobSTR adds a new layer of information that has a multitude of applications, including personal genomics, population studies, forensics analysis, and cancer genome profiling.

References

[1] Conrad, D.F., Keebler, J.E., DePristo, M.A., Lindsay, S.J., Zhang, Y., Casals, F., Idaghdour, Y., Hartl, C.L., Torroja, C., Garimella, K.V., Zilversmit, M., Cartwright, R., Rouleau, G.A., Daly, M., Stone, E.A., Hurles, M.E., Awadalla, P.: Variation in genome-wide mutation rates within and between human families. Nat. Genet. 43(7), 712–714 (2011)

[2] Frumkin, D., Wasserstrom, A., Itzkovitz, S., Stern, T., Harmelin, A., Eilam, R., Rechavi, G., Shapiro, E.: Cell lineage analysis of a mouse tumor. Cancer Res. 68(14), 5924–5931 (2008)

[3] Kayser, M., de Knijff, P.: Improving human forensics through advances in genetics, genomics and molecular biology. Nat. Rev. Genet. 12(3), 179–192 (2011)

[4] Langmead, B., Trapnell, C., Pop, M., Salzberg, S.L.: Ultrafast and memory-efficient alignment of short DNA sequences to the human genome. Genome Biol. 10(3), R25 (2009)

[5] Mirkin, S.M.: Expandable DNA repeats and human disease. Nature 447(7147), 932–940 (2007)

[6] Pearson, C.E., Nichol Edamura, K., Cleary, J.D.: Repeat instability: mechanisms of dynamic mutations. Nat. Rev. Genet. 6(10), 729–742 (2005)

[7] Skorecki, K., Selig, S., Blazer, S., Bradman, R., Bradman, N., Waburton, P.J., Ismajlowicz, M., Hammer, M.F.: Y chromosomes of Jewish priests. Nature 385(6611), 32 (1997)

[8] Walsh, B.: Estimating the time to the most recent common ancestor for the Y chromosome or mitochondrial DNA for a pair of individuals. Genetics 158(2), 897–912 (2001)

Hap-seq: An Optimal Algorithm for Haplotype Phasing with Imputation Using Sequencing Data

Dan He, Buhm Han, and Eleazar Eskin

Computer Science Dept., Univ. of California, Los Angeles, CA, 90095-1596, USA
{danhe,buhan,eeskin}@cs.ucla.edu

Abstract. Inference of haplotypes, or the sequence of alleles along each chromosome, is a fundamental problem in genetics and is important for many analyses including admixture mapping, identifying regions of identity by descent and imputation. Traditionally, haplotypes were inferred from genotype data obtained from microarrays utilizing information on population haplotype frequencies inferred from either a large sample of genotyped individuals or a reference dataset such as the HapMap. Since the availability of large reference datasets, modern approaches for haplotype phasing along these lines are closely related to imputation methods. When applied to data obtained from sequencing studies, a straightforward way to obtain haplotypes is to first infer genotypes from the sequence data and then apply an imputation method. However, this approach does not take into account that alleles on the same sequence read originate from the same chromosome. Haplotype assembly approaches take advantage of this insight and predict haplotypes by assigning the reads to chromosomes in such a way that minimizes the number of conflicts between the reads and the predicted haplotypes. Unfortunately, assembly approaches require very high sequencing coverage and are usually not able to fully reconstruct the haplotypes. In this work, we present a novel approach, Hap-seq, which is simultaneously an imputation and assembly method which combines information from a reference dataset with the information from the reads using a likelihood framework. Our method applies a dynamic programming algorithm to identify the predicted haplotype which maximizes the joint likelihood of the haplotype with respect to the reference dataset and the haplotype with respect to the observed reads. We show that our method requires only low sequencing coverage and can reconstruct haplotypes containing both common and rare alleles with higher accuracy compared to the state-of-the-art imputation methods.

Keywords: Haplotype Phasing, Imputation, Dynamic Programming, Hidden Markov Model, Genetic Variation.

1 Introduction

Haplotype, or the sequence of alleles along each chromosome, is the fundamental unit of genetic variation and inference of haplotypes is an important step in

B. Chor (Ed.): RECOMB 2012, LNBI 7262, pp. 64–78, 2012.

many analyses, including admixture mapping [20], identifying regions of identity by descent [10,7,5], imputation of uncollected genetic variation [18,17,15], and association studies [4]. It is possible to use molecular methods for obtaining haplotypes [19], but they are expensive and not amenable to high throughput technologies. For this reason, most studies collect genotype information and infer haplotypes from genotypes, referred to as haplotype inference or haplotype phasing. One strategy for haplotype phasing is to collect mother-father-child-trios and use the Mendelian inheritance patterns. However, this strategy is limited to related individuals.

Over the past two decades, there has been great interest in inference of haplotypes from genotypes of unrelated individuals [8]. Traditionally, these methods were applied to genotypes of a cohort of individuals and utilized the insight that in each region a few haplotypes accounted for the majority of genetic variation. Using this insight, these methods identified the common haplotypes and utilized these common haplotypes to infer the haplotypes for each individual in the cohort [11,9,12,21,6]. More recently, large reference datasets of genetic variation such as the HapMap [14] and 1000 Genomes [1] have become available, enabling a new class of methods, referred to as imputation methods [18,17,15]. These utilize the information in the haplotypes in these reference datasets to both predict haplotypes from genotypes as well as predict the genotypes at uncollected variants. Since these methods utilize the reference datasets, they are able to infer haplotypes from a single individual's genotypes.

Recent advances in high-throughput sequencing have dramatically reduced the cost of obtaining sequencing information from individuals [22]. One advantage of this data compared to microarray genotypes is that this information contains information on both common and rare variants. The initial approaches to inferring haplotypes from sequence data first infers genotype data from the sequencing data and then applies an imputation algorithm [17]. Unfortunately, since variants that are very rare or unique to an individual are not present in the reference datasets, phase information for these variants can not be recovered using this approach.

This commonly applied approach does not take into account an important source of information for haplotype inference. Each sequence read originates from a single chromosome and alleles spanned by that read are on the same haplotype. A set of methods referred to as haplotype assembly methods [16,3,2] have been developed around this idea. Since many reads overlap each other, these methods infer haplotypes by partitioning the reads into two sets corresponding to chromosomal origin in such a way that the number of conflicts between the reads and the predicted haplotypes is minimized. Recently, a dynamic programming method has been proposed which obtains the optimal haplotype prediction from a set of reads [13]. These methods worked well for sequencing studies where the sequencing is high and the reads are long [16], but perform very poorly for studies where the sequencing coverage is low or the reads are short.

In this paper, we propose a new method for haplotype inference from sequencing reads, Hap-seq, which combines information from a reference dataset with the information from the reads. We formulate the problem of haplotype inference using a likelihood framework. We use a hidden Markov model to represent the likelihood model of the predicted haplotypes given the reference datasets, which is similar to imputation methods [18,17,15], and we compute the posterior probability of the predicted haplotypes given the reads averaging over all possible assignments of the reads to chromosomal origin. Since consistency with the reference dataset and the read errors are independent, our joint likelihood is the product of these likelihoods. We present a dynamic programming algorithm for predicting the optimal haplotypes under this joint likelihood. This dynamic programming algorithm nests the dynamic programming algorithm over the reads with a dynamic programming algorithm over the reference haplotypes.

Through simulations generated from the HapMap data [14], we show that our method significantly outperforms traditional imputation methods. We show that using our method, haplotypes can be accurately inferred using 1X coverage assuming a 0.01 error rate. Hap-seq achieved a 2.79% switch error rate which is a 15% improvements over the number of switch errors over the state-of-the-art haplotype imputation algorithm. In addition, our method is the first practical method for predicting haplotypes for rare variants.

2   Likelihood Model

2.1   Overview

Previous methods for haplotype assembly [13] focused on finding the haplotype and the assignment of reads to each haplotype such that the number of conflicts between the reads and the predicted haplotypes is minimized. In our framework, we formulate the same intuition by computing the likelihood of a pair of haplotypes as the probability of the reads given the haplotypes using a simple error model averaged over all possible partitions of the reads into chromosomal origin. Thus, we compute the likelihood with respect to the reads

$$likelihood_{\text{reads}} = P(\text{reads}|hap_1, hap_2)$$

where hap_1 and hap_2 are the (unknown) haplotypes of an individual.

We also take into account information from the reference dataset. We compute the probability of a pair of haplotypes where each haplotype is represented as a mosaic of reference haplotypes which is the standard HMM model for imputation. Thus, we compute the likelihood with respect to the reference dataset, or the *imputation likelihood,*

$$likelihood_{\text{imputation}} = P(hap_1, hap_2|\text{reference})$$

Figure 1 shows the two types of data, sequencing reads and reference dataset.

Since these two types of data are independent, we represent the likelihood of a haplotype given the reads and the reference dataset by the joint likelihood which is the product of the two likelihoods

$$L(hap_1, hap_2) = likelihood_{\text{reads}} \times likelihood_{\text{imputation}} \propto P(hap1, hap2|\text{reads}, \text{reference})$$

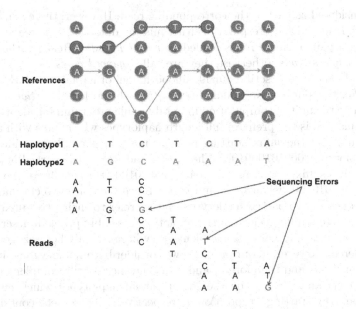

Fig. 1. An illustration of reconstructing the pair of haplotypes as well as the imputation paths for each haplotype given a set of reference sequences and a set of sequencing reads which contains errors

Our goal is to reconstruct a pair of haplotypes hap_1, hap_2 such that this likelihood is maximized. We call this objective function MIR (**M**ost likely **I**mputation based on **R**eads).

2.2   Likelihood with Respect to Sequence Reads

We will follow the notation by He et al. [13] for the haplotype assembly problem. Given a reference genome sequence and the set of reads containing sequence from both chromosomes, we align all the reads to the reference genome. However, unlike the haplotype assembly problem, where the homozygous sites (columns in the alignment with identical values) are discarded, we need to maintain the homozygous sites that are polymorphic in the reference as well as the heterozygous sites (columns in the alignment with different values). The sites are labelled as 0 or 1 arbitrarily.

A matrix X of size $m \times n$ can be built from the alignment, where m is the number of reads and n is the number of SNPs (defined as the total number of positions which are either polymorphic sites in the reference and/or heterozygous in the sample). The i-th read is described as a ternary string $X_i \in \{0, 1, -\}^n$, where '$-$' indicates a gap, namely that the allele is not covered by the fragment (again following the notation of [13] for clarity). The *start position* and *end position* of a read are the first and last positions in the corresponding row that are not '$-$', respectively. Therefore the '$-$'s in the head and tail of each row will not be considered as part of the corresponding read. However, there can be '−'s inside each read which correspond to either missing data for single reads or gaps connecting a pair of single reads (called *paired end reads*). Reads without "−" are called *gapless reads*; otherwise they are called *gapped reads*.

The goal here is to describe our likelihood model of haplotypes with respect to reads. For this purpose, we should first describe the notion of *partial haplotype* which will be the unit of computation in our dynamic programming algorithm. A partial haplotype is the prefix of full length haplotypes which ends with a suffix r of length k and the suffix starting position is i. For example, consider a full length haplotype "00010100010". When $i = 4$ and $k = 4$, the partial haplotype "0001010" has suffix $r =$ "1010". Similarly, $r =$ "0100" is the suffix of the partial haplotype "00010100" when $i = 5$ and $k = 4$. Since there are two chromosomes, we must consider two partial haplotypes and we refer to their two suffixes as r_1 and r_2. We note a difference from the haplotype assembly problem described in [13] where r_1 and r_2 are always complementary because only heterozygous sites are considered. However, in our problem, we consider both homozygous sites and heterozygous sites and therefore r_1 and r_2 are not necessarily complementary.

Let $H(i, r_1)$ and $H(i, r_2)$ be the set of partial haplotypes which end with suffix r_1 and r_2 starting at position i, respectively. These sets contain 2^{i-1} haplotypes. We define $R(i, r_1, r_2)$ as the likelihood of the reads with starting position no greater than i that are generated from the pair of partial haplotypes $h^1 \in H(i, r_1)$ and $h^2 \in H(i, r_2)$ which maximizes the likelihood of the reads. The key idea behind our approach is that we will use dynamic programming to compute this quantity for larger and larger values of i eventually allowing us to identify complete haplotypes with the highest likelihood.

Since reads are independent, we can decompose our computation of $R(i, r_1, r_2)$ into the likelihood with respect to the reads starting at each position. Let $R'(i, r_1, r_2)$ be the likelihood only for the reads starting at position i. For every position, we assume the reads span at most k sites, thus all partial haplotypes with the same suffixes r_1, r_2 staring at position i ($h^1 \in H(i, r_1)$ and $h^2 \in H(i, r_2)$) will have the same value for $R'(i, r_1, r_2)$. Each read can originate from only one of the chromosomes corresponding to either r_1 or r_2. Since we do not know the origin of the reads, we consider all possible partitions of the reads. In each partition, a read is assigned to either r_1 or r_2 but not both. We can compute the number of mismatches between the reads and r_1 and r_2 and compute a likelihood given a partition using the following:

$$R'(l, i, r_1, r_2) = e^{E_l(i, r_1, r_2)}(1 - e)^{K(i) - E_l(i, r_1, r_2)}$$

where $R'(l, i, r_1, r_2)$ is the likelihood corresponding to the l-th partition and $E_l(i, r_1, r_2)$ is the number of mismatches for the l-th partition, $K(i)$ is the total count of the alleles of all reads starting at position i, e is the sequencing error rate. $R'(i, r_1, r_2)$ is sum of the likelihoods of all the partitions weighted by the probability of each partition, namely

$$R'(i, r_1, r_2) = \sum_{l=1}^{2^{a_i}} R'(l, i, r_1, r_2) P(\text{partition } l) = \sum_{l=1}^{2^{a_i}} R'(l, i, r_1, r_2)/2^{a_i} \quad (1)$$

assuming there are totally a_i reads starting at position k, 2^{a_i} is the total number of possible partitions and the probability of each partition is equal to $\frac{1}{2^{a_i}}$.

Consider the two complete haplotypes $h^1_{\max} \in H(i, r_1)$ and $h^2_{\max} \in H(i, r_2)$ that maximize the likelihood of the reads and by definition, their likelihood is $R(i, r_1, r_2)$. For these haplotypes, R and R' have the following relationship:

$$R(i, r_1, r_2) = \prod_{f=1}^{i} R'(f, r'_1, r'_2) \qquad (2)$$

where r'_1 and r'_2 are length k suffixes of the partial haplotypes of $h^1_{\max}$ and $h^2_{\max}$ with suffix starting position f.

R' is computed for every possible suffix at each position of the dynamic programming. The dynamic programing uses the value of $R'(i, r_1, r_2)$ and the four values of $R(i - 1, x, y)$ where the suffix of x is a prefix of r_1 and the suffix of y is a prefix of r_2 to compute $R(i, r_1, r_2)$ (see [13] for details). For high sequencing coverage, we can compute just the likelihood of the most likely partition which approximates Equation (1).

2.3   Likelihood with Respect to Reference Dataset

We utilize an HMM model which is the basis of the widely used imputation methods [18,17,15]. Given a binary string r, in the HMM, for each SNP at position i, we consider the HMM state with value $S_{i,j}$ corresponding to the j-th reference haplotype h_j, where $1 \leq j \leq m$. We define the transition probability from the state $S_{i,j}$ to the state $S_{i+1,y}$ as $t_{(i,j),(i+1,y)}$ where $1 \leq y \leq m$. We assume that $t_{(i,j),(i+1,y)}$ is the same for all $y \neq j$. (Once the transition occurs, the transition probabilities to all possible states are equal.) We also define the emission probability from the state $S_{i,j}$ to the observed i-th SNP $r[i]$ as $\mu_{i,j}$. The emission probability models the mutations and we assume that the mutation rate for all the SNPs are the same. The emission probability $\mu_{i,j}$ is defined as the following:

$$\mu_{i,j} = \begin{cases} 1 - \mu & h_j[i] = r[i] \\ \mu & \text{otherwise} \end{cases}$$

where μ is the mutation rate.

In this work, we assume we know the mutation rate and the transition probabilities. An example of the HMM is shown in Figure 2.

Since we assume that we know μ as well as all $t_{(i,j),(i+1,y)}$'s, the most likely state sequence can be obtained by the Viterbi algorithm. We can also use the HMM to compute the likelihood of a partial haplotype. We define r as a length k binary string and $l(i, r, j)$ as the imputation likelihood of the partial haplotypes starting at position 0 and ending at position $i + k - 1$, with suffix r, and whose imputation at position $i + k - 1$ (the last position) is from the j-th reference sequence. $l(i, r, j)$ is the maximum value among many possible partial haplotypes having the same suffix r and the same ending HMM state $S_{i,j}$. Let $h_{l(i,r,j)}$ be the partial haplotype that maximized $l(i, r, j)$ which can usually be traced back in the algorithms.

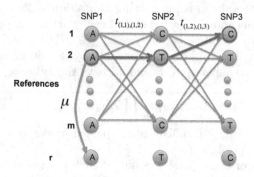

Fig. 2. Example to illustrate the HMM (Hidden Markov Model) model. There are totally m reference sequences and 3 SNP locus. The given string r is "ATC" and the optimal imputation path is highlighted as red.

3   Hap-seq

A naive algorithm to optimize MIR is to enumerate all possible pair of haplotypes and compute the likelihood, which requires time complexity $O(4^n)$, where n is the length of the haplotypes. Then for each pair, identify the number of mismatches between the reads and the haplotypes to compute the likelihood of the reads. After that, run the HMM to compute the likelihood of the imputation for both haplotypes. The complexity of this naive algorithm is obviously prohibitively large even for small n.

As the haplotype assembly problem can be optimally solved with a dynamic programming algorithm [13] and the imputation problem can be optimally solved using an HMM, we next propose a hybrid method combining a dynamic programming algorithm and an HMM and we name the method **Hap-seq**. The basic idea is to use a dynamic programming algorithm operating on partial haplotypes with suffixes r_1, r_2, which are the prefixes of full length haplotypes similar to our algorithm for reconstructing haplotypes which maximizes the likelihood of the reads. However, for each suffix, we also introduce a state encoding the current reference haplotype. Similar to the approach above, at each position we compute the likelihood of reads starting at the position. We also use an HMM to compute the likelihood of imputation for the partial haplotypes using the information on the reference state. For each position i, we store the MIRs corresponding to different partial haplotype suffixes and imputation states in the dynamic programming matrix. Then we compute the MIR values for position $i + 1$ using the previous values, and repeat this process until we obtain the full length haplotypes. To compute the MIRs for position $i + 1$ when extending the partial haplotypes, the MIR for the partial haplotypes from the previous position (i) is used in addition to the likelihood of reads for the new partial haplotypes (reads starting at position $i + 1$) and an HMM is used to compute the imputation likelihood for the extended partial haplotypes. The new MIR for the partial haplotypes is then the product of the three values (likelihood from previous step, imputation likelihood, and likelihood with respect to reads). We next discuss our Hap-seq algorithm in more details.

We define $MIR(i, r_1, r_2, j_1, j_2)$ as the value of the MIR for the set of reads with starting positions no greater than i and the pair of partial haplotypes with suffix r_1, r_2, respectively, where the current references in the imputation model at position $i + k - 1$ are the j_1-th and j_2-th reference sequences, respectively. We also define $MIR_{\max}(i, r_1, r_2)$ as the maximum MIR for $1 \leq j_1, j_2 \leq m$ given m haplotypes in the reference dataset.

We build a dynamic programming matrix and at each position i we store $MIR(i, r_1, r_2, j_1, j_2)$ for all r_1, r_2 and $1 \leq j_1, j_2 \leq m$, respectively. For each pair of r_1, r_2, we call the HMM to compute the likelihood of the imputation $l(i, r_1, j_1)$ and $l(i, r_2, j_2)$. MIR at position i can be computed as

$$MIR(i, r_1, r_2, j_1, j_2) = R(i, r_1, r_2) \times l(i, r_1, j_1) \times l(i, r_2, j_2)$$
$$\text{where } h^1_{\max} = h_{l(i, r_1, j_1)} \text{ and } h^2_{\max} = h_{l(i, r_2, j_2)}$$
$$MIR_{\max}(i, r_1, r_2) = \text{argmax}_{j_1, j_2} MIR(i, r_1, r_2, j_1, j_2) \text{ for } 1 \leq j_1, j_2 \leq m$$

The best MIR is the maximum $MIR_{\max}(n - k, r_1, r_2)$ over all r_1, r_2 where n is the full length of the haplotypes.

The objective function contains terms $R(i, r_1, r_2)$, $l(i, r_1, j_1)$, and $l(i, r_2, j_2)$, each of which is a maximum likelihood value maximized by finding the corresponding haplotypes $(h^1_{\max}, h^2_{\max})$, $h_{l(i, r_1, j_1)}$, and $h_{l(i, r_2, j_2)}$ respectively. It should be noted that in our definition, we constrain the problem using the condition $h^1_{\max} = h_{l(i, r_1, j_1)}$ and $h^2_{\max} = h_{l(i, r_2, j_2)}$, which forces the same haplotypes in each of the terms $R(i, r_1, r_2)$, $l(i, r_1, j_1)$, and $l(i, r_2, j_2)$. Thus we are searching for a single pair of haplotypes $(h^1_{\max} = h_{l(i, r_1, j_1)}, h^2_{\max} = h_{l(i, r_2, j_2)})$ which maximize the product of these three terms.

We initialize $MIR(0, r_1, r_2, j_1, j_2)$ by considering the reads starting at position 0. Given r_1, r_2, the HMM is called to compute the optimal likelihood $l(0, r_1, j_1), l(0, r_2, j_2)$ for the imputation of r_1, r_2, respectively. Then $MIR(0, r_1, r_2, j_1, j_2) = R(0, r_1, r_2) \times l(0, r_1, j_1) \times l(0, r_2, j_2)$ for $1 \leq j_1, j_2 \leq m$.

3.1   Transition Probability

The partial haplotypes at position i can be obtained by extending the partial haplotypes at position $i - 1$ with either a 0 or 1, as we consider homozygous sites and heterozygous sites. At position i, we again enumerate all length k binary strings for r_1 and r_2. Given a binary string r, $l(i, r, j)$ is obtained by calling an HMM on the j-th reference haplotype to find the likelihood for the imputation for the suffix r ending at position $i + k - 1$. In the HMM, since we need to consider both haplotypes at the same time, we build a state for each SNP which contains m^2 different values as $S_{(r_1, r_2), (i+k, j_1, j_2)}$ at position $i + k$ of the j_1-th and j_2-th reference haplotype h_{j_1}, h_{j_2}, where $1 \leq j_1, j_2 \leq m$ for all pairs of partial haplotypes with suffix r_1 and r_2, respectively. We define the transition probability from the state with value $S_{(r'_1, r'_2), (i+k-1, j_1, j_2)}$ to the state with value $S_{(r_1, r_2), (i+k, f_1, f_2)}$, where $1 \leq f_1, f_2, j_1, j_2 \leq m$ as $t_{(i+k-1, j_1, j_2), (i+k, f_1, f_2)}$.

As the assignment of the two partial haplotypes to the reference haplotypes are independent, we obtain the following formula:

$$t_{(i+k-1,j_1,j_2),(i+k,f_1,f_2)} = t_{(i+k-1,j_1),(i+k,f_1)} \times t_{(i+k-1,j_2),(i+k,f_2)}$$

where $t_{(i+k-1,j_1),(i+k,f_1)}$ and $t_{(i+k-1,j_2),(i+k,f_2)}$ are the standard transition probability used widely in the imputation methods. And we assume $t_{(i+k-1,j),(i+k,f)}$ is the same for all $1 \le f, j \le m$, $f \ne j$ for all possible r_1's and r_2's. $r_1' = (b, r_1[0, k-2])$ where b is 0 or 1. The brackets $[x, y]$ denote a substring starting from x and ending at y, both ends inclusive, where the index count starts from 0. A tuple of two strings denotes concatenation. Similarly, $r_2' = (b, r_2[0, k-2])$ where b is 0 or 1. For example, for $k=4$, $r_1 =$ "0000", $r_1' =$ "1000", we then have $r_1' = (1, r_1[0, 2])$.

3.2   Emission Probability

We also define the emission probability from the state with value $S_{(r_1,r_2),(i+k-1,j_1,j_2)}$ to the k-th observed SNPs in r_1, r_2, $r_1[k-1], r_2[k-1]$, as $\mu_{(r_1,r_2),(i+k-1,j_1,j_2)}$. Since the emissions from the two partial haplotypes are independent, we have the following formula:

$$\mu_{(r_1,r_2),(i+k-1,j_1,j_2)} = \mu_{r_1,(i+k-1,j_1)} \times \mu_{r_2,(i+k-1,j_2)}$$

The emission probability is an alternative of the mutation rate and we assume the mutation rate for all the SNPs are the same. The emission probability $\mu_{r,(i+k-1,j)}$ is defined as the following:

$$\mu_{r,(i+k-1,j)} = \begin{cases} 1 - \mu & h_j[k-1] = r[k-1] \\ \mu & \text{otherwise} \end{cases}$$

where μ is the mutation rate which is the same for all the SNPs. Therefore if the two SNPs are identical, the emission probability is 1, otherwise it is μ. We assume we know the mutation rate and the transition probabilities.

3.3   Recursions

Given the transition and emission probabilities, the MIR at position i for partial haplotypes with suffix r_1, r_2 is computed as the following:

$$MIR(i, r_1, r_2, j_1, j_2) = \text{argmax}_{b_1,b_2,y_1,y_2} \left(\mu_{r_1,(i+k-1,y_1)} \times \mu_{r_2,(i+k-1,y_2)} \right.$$
$$\times t_{(i+k-2,y_1),(i+k-1,j_1)} \times t_{(i+k-2,y_2),(i+k-1,j_2)}$$
$$\left. \times MIR\left(i-1, (b_1, r_1[0, k-2]), (b_2, r_2[0, k-2]), y_1, y_2\right) \times R'(i, r_1, r_2)\right)$$
$$\text{for } b_1 \in \{0,1\}, b_2 \in \{0,1\}, 1 \le y_1, y_2 \le m, 1 \le j_1, j_2 \le m$$

which allows us to compute $MIR_{\max}(i, r_1, r_2)$ for each i for every r_1, r_2.

3.4   Complexity

The complexity of the Hap-seq algorithm is $O(n \times m^4 \times 4^k)$ where 4^k is from the enumeration of all pairs of length-k binary strings and m^4 is from the HMM model on the m reference sequences, which is known to be reduced to m^2 by precomputing and saving the transition probabilities [17]. Therefore, the complexity of Hap-seq is $O(n \times m^2 \times 4^k)$ and typically m for imputation algorithms is around 100, for which Hap-seq is computationally feasible.

3.5   Read Length

At each position, we enumerate all possible binary strings of length k, where k is the maximum number of SNPs the reads contain. Since usually speaking the SNPs are far from each other, k is small. In our experiments, 99% of the reads contains no greater than 3 SNPs. Therefore we set a threshold as $k = 3$. For reads containing more than 3 SNPs, we simply split them. For example, for a read "0001000" starting at position 0, we split it as "000" starting at position 0, "100" starting at position 3 and "0" starting at position 6. We show later in our experiments with $k = 3$ Hap-seq can finish quickly.

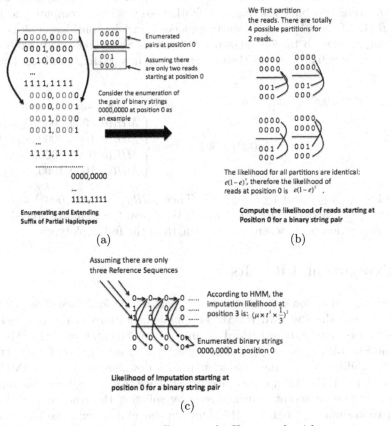

Fig. 3. Example to illustrate the Hap-seq algorithm

3.6   Illustrative Example

For illustration purpose, we show a running example of our Hap-seq algorithm in Figure 3. In Figure 3 (a), we show when $k = 4$, how we enumerate all length-4 binary strings at each position as suffixes of partial haplotypes and extend the suffix by one bit each time till the end of the haplotypes. As shown in the red color, the suffix of a binary string is identical to the prefix of the binary strings extended from it. In Figure 3 (b) and (c), we consider a pair of enumerated length-4 strings (0000, 0000) at position 0 as an example. We assume there are only two reads (001, 000) starting at position 0. In (b), we compute the likelihood of these two reads by considering all 4 possible partitions. The likelihood is computed according to Equation 1. In (c), we assume there are only three reference sequences. We apply HMM to compute the imputation likelihood for the pair of length-4 binary strings (0000,0000) at position 0. It is obvious that the best imputation path for both binary strings is through the first reference sequence without transition to other reference sequences. Then $MIR(0, 0000, 0000) = MIR(0, 0000, 0000, 1, 1)$, which is simply the product of the two likelihoods: $e(1 - e)^5 \times (\mu \times t^3 \times \frac{1}{3})^2$.

Next at position 1 (we do not show this step in the above figure), assume we enumerate two binary strings as (0001, 0001), we can compute the likelihood $R'(1, 0001, 0001)$ of reads starting at position 1 by considering all possible partitions similar to the one shown in Figure 3. Since $0000 = (0, 0001[0,2])$, $1000 = (1, 0001[0,2])$, we can then compute the new MIR at position 1 as the following:

$$MIR(1, 0001, 0001, j_1, j_2) = \mu_{1,(4,y_1)} \times \mu_{1,(4,y_2)} \times t_{(3,y_1),(4,j_1)} \times t_{(3,y_2),(4,j_2)}$$

$$\times R'(1, 0001, 0001) \times \max \begin{cases} MIR(0, 0000, 0000, y_1, y_2) \\ MIR(0, 1000, 0000, y_1, y_2) \\ MIR(0, 0000, 1000, y_1, y_2) \\ MIR(0, 1000, 1000, y_1, y_2) \end{cases}$$

where $1 \le j_1, j_2 \le 3$ and $1 \le y_1, y_2 \le 3$. Then $MIR_{\max}(1, 0001, 0001) = \mathrm{argmax}_{1 \le j_1, j_2 \le 3} MIR(1, 0001, 0001, j_1, j_2)$. We repeat the recursive procedure till we reach position $n - 4$ where n is the length of the full haplotypes.

4   Experimental Results

We perform simulation experiments to compare the performance of our Hap-seq algorithm and the standard HMM. The implementations of the two methods are as follows. The standard HMM is our own implementation of the IMPUTE v1.0 model which uses the pre-defined genetic map information for the transition probability. We use the genetic map data downloaded from the IMPUTE website. Since IMPUTE does not accept sequence read data, we allow our own implementation to accept sequence read by splitting the reads into SNP-wise information similarly to MACH [17]. Our implementation runs the Viterbi algorithm and gives the most likely haplotype phasing based on the IMPUTE model.

To make a fair comparison, for the part of Hap-seq that computes the imputation likelihood, we use the same IMPUTE v1.0 model using the same genetic map data.

We use 60 parental individuals of CEU population of HapMap Phase II data downloaded from the HapMap website. We perform a leave-one-out experiment to measure the accuracy of our method. We choose one target individual, generate simulated sequence reads, and infer the genotypes and phasing of the target individual using the 59 individuals as the reference data set. We repeat this for all 60 individuals and the results are averaged. When we generate the data, we use the per-allele sequencing error rate of 1%. We assume that both IMPUTE and Hap-seq know the true error rate. In all datasets we generate, we assume a low coverage of 1x. That is, all heterozygous sites will be covered by two reads on average, since there are two chromosomes.

The first dataset we generate is the short region of 1,000 SNP sites at the beginning of chromosome 22. This region is approximately of length 1.8 Mb. We make an unrealistic assumption that each read covers a fixed number (F) of SNPs. Although this assumption is unrealistic, in this setting the performance gain of our Hap-seq over IMPUTE will be the most prominent because every read will contain phasing information when $F > 1$. We generate data for $F = 2$, $F = 3$ and $F = 4$. The results of switch error rate, which is the proportion of consecutive heterozygote genotypes that are incorrectly phased, for both algorithms are shown in the left of Table 1. As we can see, Hap-seq makes significant improvements over IMPUTE with respect to the number of switch errors. As F increases, the improvement becomes more significant. This is because as F increases, splitting reads loses more information that the SNPs in the same reads are from the same haplotype. Therefore Hap-seq has more advantages as the reads gets longer. One thing to note is that the accuracy of both methods drops at $F = 3$ compared to $F = 2$. This is because when F increases, the total number of reads decreases that can result in some regions not effectively covered by the reads. This raises an interesting question of what are the optimal read length and coverage that maximize the imputation and phasing accuracy given a cost, which requires further investigations.

Table 1. Averaged switch error rate and the improvement of Hap-seq over IMPUTE

F	IMPUTE	Hap-seq	Improvement
2	6.7%	6.2%	7.4%
3	5.4%	4.9%	8.2%
4	7%	5.7%	12.4%

W	IMPUTE	Hap-seq	improvement
100	5.36%	5.2%	2.8%
200	6.4%	6.25%	2.5%
500	7.2%	6.9%	3.5%
1000	7.8%	7.3%	6.5%

The second dataset we generate is the same 1,000 SNP region where we generate a more realistic simulated reads that has a constant size (W) in bp. We vary W between 100, 200, 500, and 1,000 bp. Thus the number of SNPs contained in the reads is not fixed and can vary. However, as the SNPs are far from each other, over 99% of the reads contain no greater than three SNPs. The results of switch error rate for both algorithms are shown in the right of Table 1. We again observe over 2.5% improvement of Hap-seq over IMPUTE. The improvement is less significant compared with the last data set. This is because we do not fix the number of SNPs in the reads and thus there are many reads containing just one SNP. Hap-seq does not have any advantages over IMPUTE for these reads. We again observe a general increase in the improvement ratio when the read length increases.

Finally, we generate the paired-end read data of the whole chromosome 22, which contains 35,412 SNPs. We randomly select one individual and generate the paired-end sequence reads of size 1,000 bp in each end. The gap size is assumed to follow a normal distribution of mean 1,000 bp and the standard deviation of 100 bp. Therefore the number of SNPs contained in the reads can vary. We use 1X coverage and around 99% of the reads contain no greater than 3 SNPs, out of which 74% of the reads contain a single SNP. In order to parallelize our algorithm, we conduct a simple strategy. We split the chromosome 22 into chunks containing 1,000 SNPs each and run Hap-seq on each chunk in parallel only using reads in the chunk. In order to concatenate the reconstructed haplotypes from adjacent chunks seamlessly, we overlap adjacent chunks with a buffer region containing 200 SNPs. Therefore, the first chunk covers locus [0,1000], the second chunk covers locus [800, 2000], the third chunk covers locus [1800, 3000] and so on. Our intuition is that the haplotypes from adjacent chunks in the overlapped buffer region should be highly consistent with each other. Therefore, when we concatenate them, we select the best option which is able to minimize the differences between the haplotypes from adjacent chunks in the buffer region.

The program finished in 9 hours on a cluster processing the whole chromosome 22. We first compare the switch errors of Hap-seq and IMPUTE for each chunk containing 1,200 SNPs (the first and the last chunks contain different numbers of SNPs). As we can see in the left of Figure 4, it is obvious that for every chunk Hap-seq has lower switch error rate. For some chunks the improvement with respect to the number of switch errors is reduced by almost 80%. We also show the number of mismatches in the overlapping buffer regions for the haplotypes reconstructed by Hap-seq and IMPUTE in the right of Figure 4. We can see although there are certain regions with relatively high inconsistencies, most of the buffer regions are highly consistent with mismatches close to 0. We then concatenate the haplotypes from all the chunks. After the concatenation, we obtain a 2.79% overall error rate for Hap-seq on the whole chromosome 22. Compared to the 3.28% overall error rate for IMPUTE, this is a 15% improvement with respect to the number of switch errors. This again indicates that our Hap-seq algorithm is more effective in reconstructing the haplotypes using sequencing reads compared to IMPUTE.

5   Discussion

We have presented Hap-seq, a method for inferring haplotypes from sequencing data that takes into account both a population reference dataset as well as the information of alleles spanned by sequencing reads. Our method uses a

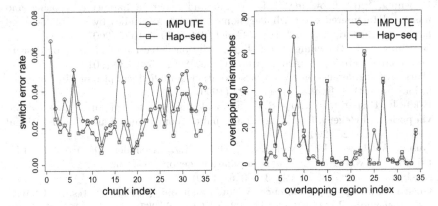

Fig. 4. (Left) The switch error rate when using IMPUTE and Hap-seq for each chunk of length 1,200 SNPs for whole chromosome 22. (Right) The number of mismatches in the overlapping buffer regions for the haplotypes reconstructed by Hap-seq and IMPUTE.

dynamic programming approach to obtain the optimal haplotype prediction under a likelihood formulation of the haplotype inference problem.

Our approach is the first practical approach that can perform haplotype inference for both common and rare variants. The current imputation based approaches that obtain genotype information from the sequencing and then apply imputation are ineffective for phasing rare variants. The reason is that reference datasets only contain common variants and since these are the only variants on the reference haplotypes, the variants which are unique to an individual can not be phased. Since our approach also utilizes read information, if a read is present in the data that spans a rare variant and a common variant, our method can recover the phase for the rare variant. To our knowledge, this is the first practical approach to phase rare variants.

Acknowledgments. D.H., B.H. and E.E. are supported by National Science Foundation grants 0513612, 0731455, 0729049, 0916676 and 1065276, and National Institutes of Health grants K25-HL080079, U01-DA024417, P01-HL30568 and PO1-HL28481. B.H. is supported by the Samsung Scholarship.

References

1. A deep catalog of human genetic variation (2010), http://www.1000genomes.org/
2. Bansal, V., Bafna, V.: HapCUT: an efficient and accurate algorithm for the haplotype assembly problem. Bioinformatics 24(16), i153 (2008)
3. Bansal, V., Halpern, A.L., Axelrod, N., Bafna, V.: An MCMC algorithm for haplotype assembly from whole-genome sequence data. Genome Research 18(8), 1336 (2008)
4. Beckmann, L.: Haplotype Sharing Methods. In: Encyclopedia of Life Sciences (ELS). John Wiley & Sons, Ltd., Chichester (2010)
5. Browning, B.L., Browning, S.R.: A fast, powerful method for detecting identity by descent. Am. J. Hum. Genet. 88(2), 173–182 (2011)

6. Browning, S.R., Browning, B.L.: Rapid and accurate haplotype phasing and missing-data inference for whole-genome association studies by use of localized haplotype clustering. Am. J. Hum. Genet. 81(5), 1084–1097 (2007)
7. Browning, S.R., Browning, B.L.: High-resolution detection of identity by descent in unrelated individuals. Am. J. Hum. Genet. 86(4), 526–539 (2010)
8. Clark, A.G.: Inference of haplotypes from pcr-amplified samples of diploid populations. Mol. Biol. Evol. 7(2), 111–122 (1990)
9. Eskin, E., Halperin, E., Karp, R.M.: Efficient reconstruction of haplotype structure via perfect phylogeny. International Journal of Bioinformatics and Computational Biology 1(1), 1–20 (2003)
10. Gusev, A., Lowe, J.K., Stoffel, M., Daly, M.J., Altshuler, D., Breslow, J.L., Friedman, J.M., Pe'er, I.: Whole population, genome-wide mapping of hidden relatedness. Genome Res. 19(2), 318–326 (2009)
11. Gusfield, D.: Haplotype Inference by Pure Parsimony. In: Baeza-Yates, R., Chávez, E., Crochemore, M. (eds.) CPM 2003. LNCS, vol. 2676, pp. 144–155. Springer, Heidelberg (2003)
12. Halperin, E., Eskin, E.: Haplotype reconstruction from genotype data using imperfect phylogeny. Bioinformatics 20(12), 1842–1849 (2004)
13. He, D., Choi, A., Pipatsrisawat, K., Darwiche, A., Eskin, E.: Optimal algorithms for haplotype assembly from whole-genome sequence data. Bioinformatics 26(12), i183 (2010)
14. International HapMap Consortium: A second generation human haplotype map of over 3.1 million snps. Nature 449(7164), 851–861 (2007)
15. Kang, H.M., Zaitlen, N.A., Eskin, E.: Eminim: An adaptive and memory-efficient algorithm for genotype imputation. Journal of Computational Biology 17(3), 547–560 (2010)
16. Levy, S., Sutton, G., Ng, P.C., Feuk, L., Halpern, A.L., Walenz, B.P., Axelrod, N., Huang, J., Kirkness, E.F., Denisov, G., et al.: The diploid genome sequence of an individual human. PLoS Biol. 5(10), e254 (2007)
17. Li, Y., Willer, C.J., Ding, J., Scheet, P., Abecasis, G.R.: Mach: using sequence and genotype data to estimate haplotypes and unobserved genotypes. Genet Epidemiol 34(8), 816–834 (2010)
18. Marchini, J., Howie, B., Myers, S., McVean, G., Donnelly, P.: A new multipoint method for genome-wide association studies by imputation of genotypes. Nature Genetics 39(7), 906–913 (2007)
19. Patil, N., Berno, A.J., Hinds, D.A., Barrett, W.A., Doshi, J.M., Hacker, C.R., Kautzer, C.R., Lee, D.H., Marjoribanks, C., McDonough, D.P., et al.: Blocks of limited haplotype diversity revealed by high-resolution scanning of human chromosome 21. Science 294(5547), 1719 (2001)
20. Patterson, N., Hattangadi, N., Lane, B., Lohmueller, K.E., Hafler, D.A., Oksenberg, J.R., Hauser, S.L., Smith, M.W., O'Brien, S.J., Altshuler, D., et al.: Methods for high-density admixture mapping of disease genes. The American Journal of Human Genetics 74(5), 979–1000 (2004)
21. Stephens, M., Smith, N.J., Donnelly, P.: A new statistical method for haplotype reconstruction from population data. The American Journal of Human Genetics 68(4), 978–989 (2001)
22. Wheeler, D.A., Srinivasan, M., Egholm, M., Shen, Y., Chen, L., McGuire, A., He, W., Chen, Y.J., Makhijani, V., Roth, G.T., et al.: The complete genome of an individual by massively parallel DNA sequencing. Nature 452(7189), 872–876 (2008)

BALLAST:
A Ball-Based Algorithm for Structural Motifs

Lu He[1], Fabio Vandin[2], Gopal Pandurangan[3], and Chris Bailey-Kellogg[1]

[1] Department of Computer Science, Dartmouth College,
6211 Sudikoff Laboratory, Hanover, NH 03755, USA
{luhe,cbk}@cs.dartmouth.edu
[2] Department of Computer Science and Center for Computational Molecular Biology,
Brown University, Providence, RI 02912, USA
vandinfa@cs.brown.edu
[3] Division of Mathematical Sciences, Nanyang Technological University,
Singapore 637371
and Department of Computer Science, Brown University, Providence, RI 02912, USA
gopal@ntu.edu.sg

Abstract. Structural motifs encapsulate *local* sequence-structure-function relationships characteristic of related proteins, enabling the prediction of functional characteristics of new proteins, providing molecular-level insights into how those functions are performed, and supporting the development of variants specifically maintaining or perturbing function in concert with other properties. Numerous computational methods have been developed to search through databases of structures for instances of specified motifs. However, it remains an open problem as to how best to leverage the local geometric and chemical constraints underlying structural motifs in order to develop motif-finding algorithms that are both theoretically and practically efficient. We present a simple, general, efficient approach, called BALLAST (Ball-based algorithm for structural motifs), to match given structural motifs to given structures. BALLAST combines the best properties of previously developed methods, exploiting the composition and local geometry of a structural motif and its possible instances in order to effectively filter candidate matches. We show that on a wide range of motif matching problems, BALLAST efficiently and effectively finds good matches, and we provide theoretical insights into why it works well. By supporting generic measures of compositional and geometric similarity, BALLAST provides a powerful substrate for the development of motif matching algorithms.

Keywords: protein structure, structural motif, sequence-structure-function relationship, geometric matching, motif matching algorithm, probabilistic analysis.

1 Introduction

With the availability of a huge and ever-increasing database of amino acid sequences, along with a smaller but also expanding and already largely representative database of three-dimensional protein structures, we are faced with the

B. Chor (Ed.): RECOMB 2012, LNBI 7262, pp. 79–93, 2012.

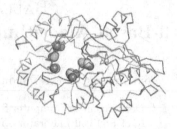

Fig. 1. Enolase superfamily motif. **(left)** Two enolase superfamily structures (backbone trace) and their instances of a motif (C$^\alpha$ spheres) common to diverse members of this superfamily [17]. Red: *E. coli* glucarate dehydratase (pdb id 1ec7D); blue: *E. coli* O-succinylbenzoate synthase 2 (pdb id 1fhv). They have 19% sequence identity and globally structurally align to $\approx$ 5 Å; the motif aligns to < 1 Å. **(right)** Motif residues from 7 templates superimposed on a mandelate racemase backbone (pdb id 2mnrA).

challenge of moving beyond characterizing what the proteins are, to what they do and how they do it. At the same time, we are presented with the opportunity to gain fundamental insights into relationships among sequence, structure, and function. Such identified relationships can further be used prospectively, e.g., to design variants whose function is specifically modified, or variants whose function is maintained while other properties (stability, solubility, etc.) are modified.

Since detailed experimental characterization of sequence-structure-function relationships is currently unable to keep pace with genomics and structural genomics efforts, computational methods are required. *Structural motifs* (Fig. 1) define patterns of amino acids that are localized within a structure and important for a particular function, and thus provide a powerful means for capturing, analyzing, and utilizing sequence-structure-function relationships. The utility of structural motifs is based on the hypothesis that, in many cases, protein function is determined not by overall fold but by a relatively small number of functionally important residues. This hypothesis is supported by convergent evolution of function, loss of function upon mutation of key residues, and the diversity of folds for some protein functions [12].

Structural motifs can better and more directly represent and utilize sequence-structure-function relationships than can alternative approaches such as sequence motifs and alignments, and global structural alignments. Typical sequence motifs may not adequately capture a compact set of key functional residues, as such residues need not be nearby in the sequence. While sequence alignment methods can often be effectively used to identify evolutionarily related proteins, and phylogenetic analysis (e.g., orthology [15]) can give further confidence in inferring related function, these techniques typically cannot help distinguish key functional residues from the overall background of evolutionarily-related amino acids. They also have a hard time dealing with cases of limited sequence identity (as in the enolases in Fig. 1). Global structure alignment techniques can identify near and remote homologs and even unrelated proteins with similar overall three-dimensional structures, but do not directly separate key functional residues from the overall scaffold. These techniques can also have difficulties distinguishing functional subclasses within a superfamily (with the enolases once again providing an example).

Motif matching is (one name for) a core problem in structural motifs; the goal is to search for instances of a motif (*query*) in a set of protein structures (*targets*). Motif matching is a complex problem, with both a compositional and a geometric component. The compositional component requires residues in the motif to be matched with compatible residues (the same or similar amino acid types, in similar chemical environments, etc.). The geometric component requires the spatial distribution of the motif residues to be similar to the spatial distribution of the matched residues. Often an additional statistical component, which is somewhat orthogonal to the actual matching problem itself, seeks to determine whether the match is likely to have occurred simply due to chance.

Numerous approaches for the motif matching problem have been proposed; see [20] for a good summary. Three approaches are fairly representative of the field, and serve to establish the key contrasts in methodology.

Geometric Hashing [23]. This is one of the most-used methods for efficiently finding three-dimensional objects represented by discrete points that have undergone an affine transformation [32]. The main idea is to preprocess the query and store its points (with labels for amino acid types, etc.) in a hash table, and to look up the targets against the hash table. The hashing and the look up are performed by choosing sets of 3 points to define coordinate systems, and for each transforming the remaining points accordingly, thereby defining rigid-body transformations that serve to align target points with query points. After its introduction in computer vision, this technique has been used in the development of many algorithms for structural biology, including motif analysis [29,6,27].

LabelHash [20]. In contrast to geometric hashing, LabelHash hashes tuples of residues (typically 3-tuples) from the target based on amino acid types rather than geometry, though each tuple does have to satisfy certain geometric constraints. Given a query, LabelHash looks up all matches to a submotif of the tuple size. It expands each partial match to a complete match using a depth-first search, a variant of the match augmentation algorithm [9]. The residues added to a match during match augmentation are not subject to the geometric constraints of reference sets, and partial matches with Root Mean Square Deviation (RMSD) greater than a certain threshold are discarded.

Graph-Based Methods [22,4]. These and other graph-based approaches (e.g., [1,11,18,30]) represent the query residues (or atoms) as vertices connected by edges for proximal pairs. In many cases, edges are defined by contact (e.g., based on a distance threshold), though Bandyopadhyay *et al.* [4] derive the graphs from almost-Delaunay triangulations. In general, graph-based methods face the subgraph isomorphism problem, a well-known NP-complete problem [28]. To tackle this, Bandyopadhyay *et al.* employ a heuristic that enables the search to be terminated when the local neighborhood of a subgraph is a witness to the impossibility of a match. Other graph-based methods formulate motif matching

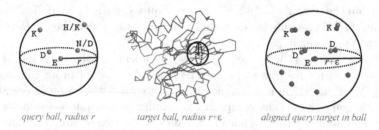

query ball, radius r   *target ball, radius r+ε*   *aligned query/target in ball*

Fig. 2. BALLAST employs *local* (ball-based) matching to find in a target structure (blue) an instance of a query motif (red) defined in terms of a set of points (geometry, e.g., C^{α} coordinates) and labels (composition, e.g., allowed amino acids). A query ball is centered on one of the query points and contains the other points. An expanded target ball, with a larger radius to account for structural variation, is scanned through the target structure, centering it at each residue. If the target ball passes some filters (e.g., it contains a sufficient number of the query labels), then possible alignments between the query and target points are evaluated. The efficiency of BALLAST stems from the fact that there are relatively few balls to consider, many of these are filtered, and the remaining ones have relatively few points to assess for matches.

in terms of clique finding, though this is also NP hard [13] and difficult to approximate [10]. In IsoCleft [22], cliques are found in a graph that has nodes for pairs of query & target residues with similar composition and edges for those pairs with similar geometry. A two-stage heuristic approach is then used to detect a match as the largest clique in this graph.

Despite the extensive amount of work on motif matching, it remains a challenge to efficiently identify all the instances of a structural motif in a database of protein structures [20]. Since the protein databank (PDB) [7] has over 76,000 structures as of Oct. 2011, efficiency is required.

Our Contribution. We focus on the *local* geometric and compositional constraints defining a structural motif, and derive a novel motif matching approach, called BALLAST (<u>Ball</u>-based <u>a</u>lgorithm for <u>s</u>tructural mo<u>t</u>ifs). Our approach combines the best properties of the previously proposed approaches: geometric hashing (geometric, but global), and subgraph matching and label hashing (local, but combinatorial). BALLAST takes advantage of the locality of the residues in a structural motif, and directly considers both geometry and composition (Fig. 2).

We provide analytical evidence of the efficiency of BALLAST, characterizing its performance under a suitable generative model for 3D structures. We derive an upper bound on the time complexity of our algorithm that holds with high probability, and is substantially better than the complexity of other algorithms. We also provide empirical evidence of the efficiency and effectiveness of BALLAST in practice. On a large and diverse set of previously studied motif matching problems, it efficiently searches a large structural database. For those searches the running time of BALLAST is comparable to what reported in Moll *et al.* [20]

for the state-of-the-art LabelHash code (though for different hardware), despite requiring no preprocessing or large index. BALLAST is relatively unaffected by the number of motif points and scales well with the motif radius.

2   Problem Statement, Algorithm, and Analysis

We represent both motifs and target structures with labeled point sets. For the points, BALLAST supports the commonly-used representations of C^α, C^β, and side-chain centroid coordinates. For the labels on the points, BALLAST currently supports amino acid types, and is readily extensible to employ other (discrete) representations of composition (e.g., physicochemical classes). Sets of allowed labels may be provided for query points (i.e., possible amino acid types, allowing for substitution). More formally, we are given a query set $Q = \{q_1, q_2, \ldots, q_k\} \subset \mathbb{R}^3$ of k points, and a target set $T \subset \mathbb{R}^3$ of n points (for a single structure at a time). We also have a function $A : Q \to 2^{\mathcal{A}}$ mapping a query point to a set of allowed amino acids (from set $\mathcal{A} = \{\text{Ala, Arg, } \ldots\}$), along with a function $a : T \to \mathcal{A}$ mapping a target point to its (single) amino acid in the structure.

Our goal is to find a subset M of T with $|M| = |Q| = k$ that *matches* Q. Many different geometric and compositional criteria have been considered in defining what constitutes a possible match, and in evaluating these to select the best. For geometric evaluation, we focus here on the common root mean squared deviation (RMSD) criterion. Let $v_M : Q \to M$ be the bijection describing a possible match between Q and M. Then, $d_{RMSD}(M, Q) = \sqrt{\frac{1}{k} \sum_{i=1}^{k} \|q_i - v_M(q_i)\|^2}$ where $\|p - q\|$ denotes the distance between points p and q. For compositional evaluation, we simply assess whether or not the target amino acids belong to the corresponding query sets. That is $d_{AA}(M, Q) = \sum_{i=1}^{k} I\{a(v_M(q_i)) \notin A(q_i)\}$ where $I\{\cdot\}$ is the indicator function. BALLAST readily supports variations of these criteria (including distance differences and substitution scores), so we will continue to refer to geometric and compositional criteria generically. We consider a match to be a *candidate* if it satisfies *constraints* on the geometric and compositional criteria, namely that d_{RMSD} is at most a user-specified threshold θ and d_{AA} is zero (all amino acid types match).

BALLAST assumes that a motif is both compact and relatively similar to the query. For compactness (see Fig. 2), we assume that there is a ball of radius r, centered on one of the points in Q and containing all of the points, such that r is "small" compared to the overall structure. For geometric similarity, we assume that for each pair of points in Q, the corresponding pair of points in T has about the same distance, within a user-specified parameter $\varepsilon \geq 0$. Note that this is a local geometric constraint, somewhat complementary to the global RMSD constraint above; a candidate must satisfy both constraints. This further implies that the instance in the target fits within a ball of radius of at most $r + \varepsilon$ when centered on one of the points in T. These assumptions, which also underlie graph-based methods, generally hold for structural motifs (particularly those defining catalytic sites), as we demonstrate in the results. We also show in the theoretical analysis below that they directly lead to the efficiency of BALLAST.

Algorithm 1. Pseudocode for algorithm BALLAST.

Input: Query set Q, target set T, radius expansion $\varepsilon > 0$, RMSD threshold θ
Output: Candidate matches $C \subset 2^T$

1 $\hat{q} \leftarrow \arg\min_{q \in Q} \max\{\|q - q_i\| : q_i \in Q \setminus \{q\}\}$;
2 $r \leftarrow \max\{\|\hat{q} - q_i\| : q_i \in Q \setminus \{\hat{q}\}\}$;
3 $C \leftarrow \emptyset$;
4 **for** $p \in T$ **do**
5   **if** $a(p) \in A(\hat{q})$ **then**
6          $B_p(r + \varepsilon) \leftarrow \{p' \in T \setminus \{p\} : \|p - p'\| \leq r + \varepsilon\}$;
7   **for** $q_i \in Q \setminus \{\hat{q}\}$ **do**
8              $d_i \leftarrow \|\hat{q} - q_i\|$;
9              $B_p^{(i)}(r + \varepsilon) \leftarrow \{p' \in B_p(r + \varepsilon) : a(p') \in A(q_i), \|p - p'\| \in [d_i - \varepsilon, d_i + \varepsilon]\}$;
10         $C \leftarrow C \cup \{M \in \prod_i B_p^{(i)}(r + \varepsilon) : M \text{ has no repeats}, d_{RMSD}(M, Q) < \theta,$
            $\forall q \neq q' \in Q : |\|q - q'\| - \|v_M(q) - v_M(q')\|| \leq \varepsilon\}$;
11 Sort C by geometric and compositional criteria;
12 **return** C ;

We note that r is part of the definition of a motif and follows from the given points, while ε is part of the definition of a match and is set by the user. In the results, we study the effects of ε on the output and efficiency.

The basic idea of our algorithm is straightforward; see Algorithm 1 and Fig. 2. We find the ball of minimum radius r, centered at some $\hat{q} \in Q$, that contains all the points in query Q. Then we separately consider each point p in target T, and examine the set of points $B_p(r + \varepsilon)$ within the ball centered at p and of radius $r + \varepsilon$. We generate as candidate matches all the subsets of size k of $B_p(r + \varepsilon)$ that contain p and satisfy the geometric and compositional constraints. While candidate generation could be done in a brute force fashion, we instead filter the possible matched target points for each query point to those correspondingly close to the center and of the corresponding amino acid type. We then take one point from each set, avoiding repetitions and ensuring satisfaction of the constraints. We show below that, while this generation step could be expensive, it is likely to be cheap due to the filtering, the locality and compactness of the ball, and the physical nature of protein packing. Finally, we rank the candidates.

We now analyze the efficiency of BALLAST. The ball of minimum radius r that contains all the points in Q (Lines 1, 2) can be found in time $O(k^2)$ by computing the $O(k^2)$ distances for all pairs of points in Q, and then finding for each point q in Q the maximum of the $k - 1$ distances between q and the other points in Q. The naive way to find all the points in $B_p(r + \varepsilon)$ (Line 6) requires time $O(n)$. This naive implementation is very efficient in practice and has been used in our experiments. However, we note that the complexity of this part can be improved employing a range tree [16,31]. A range tree is a data structure on n points (in 3D space) that can be built in time $O(n \log^2 n)$. It allows orthogonal range queries to be answered in time $O(\log^2 n + w)$, where w is the number of points reported. Since we want to find all the points in $B_p(r + \varepsilon)$, we can first perform a orthogonal range query to retrieve all the points in the cube with edge length $2(r + \varepsilon)$ centered at p. Assuming there are w such points, then in time $O(w)$ we can then find the ones in $B_p(r + \varepsilon)$. For the generation of candidate matches (Lines 7-10), if we denote by m the number of points in $B_p(r+\varepsilon)$, then in the worst case there are $\binom{m}{k}$ candidates; we tighten this in the corollary below, based on our geometric and compositional constraints. Thus the generation of candidate matches requires $O\left(\binom{m}{k} f_c(k)\right)$ time, where $f_c(k)$ is the time required to evaluate a subset of k points for the constraints (instantiated for our constraints below). Therefore the time complexity of our algorithm is $O\left(k^2 + n\left(\log^2 n + w + \binom{m}{k} f_c(k)\right)\right)$.

The efficiency of our algorithm strongly depends on the number m of points that are found in $B_p(r+\varepsilon)$. In the worst case m could be as large as n, and thus our algorithm could in the worst case require $\Omega(n^k)$ time, but in practice our method is extremely efficient. To understand why, we analyze the performance of our algorithm when the input is not adversarially chosen, but when the points in T are drawn from a probability distribution. This distribution is the same considered for the $G(n, r, \ell)$ random geometric graph model [21], a generalization of the $G(n, r)$ random geometric graph model [25] that scales to arbitrary sizes. In the $G(n, r, \ell)$ model, the vertices are points placed uniformly at random in $[0, \ell]^3$. We present a probabilistic analysis to show that the average case performance of the algorithm is good; this provides a theoretical insight why the algorithm works efficiently in practice.

We now prove (full proofs are given in Supplemental Material) that if the points in T are drawn uniformly at random in $[0, \ell]^3$, and for reasonable values of the parameters r and ε, the number of points inside $B_p(r+\varepsilon)$ is small whp[1].

Lemma 1. *Let T be a set of n points drawn uniformly at random from $[0, \ell]^3$ and $r, \varepsilon \geq 0$ such that $r + \varepsilon \in O\left(\ell\left(\frac{\log n}{n}\right)^{1/3}\right)$. Then $m = \max_{p \in T}\{|B_p(r + \varepsilon)|\} \in O(\log n)$ whp.*

Proof (Sketch). The expected number of points in $B_p(r + \varepsilon)$ is bounded by $O(\log n)$. Thus by Chernoff bound [19] the number of points in $B_p(r + \varepsilon)$ is $O(\log n)$ with probability $\geq 1 - \frac{1}{n^d}$, for a constant $d > 1$. By union bound on all points $p \in T$, we have that $m \in O(\log n)$ whp. □

Lemma 1 gives theoretical evidence of why the use of a ball results in an efficient approach: since "few" residues are found in a ball whp. when the residues are placed randomly, few subsets are considered for candidate generation and hence few candidates are explicitly examined. We can use this to bound the overall time complexity.

Theorem 1. *Let T be a set of n points drawn uniformly at random from $[0, \ell]^3$, Q a set of $k \in o(\log n)$ points and $r, \varepsilon \geq 0$ such that $r + \varepsilon \in O\left(\ell\left(\frac{\log n}{n}\right)^{1/3}\right)$. Then for any (small) constant $d > 0$ the time complexity of our algorithm is bounded by $O\left(n^{1+d} f_c(k)\right)$ whp.*

---

[1] We say that an event holds *with high probability*, abbreviated whp., if it holds with probability at least $1 - n^{-c}$ for some constant $c > 0$, for sufficiently large n.

Proof (Sketch). From Lemma 1, $m \in O(\log n)$ whp. Thus for n sufficiently large we have $\binom{m}{k} \in O(n^d)$ for any (small) constant $d > 0$. The analysis of Lemma 1 can be adapted to consider the event $E =$ "a point drawn uniformly at random from $[0, \ell]^3$ is in the cube of edge $2(r + \varepsilon)$ centered at p". In this way we prove that w is $O(\log n)$, and the theorem follows.   □

The complexity of our approach depends on the effectiveness in the constraints used to filter candidates (through $f_c(k)$). We now obtain a more precise analysis by incorporating the constraints in our current implementation. Note that the rigid superposition to evaluate $d_{RMSD}(M, Q)$ can be computed in time $O(k)$ [2], and that the time complexity to check if M satisfies pairwise distances and allowed amino acids substitutions constraints is $O(k^2)$. Thus in this case we have $f_c(k) \in O(k^2)$.

Corollary 1. *Let T be a set of n points drawn uniformly at random from $[0, 1]^3$, Q a set of $k \in o(\log n)$ points and $\varepsilon \geq 0$ such that $r + \varepsilon \in O(n^{-1/3})$. If we look for matches satisfying our constraints (maximum RMSD of θ, matching amino acid types, and maximum pairwise distance expansion of ϵ), the time complexity of algorithm is bounded by $o(n^{1+d} \log^2 n)$ whp. for any (small) constant $d > 0$.*

It follows that BALLAST is more efficient than previous motif matching approaches. **Geometric hashing.** All possible bases of 3 points in T are considered, and the points in T are transformed to each such basis. Since there are $\Theta(n^3)$ such bases and each transformation requires time $\Theta(n)$, the total complexity is $\Theta(n^4)$. **LabelHash.** For a new protein target T. LabelHash starts with "reference sets", all 3-tuples from T, as possible seeds for matching, and then augments them to full-size matches. Thus the time complexity is at least $\Theta(n^3)$. This is of course a loose characterization, since it does not take into account the augmentation phase. (For a fixed target, the LabelHash index avoids the recomputation of the reference sets of the target. A similar strategy could be used with our approach, e.g., precomputing the points inside $B_p(r + \varepsilon)$ for each point p in each target structure, for different values of r and ε.) **Graph-based methods.** Even before tackling the NP-hard subgraph isomorphism, Bandyopadhyay *et al.* [4] compute an almost-Delaunay triangulation. The proposed algorithm [5] requires time $O(n^5 \log n)$ in the worst case, but runs in $O(n^2 \log n)$ expected time (no result whp. is proved). IsoCleft [22] uses the Bron and Kerbosch algorithm [8] to detect the largest clique, which can take up to $O(3^{n/3})$ time.

3   Results

In order to assess the practical utility of BALLAST, we applied it to a wide range of matching problems previously studied by LabelHash [20]. Two case studies enable us to explore the matching of structural motifs initially defined from structural analysis (enolases [17]) and sequence analysis (SOIPPA [33]). A set of 147 motifs derived from the catalytic site atlas (CSA [26]) are not so rigorously characterized but serve as a large-scale benchmark.

Since BALLAST addresses the motif matching problem rather than the motif discovery problem, we focus on its performance in finding motifs, not on the significance of the motifs themselves (e.g., by assessing the p-values under some null model). BALLAST is guaranteed to find all instances satisfying the definition of a motif and the settings of the ε (local distance expansion) and θ (global RMSD) parameters. In order to analyze the structural variability underlying the motif and the effects of the geometric constraints, we vary these parameters and characterize the numbers of matches in the "foreground" dataset used to develop the motif (essentially a sensitivity measure) as well as in a large "background" dataset of structures (specificity). The case study-specific foregrounds are introduced below. For the background, we used 30111 non-redundant protein structures from the PDB, clustered by BLASTClust at 95% sequence identity.

Moll et al. [20] performed an extensive evaluation of the performance of Label-Hash, including its scalability with multiple cores. Our current implementation of BALLAST is in single-threaded Java code, but is embarrassingly parallel and could easily be extended to distribute different subsets of the target database to different cores. For now we simply study the single-core performance of BALLAST in searching our 30111-member background database, analyzing the dependence on ε (but not on θ, as it is just a post-processing filter). Results provided in terms of wall-clock time on a Linux machine with an AMD Opteron 2435 processor and 32 GB memory (though we do not require or utilize large memory).

We do not present a direct performance comparison against LabelHash since we run BALLAST on different hardware and a different background dataset from that reported for LabelHash [20] (21745 structures, roughly 2/3 our background). Indeed, our purpose is only to show that the current straightforward implementation of BALLAST has reasonable performance, comparable to other, highly-optimized tools. We do see in all our test cases that the wall-clock times are fairly comparable, on the same order of magnitude (though again with different hardware). We have an advantage when motifs have more points; they have an advantage when the amino acid labels are unambiguous. As we have emphasized throughout, our key contribution is a new algorithmic framework that combines locality and geometry, and provides a strong theoretical rationale for efficient performance. We also note that, as part of its simplicity, BALLAST requires no extra data structures and only preprocesses the PDB files into binary files in order to speed the loading time (a few MB extra). In contrast, LabelHash employs a preprocessed 9.5GB hash table for a 21745-structure background, which increases to 65GB when considering the entire PDB; as reported [20] this number would grow to approximately 5TB with reference sets of size 4.

3.1 Enolase Superfamily

The enolase superfamily (ES) includes 7 major sub-groups that share core catalytic sites supporting the abstraction of a proton from a carbon adjacent to a carboxylic acid, in order to form an enolate anion intermediate [3]. Enzymes in the superfamily have in common two domains, an N-terminal capping domain for substrate specificity and a C-terminal TIM beta/alpha-barrel domain

Enolases

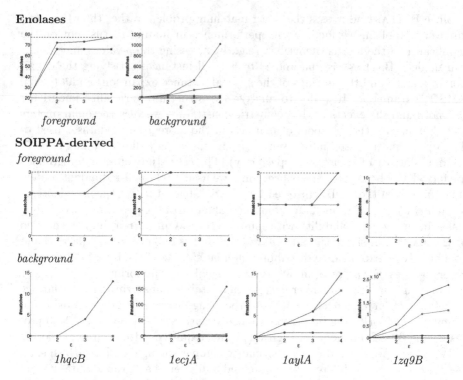

Fig. 3. Matches found by BALLAST in the foreground and background for enolases and SOIPPA-derived motifs (identified by PDB id), under different values for ε (x-axis) and RMSD threshold (different lines, red: 0.5; blue: 1.0; magenta: 1.5; green: 2.0). The foreground sizes are indicated by black dashed horizontal lines; the background size, not shown, is roughly 30111, though we exclude foreground structures.

containing key catalytic residues at the ends of the beta strands. A five-residue structural motif common to the superfamily was developed by [17] based on 7 representative structure templates (Fig. 1). Using residue numbering based on mandelate racemase (PDB id 2MNR) and listing multiple allowed amino acid types where appropriate, the ES motif includes KH164, D195, E221, EDN247, and HK297; we note that according to the structure-function linkage database [24] there is some ambiguity in the first position. As the superfamily is known to have particularly diverse structures in terms of C^α RMSDs, side-chain centroids were instead used to define the motif geometry. While Meng et al. originally used SPASM [14] for motif matching, Moll et al. [20] demonstrated that LabelHash could also successfully match it against the superfamily members.

We used BALLAST to search for instances of the ES motif in a foreground ES family benchmark of 77 chains provided by Meng et al. (excluding the one with no PDB code). The top part of Fig. 3 illustrates the number of matches found in both the superfamily and background under different settings of the parameters. We reemphasize that BALLAST is complete with respect to the parameter values, so this analysis is a characterization of the quality of the motif, rather than the quality of the algorithm. That is, the efficiency of BALLAST enables us to characterize the structural variability over the instances of a motif and the effects on sensitivity and specificity in trying to account for that variability by allowing some local and global geometric "slop". We find that the enolase motif is rather robust. The foreground requires an RMSD of 1 Å or more to allow for variation among instances of the motif. The pairwise distances can vary by around 2 Å but are relatively stable with any ε at least that large. Looser settings also lead to the identification of multiple instances in some of the foreground structures, though we only report one match per structure in the figure. With RMSD at 1 Å and ε at 2 Å, we find 66 of the 77 foreground matches. Six of the missing ones require an RMSD of 1.1 Å and one also requires ε of 3 Å. The remaining four all belong to the subfamily of methylaspartate ammonialyase, which is apparently more structurally variable, requiring RMSD of 1.5 Å and ε of 4 Å.

There are relatively few instances of the ES motif in the background. With the basic settings (RMSD: 1 Å, ε: 2 Å) required to achieve reasonable foreground coverage (66/77, 86%), there are only 68 non-foreground matches among the 30100 background structures (0.23%), excluding 11 structures in the foreground. Of those, 22 were not in the original foreground dataset but actually do belong to the enolase superfamily according to the structure-function linkage database [24]. The remaining 46 are not known to be ES members although many have very good instances of the motif (19 with RMSDs $\leq$ 0.5 Å). Bumping the RMSD up to 1.5 Å while holding ε at 2 Å yields better sensitivity (72/77, 93.5%) at the price of some additional background hits (total of 82, 0.27%, with 23 not in the foreground but in the superfamily). Increasing ε at the higher RMSD threshold has detrimental effects on specificity. Thus BALLAST enables us to conclude that the ES motif provides a fairly "tight" specification of the structural pattern common to the superfamily and distinct from other structures.

The dashed line in Fig. 4(left) characterizes the running time of BALLAST for the background search at different ε values and a fixed RMSD threshold of 1.5 Å. While increasing the distance expansion parameter results in a larger ball size and more potential matches to assess, even the 4 Å setting only requires an additional 197 seconds beyond that for the baseline 1 Å. In terms of a rough comparison of wall-clock times (on different hardware; see the start of the Results for a discussion), we note that LabelHash [20] reported roughly 1000 seconds for a background search, about twice as long as BALLAST.

3.2 SOIPPA-Derived Motifs

Xie and Bourne [33] developed the Sequence Order-Independent Profile-Profile Alignment (SOIPPA) method to align protein structures independent of the sequential order of the residues, and identify motifs with similar local structures but distinct sequences. Moll et al. [20] derived structural motifs from SOIPPA motifs by, for each SOIPPA motif, using the C^α coordinates from one template structure, along with all SOIPPA-identified alternative amino acid types. Motif details are provided in an appendix for the interested reader.

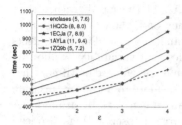 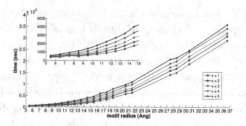

Fig. 4. Wall-clock timing results for BALLAST background searches. **(left)** ES motif (dashed lines) and SOIPPA-derived motifs (solid), with varying ε (x-axis) and fixed RMSD (1.5 Å) . In the legend, each motif is characterized by (# points, radius). **(right)** Averages over motifs in CSA database, grouped by radius (within ±0.5 Å, at different ε values (lines) and fixed RMSD (1.5 Å).

As discussed, BALLAST enables us to evaluate the robustness of a motif by performing searches at different threshold values and thereby assessing structural variability in terms of these local (ε) and global (RMSD) parameters. We matched each motif against a foreground dataset consisting of the original SOIPPA-aligned structures, as well the entire background database; the bottom part of Fig. 3 summarizes the numbers of matches. Note that the motifs show up multiple times in some of the foreground structures (details in the Supplemental Material); each is counted separately in these figures. The different foreground datasets clearly have different levels of structural diversity, as might be expected from motifs initially derived from sequence profiles. For example, 1zq9B is very "tight", with the entire foreground covered at any setting of the parameters, and hitting only 19 background structures with, e.g., ε of 1 Å and RMSD of 0.5 Å. 1ecjA is also quite tight, with ε of 2 Å and RMSD of 1 Å covering the entire foreground and only 3 members of the background. It also remains quite stable to background hits with increasing ε under the smaller RMSD thresholds. On the other hand, we need ε of 4 Å and RMSD of 2 Å to cover the foregrounds for 1hqcB and 1aylA, hitting respectively 16 matches in 13 unique chains (1hqcB) and 36 in 15 (1aylA). In the case of 1hqcB, the two chains of 1hqc are covered before the other foreground chain. Again we see the power of BALLAST in performing a range of motif searches and helping characterize the trade-offs required to account for structural variability.

Efficiency-wise, our implementation took less than 1100 seconds of wall-clock time to match each motif against the background database even with ε set to 4 Å (Fig. 4, left). The motifs range from 5 residues up to 11 residues and about 7.2 Å to 9.4 Å in ball radius, with the larger ones taking a bit longer. We see good scalability over the ε range. 1zq9b suffers the largest loss due to the extra time for outputting the large number of matches. These numbers again compare very favorably to those reported by LabelHash (though again on different hardware with a different background), which exceed 5000 seconds. This is because the LabelHash hash keys are typically for only 3 residues, and the extension from the quick identification of those "core" sets to an entire motif (of up to 11 points) is relatively expensive. In contrast, BALLAST simultaneously filters on both geometry and amino acid content. This same contrast holds for graph-based approaches, as subgraph isomorphism scales poorly with subgraph size.

3.3 Catalytic Site Atlas

The Catalytic Site Atlas (CSA) [26] defines residues implicated as comprising catalytic sites for a range of families. Moll *et al.* [20] constructed 147 motifs from 147 CSA sites within 118 unique EC classes spanning 6 top-level EC classifications (oxidoreductases, transferases, hydrolases, lyases, iosmerases, and ligases). Each motif was defined using the C^α geometry and amino acid types for a single representative structure, due to the lack of characterized substitutions and alignments. We followed the same procedure to generate an analogous dataset, though note that the actual members may be different from those used by [20] (and we do have different sizes of the foreground sets), due to changing databases and so forth. The task is then to search each motif against a foreground comprising the members of the corresponding EC family, as well as the background.

Unlike the enolase and SOIPPA-derived motifs, these are not rigorously defined or assessed as "motifs" per se, so they vary widely in their ability to capture the foreground and not the background. In fact, [20] found that while the motifs are quite specific, covering only 0.1–0.2% of the background, their sensitivity ranges from 0% to 100%. They discussed a number of reasons, including the fact that no amino acid substitutions were allowed, as well as the construction of the motifs based on CSA rather than EC classes. We found similar lack of specificity and sensitivity (illustrated in the Supplemental Material), but still use this dataset as a large-scale study of the effects of different motif definitions, with the number of points ranging from 4 to 8 and the radius from about 5 Å to about 37 Å [!].

Fig. 4 (right panel) summarizes the wall-clock times required for background searches, aggregated by the radius, with different lines for different values of ε. The performance does depend on the motif radius, though the larger radii aren't really appropriate structural motifs. For motifs with a radius of at most 9 Å (in line with the case study motifs), the average running time was 600 seconds, while for larger motifs of radius 15 Å, it degraded smoothly to 1300 seconds. LabelHash did quite a bit better on these searches, averaging about 150 seconds for the (different) background search, since most of the motifs have a small number of points (4 or 5) and unique amino acid labels, so that indeed most of the effort is handled by hashing. We also aggregated the times by the number of points (figures provided in Supplemental Material). As we observed for SOIPPA motifs, BALLAST is relatively insensitive to the number of motif points, in contrast to LabelHash and graph-based methods.

4   Conclusion

We have presented a new approach to structural motif matching, making use of balls to localize computations, and directly utilizing both local geometry and chemical composition to find motif instances. We showed that our algorithm is efficient and effective in both theory and in practice. BALLAST's efficiency and its interpretable, tweakable parameters enable the analysis of structural variability inherent in the definition of a motif, and implications for specificity and sensitivity. It can thus also be quite useful in motif discovery. To a large extent, the BALLAST approach is generic to assessments of geometric and compositional similarity, and while we instantiated it with common choices, it can readily support a variety of alternatives (distance difference, weighted metrics; solvent accessibility, local chemical environment). BALLAST provides a powerful and efficient substrate for exploring fundamental questions in defining and developing motifs and characterizing sequence-structure-function relationships, and we look forward to further extending and applying it in a wide range of such contexts.

Acknowledgement. This work was supported in part by NSF grant CCF-0915388 / 1023160, collaborative among CBK, GP, and FV.

Supplemental Material. A document with full proofs and additional results is available at `www.cs.dartmouth.edu/~cbk/ballast/recomb12-suppl.pdf`. A Java implementation of BALLAST is freely available upon request.

References

1. Artymiuk, P.J., Poirrette, A.R., Grindley, H.M., Rice, D.W., Willett, P.: A graph-theoretic approach to the identification of three-dimensional patterns of amino acid side-chains in protein structures. J. Mol. Biol. 243, 327–344 (1994)
2. Arun, K.S., Huang, T.S., Blostein, S.D.: Least-squares fitting of two 3-d point sets. IEEE Trans. Pattern Anal. Mach. Intell. 9, 698–700 (1987)
3. Babbitt, P.C., Hasson, M.S., et al.: The enolase superfamily: A general strategy for enzyme-catalyzed abstraction of the α-protons of carboxylic acids. Biochemistry 35(51), 16489–16501 (1996)
4. Bandyopadhyay, D., Huan, J., et al.: Identification of family-specific residue packing motifs and their use for structure-based protein function prediction: I. Method development. J. Comput. Aided Mol. Des. 23, 773–784 (2009)
5. Bandyopadhyay, D., Snoeyink, J.: Almost-delaunay simplices: nearest neighbor relations for imprecise points. In: Proc. SODA, pp. 410–419 (2004)
6. Barker, J.A., Thornton, J.M.: An algorithm for constraint-based structural template matching: application to 3D templates with statistical analysis. Bioinformatics 19, 1644–1649 (2003)
7. Bernstein, F.C., Koetzle, T.F., et al.: The Protein Data Bank: a computer-based archival file for macromolecular structures. J. Mol. Biol. 112, 535–542 (1977)
8. Bron, C., Kerbosch, J.: Algorithm 457: finding all cliques of an undirected graph. Commun. ACM 16, 575–577 (1973)
9. Chen, B.Y., Fofanov, V.Y., et al.: The MASH pipeline for protein function prediction and an algorithm for the geometric refinement of 3D motifs. J. Comput. Biol. 14, 791–816 (2007)
10. Feige, U., Goldwasser, S., Lovász, L., Safra, S., Szegedy, M.: Interactive proofs and the hardness of approximating cliques. J. ACM 43, 268–292 (1996)

11. Gardiner, E.J., Artymiuk, P.J., et al.: Clique-detection algorithms for matching three-dimensional molecular structures. J. Mol. Graph. Model. 15, 245–253 (1997)

12. Hegyi, H., Gerstein, M.: The relationship between protein structure and function: a comprehensive survey with application to the yeast genome. J. Mol. Biol. 288, 147–164 (1999)

13. Karp, R.M.: Reducibility among combinatorial problems. Complexity of Computer Computations 40(4), 85–103 (1972)

14. Kleywegt, G.J.: Recognition of spatial motifs in protein structures. J. Mol. Biol. 285, 1887–1897 (1999)

15. Loewenstein, Y., Raimondo, D., et al.: Protein function annotation by homology-based inference. Genome Biol. 10, 207 (2009)

16. Lueker, G.S.: A data structure for orthogonal range queries. In: Proc. FOCS, pp. 28–34. IEEE Computer Society, Washington, DC (1978)

17. Meng, E.C., et al.: Superfamily active site templates. Proteins 55, 962–976 (2004)

18. Milik, M., Szalma, S., Olszewski, K.A.: Common Structural Cliques: a tool for protein structure and function analysis. Protein Eng. 16, 543–552 (2003)

19. Mitzenmacher, M., Upfal, E.: Probability and Computing: Randomized Algorithms and Probabilistic Analysis. Cambridge Univ. Press, New York (2005)

20. Moll, M., Bryant, D.H., Kavraki, L.E.: The labelhash algorithm for substructure matching. BMC Bioinformatics 11, 555 (2010)

21. Muthukrishnan, S., Pandurangan, G.: The bin-covering technique for thresholding random geometric graph properties. In: Proc. SODA, pp. 989–998 (2005)

22. Najmanovich, R., Kurbatova, N., Thornton, J.: Detection of 3D atomic similarities and their use in the discrimination of small molecule protein-binding sites. Bioinformatics 24, i105–i111 (2008)

23. Nussinov, R., Wolfson, H.J.: Efficient detection of three-dimensional structural motifs in biological macromolecules by computer vision techniques. PNAS 88, 10495–10499 (1991)

24. Pegg, S.C., Brown, S.D., et al.: Leveraging enzyme structure-function relationships for functional inference and experimental design: the structure-function linkage database. Biochemistry 45, 2545–2555 (2006)

25. Penrose, M.D.: Random Geometric Graphs. Oxford University Press (2003)

26. Porter, C.T., Bartlett, G.J., Thornton, J.M.: The Catalytic Site Atlas: a resource of catalytic sites and residues identified in enzymes using structural data. Nucleic Acids Res. 32, D129–D133 (2004)

27. Shulman-Peleg, A., Nussinov, R., Wolfson, H.J.: Recognition of functional sites in protein structures. J. Mol. Biol. 339, 607–633 (2004)

28. Ullmann, J.R.: An algorithm for subgraph isomorphism. J. ACM 23, 31–42 (1976)

29. Wallace, A.C., Borkakoti, N., Thornton, J.M.: TESS: a geometric hashing algorithm for deriving 3D coordinate templates for searching structural databases. Application to enzyme active sites. Protein Sci. 6, 2308–2323 (1997)

30. Wangikar, P.P., et al.: Functional sites in protein families uncovered via an objective and automated graph theoretic approach. J. Mol. Biol. 326, 955–978 (2003)

31. Willard, D.E.: Predicate-Oriented Database Search Algorithms. Outstanding Dissertations in the Computer Sciences. Garland Publishing, New York (1978)

32. Wolfson, H.J., Rigoutsos, I.: Geometric hashing: An overview. Computing in Science and Engineering 4, 10–21 (1997)

33. Xie, L., Bourne, P.E.: Detecting evolutionary relationships across existing fold space, using sequence order-independent profile-profile alignments. PNAS 105, 5441–5446 (2008)

Evolution of Genome Organization by Duplication and Loss: An Alignment Approach

Patrick Holloway[1], Krister Swenson[2], David Ardell[3], and Nadia El-Mabrouk[4]

[1] Département d'Informatique et de Recherche Opérationnelle (DIRO),
Université de Montréal, H3C 3J7, Canada
patrick.holloway@umontreal.ca
[2] DIRO and University of McGill Computer Science
swensonk@iro.umontreal.ca
[3] Center for Computational Biology, School of Natural Sciences,
5200 North Lake Road, University of California, Merced, CA 95343
dardell@ucmerced.edu
[4] DIRO
mabrouk@iro.umontreal.ca

Abstract. We present a comparative genomics approach for inferring ancestral genome organization and evolutionary scenarios, based on a model accounting for content-modifying operations. More precisely, we focus on comparing two ordered gene sequences with duplicated genes that have evolved from a common ancestor through duplications and losses; our model can be grouped in the class of "Block Edit" models. From a combinatorial point of view, the main consequence is the possibility of formulating the problem as an alignment problem. On the other hand, in contrast to symmetrical metrics such as the inversion distance, duplications and losses are asymmetrical operations that are applicable to one of the two aligned sequences. Consequently, an ancestral genome can directly be inferred from a duplication-loss scenario attached to a given alignment. Although alignments are *a priori* simpler to handle than rearrangements, we show that a direct approach based on dynamic programming leads, at best, to an efficient heuristic. We present an exact pseudo-boolean linear programming algorithm to search for the optimal alignment along with an optimal scenario of duplications and losses. Although exponential in the worst case, we show low running times on real datasets as well as synthetic data. We apply our algorithm in a phylogenetic context to the evolution of stable RNA (tRNA and rRNA) gene content and organization in *Bacillus* genomes. Our results lead to various biological insights, such as rates of ribosomal RNA proliferation among lineages, their role in altering tRNA gene content, and evidence of tRNA class conversion.

Keywords: Comparative Genomics, Gene order, Duplication, Loss, Linear Programming, Alignment, Bacillus, tRNA.

B. Chor (Ed.): RECOMB 2012, LNBI 7262, pp. 94–112, 2012.
© Springer-Verlag Berlin Heidelberg 2012

1   Introduction

During evolution, genomes continually accumulate mutations. In addition to base mutations and short insertions or deletions, genome-scale changes affect the overall gene content and organization of a genome. Evidence of these latter kinds of changes are observed by comparing the completely sequenced and annotated genomes of related species. Genome-scale changes can be subdivided into two categories: (1) the *rearrangement operations* that shuffle gene orders (inversions, transpositions, and translocations), and (2) the *content-modifying operations* that affect the number of gene copies (gene insertions, losses, and duplications). In particular, gene duplication is a fundamental process in the evolution of species [25], especially in eukaryotes [5,9,12,16,22,34], where it is believed to play a leading role for the creation of novel gene function. In parallel, gene losses through pseudogenization and segmental deletions, appear generally to maintain a minimum number of functional gene copies [5,9,10,12,16,22,25]. Transfer RNAs (tRNAs) are typical examples of gene families that are continually duplicated and lost [3,28,31,35]. Indeed, tRNA clusters (or operons in microbial genomes) are highly dynamic and unstable genomic regions. In *Escherichia coli* for example, the rate of tRNA gene duplication/loss events has been estimated to be about one event every 1.5 million years [3,35].

One of the main goals of comparative genomics is to infer evolutionary histories of gene families, based on the comparison of the genomic organization of extant species. Having an evolutionary perspective of gene families is a key step towards answering many fundamental biological questions. For example, tRNAs are essential to establishing a direct link between codons and their translation into amino-acids. Understanding how the content and organization of tRNAs evolve is essential to the understanding of the translational machinery, and in particular, the variation in codon usage among species [20,11].

In the genome rearrangement approach to comparative genomics, a genome is modeled as one or many (in case of many chromosomes) linear or circular sequences of genes (or other building blocks of a genome). When each gene is present exactly once in a genome, sequences can be represented as permutations. In the most realistic version of the rearrangement problem, a sign (+ or -) is associated with a gene, representing its transcriptional orientation. Most genome rearrangement studies have focused on signed permutations. The pioneering work of Hannenhalli and Pevzner in 1995 [17,18], has led to efficient algorithms for computing the inversion and/or translocation distance between two signed permutations. Since then, many other algorithms have been developed to compare permutations subject to various rearrangement operations and based on different distance measures. These algorithms have then been used from a phylogenetic perspective to infer ancestral permutations [24,6,8,23,30] and evolutionary scenarios on a species tree. An extra degree of difficulty is introduced in the case of sequences containing multiple copies of the same gene, as the one-to-one correspondence between copies is not established in advance. A review of the methods used for comparing two ordered gene sequences with duplicates can be found in [14,15]. They can be grouped into two main classes.

The "Match-and-Prune" model aims at transforming strings into permutations, so as to minimize a rearrangement distance between the resulting permutations. On the other hand, the "Block Edit" model consists of performing the minimum number of "allowed" rearrangement and content-modifying operations required to transform one string into the other. Most studied distances and ancestral inference problems in this category are NP-complete [15].

In this paper, we focus on comparing two ordered gene sequences with duplicates that have evolved from a common ancestor through duplications and losses. In contrast to the approaches cited above and reviewed in [15], only content-modifying operations are considered. Such a simplified model is required to study the evolution of gene families mainly affected by duplications and losses, for which a general model involving rearrangement events may be misleading. From a combinatorial point of view, the main consequence of removing rearrangement operations is the fact that gene organization is preserved, which allows us to reformulate the problem of comparing two gene orders as an alignment problem. On the other hand, in contrast to symmetrical metrics such as the Hamming distance for nucleotide sequences, or the inversion distance for ordered sequences, duplication and loss are asymmetrical operations that are applicable to one of the two aligned sequences. Consequently an ancestral genome can directly be inferred from a duplication-loss scenario attached to a given alignment.

Although alignments are *a priori* simpler to handle than rearrangements, there is no direct way of inferring optimal alignments together with a related duplication-loss scenario for two gene orders, as detailed in Section 4. Even our simpler goal of finding an alignment is fraught with difficulty as a naive branch-and-bound approach to compute such an alignment is non-trivial; trying all possible alignments with all possible duplication and loss scenarios for each alignment is hardly practicable. As it is not even clear how, given an alignment, we can assign duplications and losses in a parsimonious manner, we present in Section 4.1 a pseudo-boolean linear programming (PBLP) approach to search for the optimal alignment along with an optimal scenario of duplications and losses. The disadvantage of the approach is that, in the worst case, an exponential number of steps could be used by our algorithm. On the other hand, we show in Section 5.2 that for real data, and larger simulated genomes, the running times are quite reasonable. Further, the PBLP is flexible in that a multitude of weighting schemes for losses and duplications could be employed to, for example, favor certain duplications over others, or allow for gene conversion. In Section 5.1, we apply our algorithm in a phylogenetic context to infer the evolution of stable RNA (tRNA and rRNA) gene content and organization in various genomes from the genus *Bacillus*, a so-called "low G+C" gram-positive clade of Firmicutes that includes the model bacterium *B. subtilis* as well as the agent of anthrax. Stable RNA operon organization in this group is interesting because it has relatively fewer operons that are much larger and contain more segmental duplicates than other bacterial groups. We obtained results leading to various biological insights, such as more accurate quantification of ribosomal

RNA operon proliferation, their role in altering tRNA gene content, and evidence of tRNA gene class conversion.

2   Research Context

The evolution of g genomes is often represented by a phylogenetic (or species) tree T, binary or not, with exactly g leaves, each representing a different genome. When such a species tree T is known for a set of species, then we can use the gene order information of the present-day genomes to infer gene order information of ancestral genomes identified with each of the internal nodes of the tree. This problem is known in the literature as the "small" phylogeny problem, in contrast to the "large" phylogeny problem which is one of finding the actual phylogenetic tree T.

Although our methods may be extended to arbitrary genomes, we consider single chromosomal (circular or linear) genomes, represented as gene orders with duplicates. More precisely, given an alphabet Σ where each character represents a specific gene family, a **genome** or **string** is a sequence of characters from Σ where each character may appear many times. As the content-modifying operations considered in this paper do not change gene orientation, we can assume, w.l.o.g. that genes are unsigned. For example, given $\Sigma = \{a, b, c, d, e\}$, $A = $ "ababcd" is a genome containing two gene copies from the gene family identified by a, two genes from the gene family b, and a single gene from each family c and d. A gene in a genome A is a **singleton** if it appears exactly once in A (for example c and d in A), and a **duplicate** otherwise (a and b in A above).

Let $\mathcal{O}$ be a set of "allowed" evolutionary operations. The set $\mathcal{O}$ may include organizational operations such as Reversals (R) and Transpositions (T), and content-modifying operations such as Duplications (D), Losses (L) or Insertions (I). For example, $\mathcal{O} = \{R, D, L\}$ is the set of operations in an evolutionary model involving reversals, duplications and losses. In the next section, we will formally define the operations involved in our model of evolution.

Given a genome A, a **mutation** on A is characterized by an operation O from $\mathcal{O}$, the substring of A that is affected by the mutation, as well as possibly other characteristics such as the position of the re-inserted, removed (in case of transposition), or duplicated substring. For simplicity, consider a mutation $O(k)$ to be characterized solely by the operation O from $\mathcal{O}$, and the size k of the substring affected by the mutation. Consider $c(O(k))$ to be a cost function defined on mutations. Finally, given two genomes A and X, an **evolutionary history** $O_{A \to X}$ from A to X is a sequence of mutations (possible of length 0) transforming A into X.

Let A, X be two strings on Σ with A being a **potential ancestor** of X, meaning that there is at least one evolutionary history $O_{A \to X} = \{O_1(k_1), \cdots, O_l(k_l)\}$ from A to X. Then the cost of $O_{A \to X}$ is:

$$C(O_{A \to X}) = \sum_{i=1}^{l} c(O_i(k_i))$$

Now let $\mathcal{O}_{A \to X}$ be the set of possible histories transforming A into X. Then we define:

$$C(A \to X) = \min_{O_{A \to X} \in \mathcal{O}_{A \to X}} C(O_{A \to X})$$

Then, the small phylogeny problem can be formulated as one of finding strings at internal nodes of a given tree T that minimize the total cost:

$$C(T) = \sum_{\text{all branches } b_i \text{ of } T} C(X_{i,1} \to X_{i,2})$$

where $X_{i,1}, X_{i,2}$ are the strings labeling the nodes of T adjacent to the branch b_i, with the node labeled $X_{i,1}$ being the parent of the node labeled $X_{i,2}$.

For most restrictions on genome structure and models of evolution, the simplest version of the small phylogeny problem — the median of three genomes — is NP-hard [7,26,32]. When duplicate genes are present in the genomes, even finding minimum distances between two genomes is almost always an NP-Hard task [19]. In this paper, we focus on **cherries of a species tree** (*i.e.* on subtrees with two leaves). The optimization problem we consider can be formulated as follows:

Two Species Small Phylogeny Problem:
INPUT: Two genomes X and Y.
OUTPUT: A potential common ancestor A of X and Y minimizing $C(A \to X) + C(A \to Y)$.

Solving the TWO SPECIES SMALL PHYLOGENY PROBLEM (2-SPP) can be seen as a first step towards solving the problem on a given phylogenetic tree T. The most natural heuristic to the Small Phylogeny Problem, that we will call the SPP-HEURISTIC, is to traverse T depth-first, and to compute successive ancestors of pairs of nodes. Such a heuristic can be used as the initialization step of the *steinerization* method for SPP [30,4]. The sets of all optimal solutions output by an algorithm for the 2-SPP applied to all pairs of nodes of T (in a depth-first traversal) can alternatively be used in an iterative local optimization method, such as the dynamic programming method developed in [21].

3   The Duplication and Loss Model of Evolution

Our evolutionary model accounts for two operations, Duplication (denoted D) and Loss (denoted L). In other words $\mathcal{O} = \{D, L\}$, where D and L are defined as follows. Let $X[i \ldots i + k]$ denote the substring $X_i X_{i+1} \cdots X_{i+k}$ of X.

- D: A **Duplication** of size $k+1$ on $X = X_1 \cdots X_i \cdots X_{i+k} \cdots X_j X_{j+1} \cdots X_n$ is an operation that copies the substring $X[i \ldots i + k]$ to a location j of X outside the interval $[i, i+k]$ (*i.e.* preceding i or following $i+k$). In the latter case, D transforms X into

$$X' = X_1 \cdots X_i \cdots X_{i+k} \cdots X_{j-1} \underline{X_i \cdots X_{i+k}} X_{j+1} \cdots X_n$$

We call the original copy $X[i \ldots i+k]$ the **origin**, and the copied string the **product** of the duplication D.

– L: A **Loss** of size k is an operation that removes a substring of size k from X.

Notice that gene insertions could be considered in our model as well. In particular, our linear programming solution is applicable to an evolutionary model involving insertions, in addition to duplications and losses. We ignore insertions for two main reasons: (1) insertions and losses are two symmetrical operations that can be interchanged in an evolutionary scenario. Distinguishing between insertions and losses may be possible on a phylogeny, but cannot be done by comparing two genomes; (2) gene insertions are usually due to lateral gene transfer, which may be rare events compared to nucleotide-level mutations that eventually transform a gene into a pseudogene.

As duplication and loss are content-modifying operations that do not shuffle gene order, the TWO SPECIES SMALL PHYLOGENY PROBLEM can be posed as an alignment problem. However, the only operations that are "visible" on an alignment are the events on an evolutionary history that are not obscured by subsequent events. Moreover, as duplications and losses are asymmetrical operations, an alignment of two genomes X and Y does not reflect an evolutionary path from X to Y (as operations going back to a common ancestor are not defined), but rather two paths going from a common ancestor to both X and Y. A precise definition follows.

Definition 1. *Let X and Y be two genomes. A **visible history** of X and Y is a triplet $(A, O_{A \to X}, O_{A \to Y})$ where A is a potential ancestor of both X and Y, and $O_{A \to X}$ (respectively $O_{A \to Y}$) are evolutionary histories from A to X (respectively from A to Y) verifying the following property: Let D be a duplication in $O_{A \to X}$ or $O_{A \to Y}$ copying a substring S. Let S_1 be the origin and S_2 be the product of D. Then D is not followed by any other operation inserting (by duplication) genes inside S_1 or S_2, or removing (by loss) genes from S_1 or S_2. We call a **visible ancestor** of X and Y a genome A belonging to a visible history $(A, O_{A \to X}, O_{A \to Y})$ of X and Y.*

We now define an alignment of two genomes.

Definition 2. *Let X be a string on Σ, and let Σ^- be the alphabet Σ augmented with an additional character " $-$ ". An **extension of** A is a string A^- on Σ^- such that removing all occurrences of the character " $-$ " from A^- leads to the string A.*

Definition 3. *Let X and Y be two strings on Σ. An **alignment** of size α of X and Y is a pair (X^-, Y^-) extending (X, Y) such that $|X^-| = |Y^-| = \alpha$, and for each i, $1 \leq i \leq \alpha$, the two following properties hold:*

– *If $X_i^- \neq$ " $-$ " and $Y_i^- \neq$ " $-$ " then $X_i^- = Y_i^-$;*
– *X_i^- and Y_i^- cannot be both equal to " $-$ ".*

Let $\mathcal{A} = (X^-, Y^-)$ be an alignment of X and Y of size α. It can be seen as a $2 \times \alpha$ matrix, where the ith column $\mathcal{A}_i$ of the alignment is just the ith column of the matrix. A column is a **match** iff it does not contain the character '-', and a **gap** otherwise. A gap $\begin{bmatrix} X_i \\ - \end{bmatrix}$ is either part of a loss in Y, or part of a duplication in X (only possible if the character X_i is a duplicate in X). The same holds for the column $\begin{bmatrix} - \\ Y_j \end{bmatrix}$. An interpretation of $\mathcal{A}$ as a sequence of duplications and losses is called a **labeling** of $\mathcal{A}$. The **cost of a labeled alignment** is the sum of costs of all underlying operations.

As duplications and losses are asymmetric operations that are applied explicitly to one of the two strings, each labeled alignment $\mathcal{A}$ of X and Y leads to a unique common ancestor A for X and Y. The following theorem (whose proof is left to the appendix) shows that this, and the converse is true.

Theorem 1. *Given two genomes X and Y, there is a one-to-one correspondence between labeled alignments of X and Y and visible ancestors of X and Y.*

See Figure 1 for an example.

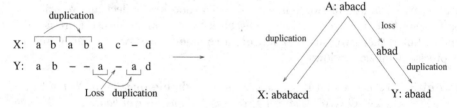

Fig. 1. Left: a labeled alignment between two strings $X = ababacd$ and $Y = abaad$. Right: the ancestor A and two histories respectively from A to X and from A to Y obtained from this alignment. The order of operations in the history from A to Y is arbitrary.

In other words, Theorem 1 states that the TWO SPECIES SMALL PHYLOGENY PROBLEM reduces in the case of the Duplication-Loss model of evolution to the following optimization problem.

Duplication-Loss Alignment Problem:
INPUT: Two genomes X and Y on Σ.
OUTPUT: A labeled alignment of X and Y of minimum cost.

4   Method

Although alignments are *a priori* simpler to handle than rearrangements, a straightforward way to solve the DUPLICATION-LOSS ALIGNMENT PROBLEM is not known. We show in the following paragraphs, that a direct approach based on dynamic programming leads, at best, to an efficient heuristic, with no guarantee of optimality.

Let X be a genome of size n and Y be a genome of size m. Denote by $X[1\ldots i]$ the prefix of size i of X, and by $Y[1\ldots j]$ the prefix of size j of Y. Let $C(i,j)$ be the minimum cost of a labeled alignment of $X[1\ldots i]$ and $Y[1\ldots j]$. Then the problem is to compute $C(m,n)$.

DP: A natural idea would be to consider a dynamic programming approach (DP), computing $C(i,j)$, for all $1 \leq i \leq n$ and all $1 \leq j \leq m$. Consider the variables $M(i,j)$, $D_X(i,j)$, $D_Y(i,j)$, $L_X(i,j)$ and $L_Y(i,j)$ which reflect the minimum cost of an alignment $\mathcal{A}_{i,j}$ of $X[1\ldots i]$ and $Y[1\ldots j]$ satisfying respectively, the constraint that the last column of $\mathcal{A}_{i,j}$ is a match, a duplication in X, a duplication in Y, a loss in X, or a loss in Y. Consider the following recursive formulae.

- $M(i,j) = \begin{cases} C(i-1, j-1) \text{ if } X[i] = Y[j] \\ +\infty \qquad\qquad \text{ otherwise} \end{cases}$

- $L_X(i,j) = \min_{0 \leq k \leq i-1}[C(k,j) + c(L(i-k))]$
  (the corresponding formula holds for $L_Y(i,j)$)

- $D_X(i,j) = \begin{cases} +\infty & \text{if } X[i] \text{ is a singleton} \\ \min_{l \leq k \leq i-1}[C(k,j) + c(D(i-k))] & \text{otherwise,} \end{cases}$
  where $X[l\ldots i]$ is the longest suffix of $X[1\ldots i]$ that is a duplication
  (the corresponding formula holds for $D_Y(i,j)$).

The recursions for D_X and D_Y imply that duplicated segments are always inserted to the right of the origin. Unfortunately, such an assumption cannot be made while maintaining optimality of the alignment. For example, given the cost $c(D(k)) = 1$ and $c(L(k)) = k$, the optimal labeled alignment of $S_1 = abxabxab$ and $S_2 = xabx$ aligns ab of S_2 with the second ab of S_1, leading to an optimal history with two duplications inserting the second ab of S_1 to its left and to its right. Such an optimal scenario cannot be recovered by **DP**.

DP-2WAY: As a consequence of the last paragraph, consider the two-way dynamic programming approach DP-2WAY that computes $D_X(i,j)$ (resp. $D_Y(i,j)$) by looking for the longest suffix of $X[1\ldots i]$ (resp. $Y[1\ldots j]$) that is a duplication in the whole genome X (resp. Y). Unfortunately, DP-2WAY may lead to invalid cyclic evolutionary scenarios, as the same scenario may involve two duplications: one with origin S_1 and product S_2, and one with origin S_2 and product S_1, where S_1 and S_2 are two duplicated strings. This is described in more detail in Section 4.1.

DP-2WAY-UNLABELED: The problem mentioned above with the output of DP-2WAY is not necessarily the alignment itself, but rather the label of the alignment. As a consequence, one may think about a method, DP-2WAY-UNLABELED, that would consider the unlabeled alignment output by DP-2WAY, and label it in an optimal way (*e.g.* find an evolutionary scenario of minimum cost that is in agreement with the alignment). Notice first that the problem of finding a most parsimonious labeling of a given alignment, is not *a priori* an easy problem, and there is no direct and simple way to do it. Moreover, although DP-2WAY-UNLABELED is likely to be a good heuristic algorithm to the DUPLICATION-LOSS ALIGNMENT PROBLEM, it would not be an exact algorithm, as an optimal cyclic alignment is not guaranteed to have a valid labeling leading to an optimal labeled alignment. Figure 2 shows such an example; an optimal cyclic duplication and loss scenario can be achieved by both alignments (5 operations), while the optimal acyclic scenario can only be achieved by the alignment of Figure 2b.

4.1 The Pseudo-Boolean Linear Program

Consider genome X of length n and genome Y of length m. We show how to compute a labeled alignment of X and Y by use of pseudo-boolean linear programming (PBLP). The alignment that we compute is guaranteed to be optimal. While in the worst case our program could take an exponential number (in the length of the strings) of steps to find the alignment, our formulation has a cubic number of equations variables, and is far more efficient than scoring all possible alignments along with all possible duplication/loss scenarios. We show that practical running times can be achieved on real data in Section 5.2.

For any alignment, an element of the string X could be considered a loss (this corresponds to gaps in the alignment), a match with an element of Y, or a duplication from another element in X (these also appear as gaps in the alignment). Thus, in a feasible solution, every element must be "covered" by one of those three possibilities. The same holds for elements of Y. Figure 2 shows two

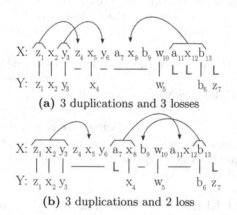

(a) 3 duplications and 3 losses

(b) 3 duplications and 2 loss

Fig. 2. Alignments for strings X = "zxyzxyaxbwaxb" and Y = "zxyxwb". We consider the following cost: $c(D(k)) = 1$ and $c(L(k)) = k$ for any integer k. Matches are denoted by a vertical bar, losses denoted by an "L", and duplications denoted by bars, brackets, and arrows. Alignment (a) yields 6 operations and implies ancestral sequence "zxyxwaxbz", while (b) yields 5 operations and implies ancestral sequence "zxyaxwbz".

possible alignments for a given pair of strings, along with the corresponding set of duplications and losses. In the alignment of Figure 2a, character x_8 is covered by the duplication of x_{12}, character a_{11} is covered by a loss, and character x_5 is covered by a match with character x_4 in Y.

Let M_j^i signify the match of character X_i to character Y_j in the alignment. Say character X_i could be covered by matches $M_1^i, M_2^i, \ldots, M_{p_i}^i$ or by duplications $DX_1^i, DX_2^i, \ldots, DX_{s_i}^i$. If we consider each of those to be a binary variable (can take value 0 or 1) and take the binary variable LX^i as corresponding to the possibility that X_i is a loss, then we have the following equation to ensure that character X_i is covered by exactly one operation:

$$LX^i + M_1^i + M_2^i + \cdots + M_{p_i}^i + DX_1^i + DX_2^i + \cdots + DX_{s_i}^i = 1, \qquad (1)$$

where p_i and s_i are the number of matches and duplications that could cover character X_i. A potential duplication in X (DX_l^i for some l) corresponds to a pair of distinct, but identical, substrings in X. Each pair of such substrings yields two potential duplications (substring A was duplicated from substring B or substring B was duplicated from substring A). Each of the possible $O(n^3)$ duplications gets a variable. Each position in Y gets a similar equation to Equation 1.

The order of the matches with respect to the order of the strings must be enforced. For example, in Figure 2 it is impossible to simultaneously match w_{10} and x_{12} from X with w_5 and x_4 of string Y; the assignment of variables corresponding to this case must be forbidden in the program. Thus, we introduce equations enforcing the order of the matches. Recall that M_j^i is the variable corresponding to the match of the ith character from X with the jth character from Y. The existence of match M_j^i implies that any match M_l^k where $k \geq i$ and $l \leq j$ (or $k \leq i$ and $l \geq j$) is impossible, so must be forbidden by the program. The constraints can be written as:

$$M_j^i + M_{l_1}^{k_1} \leq 1, \ M_j^i + M_{l_2}^{k_2} \leq 1, \ldots, \ M_j^i + M_{l_{t_i}}^{k_{t_i}} \leq 1 \qquad (2)$$

where t_i is the number of matches conflicting with M_j^i, and for any $M_{l_u}^{k_u}$ we have either $k_u \geq i$ and $l_u \leq j$, or $k_u \leq i$ and $l_u \geq j$. There are at most a linear number of inequalities for each of the possible $O(n^2)$ matches.

Equality 1 ensures that each character will be covered by exactly one D, M, or L. Our objective function minimizes some linear combination of all Ls and all Ds:

$$min \ c_1 LX^1 + \cdots + c_n LX^n + c_{n+1} LY^1 + \cdots + c_{n+m} LY^m + c_{n+m+1} D_1 + \cdots + c_{n+m+q} D_q \qquad (3)$$

where c_l is a cost of the lth operation and q is the total number of duplications for all positions of X and Y (i.e. $D_l = DX_s^i$ or $D_l = DY_r^j$ for some $i, j, s,$ and r). The full PBLP (excluding the trivial integrality constraints) is then:

$$\min \quad c_1 LX^1 + \cdots + c_n LX^n + c_{n+1} LY^1 + \cdots + c_{n+m} LY^m +$$
$$c_{n+m+1} D_1 + c_{n+m+2} D_2 + \cdots + c_{n+m+q} D_q$$

$$\text{s.t.} \quad LX^i + M_1^i + M_2^i + \cdots + M_{p_i}^i + DX_1^i + \cdots + DX_{s_i}^i = 1, \quad 0 \le i \le n$$
$$LY^j + M_j^1 + M_j^2 + \cdots + M_j^{q_j} + DY_1^j + \cdots + DY_{r_j}^j = 1, \quad 0 \le j \le m$$
$$M_j^i + M_{l_1}^{k_1} \le 1, \; M_j^i + M_{l_2}^{k_2} \le 1, \ldots, M_j^i + M_{l_{t_i}}^{k_{t_i}} \le 1 \quad , \quad \forall i,j \text{ s.t. } X_i = Y_j \text{ and}$$
$$(k_m \ge i \text{ and } l_m \le j), \text{ or}$$
$$(k_m \le i \text{ and } l_m \ge j)$$

In the example illustrated in Figure 2 — where $X = $ "zxyzxyaxbwaxb" and $Y = $ "zxyxwb" — there are 17 variables corresponding to matches, 20 variables corresponding to losses, and 24 variables corresponding to duplications.

Cyclic Duplications. Recall the definition of the product of a duplication; in Figure 2b, the product of the leftmost duplication is z_4, x_5, and y_6. Consider a sequence of duplications $D_1, D_2, \ldots, D_l$ and characters $a_1, a_2, \ldots, a_l$ such that character a_i is in the product of D_i and is the duplication of the character a_{i-1} (a_1 is the duplication of some character a_0). We call this set of duplications **cyclic** if $D_1 = D_l$. Consider the set of duplications $\{D_1, D_2\}$ where D_1 duplicates the substring $X_1 X_2$ to produce the substring $X_3 X_4$ and D_2 duplicates the substring $X_4 X_5$ to produce the substring $X_1 X_2$. This implies the sequence of characters X_2, X_4, X_1, X_3 corresponding to the cyclic duplication D_1, D_2, D_1.

Theorem 2. *A solution to the PBLP of Section 4.1 that has no cyclic set of duplications is an optimal solution to the Duplication-Loss Alignment problem.*

Proof. Equation 1 ensures that each character of X is either aligned to a character of Y, aligned to a loss in Y, or the product of a duplication. The similar holds for each character of Y. Since there exists no cyclic set of duplications, then the solution given by the PBLP is a feasible solution to the Duplication-Loss Alignment problem. The minimization of Formula 3 guarantees optimality. $\square$

However, if there does exist a cyclic duplication set, the solution given by the PBLP is not a feasible solution since the cycle implies a scenario that is impossible; the cycle implies a character that does not exist in the ancestor but does appear in X. A cyclic duplication set $\{D_1, D_2, \ldots, D_l\}$ can be forbidden from a solution of the PBLP by the inclusion of the following inequality:

$$D_1 + D_2 + \cdots + D_l \le l - 1. \tag{4}$$

The following algorithm guarantees an acyclic solution to the PBLP.

It simply runs the PBLP and each time it finds a cyclic set of duplications, it adds the corresponding constraint to forbid the set and reruns the PBLP. It is clear that the algorithm of Figure 1 guarantees an acyclic solution:

Algorithm 1 Pairwise-Alignment(PBLP)

---

get solution S to the PBLP
while S has a cycle **do**
  for each cycle $D_1 + D_2 + \cdots + D_l$ in S **do**
    PBLP $\leftarrow$ PBLP plus constraint $D_1 + D_2 + \cdots + D_l \leq l - 1$
  end for
  get solution S to the PBLP
end while
return S

Theorem 3. *Algorithm 1 returns an optimal solution to the Duplication-Loss Alignment problem.*

Note that the constraints to forbid all possible cyclic sets of duplications, given a particular X and Y, could be added to the PBLP from the start, but in the worst case there is an exponential number of such constraints. We will see in Section 5.2 that in practice we do not have to rerun the PBLP many times to find an acyclic solution.

5   Applications

5.1   Evolution of Stable RNA Gene Content and Organization in *Bacillus*

The stable RNAs are chiefly transfer RNAs (tRNAs) and ribosomal RNAs (rRNAs), which are essential in the process of translating messenger RNAs (mR-NAs) into protein sequences. They are usually grouped in the genome within clusters (or operons in the case of microbial genomes), representing highly repetitive regions, causing genomic instability through illegitimate homologous recombination. In consequence, stable RNA families are rapidly evolving by duplication and loss [35,3,28].

We applied Algorithm 1 in a phylogenetic context, using the SPP-HEURISTIC described at the end of Section 2, to analyze the stable RNA content and organization of 5 *Bacillus* lineages: *Bacillus cereus ATCC 14579 (NC4722)*, *Bacillus cereus E33L (NC6274)*, *Bacillus anthracis (NC7530)*, *Bacillus licheniformis ATCC 14580 (NC6322)* and *Bacillus subtilis (NC964)*. The overall number of represented RNA families in these genomes is around 40, and the total number of RNAs in each genome is around 120. Our PBLP algorithm processes each pair of these genomes in a few seconds. We used the following cost for duplications and losses: $c(D(k)) = 1$ and $c(L(k)) = k$, for any integer k representing the size of an operation. The phylogeny in Figure 3 reflects the NCBI taxonomy. Each leaf is labeled by a block representation of the corresponding genome. Details on each colored block is given in Figure 4 of the appendix.

The costs and evolutionary scenarios given in Figure 3 are those output by our algorithm after interpretation. In particular, the five *Bacillus* genomes all show a large inverted segment in the region to the left of the origin of replication (the right part of each linearized representation in Figure 3). As our algorithm does not handle inversions, we preprocessed the genomes by inverting this segment. The genome representations given in Figure 3 are, however, the true ones. Signs given below the red bars represent their true orientations. Consequently, the duplication of the right-most red bar to the left-most position should be interpreted as an inverted block duplication that occurred around the origin of replication. On the other hand, some duplications of the red bar have been reported by our algorithm as two separate duplications of two segments separated by a single gene. After careful consideration (see Figure 5 of the appendix), these pairs of duplications are more likely a single duplication obscured by subsequent substitution, functional shift or loss of a single tRNA gene. Also, when appropriate (*e.g.* when a lone gene is positioned in a lone genome), we interpreted some of the losses in our alignments as insertions.

The ancestral genomes given in Figure 3 are those output by our algorithm. They reflect two separate inverted duplications that would have occurred independently in each of the two groups (*cereus, anthracis*) and (*licheniformis, subtilis*). We could alternatively infer that the ancestral *Bacillus* genome already contained both the origin and product of the inverted duplication. The consequence would be the simultaneous loss of the leftmost red bar in three of the five considered genomes. Moreover, nucleotide sequence alignment of the red bars in *subtilis* and *cereus ATCC* reveal a higher conservation of pairs of paralogous bars versus orthologous ones, which may indicate that inverted duplications are recent. Whatever the situation is, our inference of a duplication around the origin of replication is in agreement with the observation that has been largely reported in the literature that bacterial genomes have a tendency to preserve a symmetry around the replication origin and terminus [13,33,1]. The results also show independent proliferation of ribosomal RNA gene-containing operons in *Bacillus*, which has been associated to selection for increased growth rate [2]. They also show that in *Bacillus*, growth-selection on ribosomal RNA operon expansions may significantly alter tRNA gene content as well. The results given in Figure 5 of the appendix also suggest that some tRNA genes may have been affected by substitutions leading to conversions of function. Such tRNA functional shifts have been detected in metazoan mitochondrial genomes [27] and bacteria [29].

5.2   Execution Time

Running times were recorded using a 12-core AMD 2.1GHZ processor, with 256GB of RAM, and the (multithreaded) IBM CPLEX solver under the default settings. Note that a significant amount of memory (> 1GB) was required only for sequences of several thousand genes; all tests reported here could be run on a standard laptop with 2 GB of memory. Alignments of all pairs of genomes for each of three sets of stable RNA gene orders were computed. The average

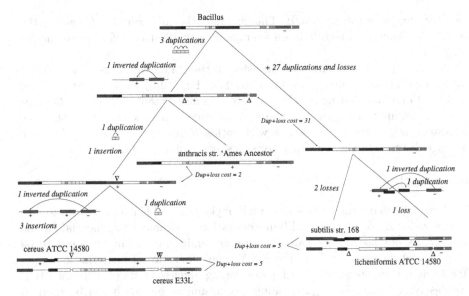

Fig. 3. An inferred evolutionary history for the five *Bacillus* lineages identified with each of the five leaves of the tree. Circular bacterial genomes have been linearized according to their origin of replication (*e.g.* the endpoints of each genome is its origin of replication). Bar length is proportional to the number of genes in the corresponding cluster. A key for the bars is given in Figure 4, except for white bars that represent regions that are perfectly aligned inside the two groups (*cereus, anthracis*) and (*licheniformis, subtilis*), but not between the two groups. More specifically, the 27 duplications and losses reported on the top of the tree are obtained from the alignment of these white regions. Finally, each Δ represents a loss and each ∇ is an insertion.

computation time for the *Bacillus* pairs was under thirty seconds. The average computation time for pairs from 13 *Staphylococcus* was under a second. Pairs from a dataset of *Vibrionaceae* which had a very high number of paralogs and a large number of rearrangements took a couple of days.

5.3   Simulations

Simulations were run in order to explore the limits of our method (full results not shown due to space limitations). A random sequence R was drawn from the set of all sequences of length n and alphabet size a. l moves were then applied to R to obtain the ancestral sequence A. To obtain the extant sequences X and Y, l more moves were applied to A for each. The set of moves were segmental duplications and single gene losses. The length of a duplication was drawn from a Gaussian distribution with mean 5 and standard deviation 2; these lengths were consistent with observations on *Bacillus* and *Staphylococcus*. Average running times for sequences with a fixed ratio of $2l/n = 1/5$ and $a/n = 1/2$ (statistics similar to those observed in *Bacillus*) were always below 6 minutes for $n < 800$. Sequences of length 2000 took less than 2 hours and sequences of length 5000 took a couple of days. When varying l, n, and a the most telling factor for running time was the ratio a/n. This explains the high running times for the set of *Vibrionaceae* which had, on average, nearly 100 moves for a sequence of length 140.

The distance of our computed ancestor to the simulated ancestor was found by computing an alignment between the two. For values of $n = 120$, $a = 40$, and $l = 15$ (values that mimic the statistics of more distant pairs of the *Bacillus* data) we compute ancestors that are, on average, 5 moves away from the true ancestor. In general, for sequences with ratios $2l/n = 1/5$ and $a/n = 1/2$, the average distance to the true ancestor stays at about 15% of l.

6 Conclusion

We have considered the two species small phylogeny problem for an evolutionary model reduced to content-modifying operations. Although exponential in the worst case, our pseudo-boolean linear programming algorithm turns out to be fast on real datasets, such as the RNA gene repertoire of bacterial genomes. We have also explored avenues for developing efficient non-optimal heuristics. As described in Section 4, a dynamic programming approach can be used to infer a reasonable, though not necessarily optimal unlabeled alignment of two genomes. An important open problem is then to discern the complexity of finding a minimum labeling for a given alignment. The implication being that if this problem is NP-hard, then the two species small phylogeny problem is not even in NP.

Application to the *Bacillus* lineages has pointed out a number of generalizations that could be introduced to the evolutionary model, such as inversions, inverted duplications, gene conversion (substitutions), and insertions. Substitutions can easily be accommodated in our PBLP by allowing matching of different genes, and attributing a cost according to the likelihood of that gene conversion. Our boolean programming model is also likely to handle "visible", in term of non-crossing inversions and inverted duplications. As for insertions, they can not be distinguished from losses by the comparison of pairs of genomes. A deterministic methodology could, however, be established to distinguish between them based on results obtained on a complete phylogenetic tree.

References

1. Ajana, Y., Lefebvre, J.F., Tillier, E., El-Mabrouk, N.: Exploring the Set of All Minimal Sequences of Reversals - An Application to Test the Replication-Directed Reversal Hypothesis. In: Guigó, R., Gusfield, D. (eds.) WABI 2002. LNCS, vol. 2452, pp. 300–315. Springer, Heidelberg (2002)
2. Ardell, D.H., Kirsebom, L.A.: The genomic pattern of tDNA operon expression in E. *coli*. PLoS Comp. Biol. 1(1:e12) (2005)
3. Bermudez-Santana, C., Attolini, C.S., Kirsten, T., Engelhardt, J., Prohaska, S.J., Steigele, S., Stadler, P.: Genomic organization of eukaryotic tRNAs. BMC Genomics 11(270) (2010)
4. Blanchette, M., Bourque, G., Sankoff, D.: Breakpoint phylogenies. In: Genome Informatics Workshop (GIW), pp. 25–34 (1997)

5. Blomme, T., Vandepoele, K., De Bodt, S., Silmillion, C., Maere, S., van de Peer, Y.: The gain and loss of genes during 600 millions years of vertebrate evolution. Genome Biology 7, R43 (2006)
6. Bourque, G., Pevzner, P.A.: Genome-scale evolution: Reconstructing gene orders in the ancestral species. Genome Research 12, 26–36 (2002)
7. Caprara, A.: Formulations and hardness of multiple sorting by reversals. In: RECOMB, pp. 84–94 (1999)
8. Chauve, C., Tannier, E.: A methodological framework for the reconstruction of contiguous regions of ancestral genomes and its application to mammalian genomes. PloS Computational Biology 4, e1000234 (2008)
9. Cotton, J.A., Page, R.D.M.: Rates and patterns of gene duplication and loss in the human genome. Proceedings of the Royal Society of London. Series B 272, 277–283 (2005)
10. Demuth, J.P., De Bie, T., Stajich, J., Cristianini, N., Hahn, M.W.: The evolution of mammalian gene families. PLoS ONE 1, e85 (2006)
11. Dong, H., Nilsson, L., Kurland, C.G.: Co-variation of tRNA abundance and codon usage in Escherichia coli at different growth rates. Journal of Molecular Biology 260, 649–663 (2006)
12. Eichler, E.E., Sankoff, D.: Structural dynamics of eukaryotic chromosome evolution. Science 301, 793–797 (2003)
13. Eisen, J.A., Heidelberg, J.F., White, O., Salzberg, S.L.: Evidence for symmetric chromosomal inversions around the replication origin in bacteria. Genome Biology 1(6) (2000)
14. El-Mabrouk, N.: Genome rearrangement with gene families. In: Mathematics of Evolution and Phylogeny, pp. 291–320. Oxford University Press (2005)
15. Fertin, G., Labarre, A., Rusu, I., Tannier, E., Vialette, S.: Combinatorics of genome rearrangements. The MIT Press, Cambridge (2009)
16. Hahn, M.W., Han, M.V., Han, S.-G.: Gene family evolution across 12 drosophilia genomes. PLoS Genetics 3, e197 (2007)
17. Hannenhalli, S., Pevzner, P.A.: Transforming men into mice (polynomial algorithm for genomic distance problem). In: Proceedings of the IEEE 36th Annual Symposium on Foundations of Computer Science, pp. 581–592 (1995)
18. Hannenhalli, S., Pevzner, P.A.: Transforming cabbage into turnip (polynomial algorithm for sorting signed permutations by reversals). JACM 48, 1–27 (1999)
19. Jiang, M.: The Zero Exemplar Distance Problem. In: Tannier, E. (ed.) RECOMB-CG 2010. LNCS, vol. 6398, pp. 74–82. Springer, Heidelberg (2010)
20. Kanaya, S., Yamada, Y., Kudo, Y., Ikemura, T.: Studies of codon usage and tRNA genes of 18 unicellular organisms and quantification of Bacillus subtilis tRNAs: Gene expression level and species-specific diversity of codon usage based on multivariate analysis. Gene 238, 143–155 (1999)
21. Kováč, J., Brejová, B., Vinař, T.: A Practical Algorithm for Ancestral Rearrangement Reconstruction. In: Przytycka, T.M., Sagot, M.-F. (eds.) WABI 2011. LNCS (LNAI), vol. 6833, pp. 163–174. Springer, Heidelberg (2011)
22. Lynch, M., Conery, J.S.: The evolutionary fate and consequences of duplicate genes. Science 290, 1151–1155 (2000)
23. Ma, J., Zhang, L., Suh, B.B., Raney, B.J., Burhans, R.C., Kent, W.J., Blanchette, M., Haussler, D., Miller, W.: Reconstructing contiguous regions of an ancestral genome. Genome Research 16, 1557–1565 (2007)
24. Moret, B., Wang, L., Warnow, T., Wyman, S.: New approaches for reconstructing phylogenies from gene order data. Bioinformatics 173, S165–S173 (2001)

25. Ohno, S.: Evolution by gene duplication. Springer, Berlin (1970)
26. Pe'er, I., Shamir, R.: The median problems for breakpoints are NP-complete. Elec. Colloq. on Comput. Complexity 71 (1998)
27. Rawlings, T.A., Collins, T.M., Bieler, R.: Changing identities: tRNA duplication and remolding within animal mitochondrial genomes. Proceedings of the National Academy of Sciences USA 100, 15700–15705 (2003)
28. Rogers, H.H., Bergman, C.M., Griffiths-Jones, S.: The evolution of tRNA genes in Drosophila. Genome Biol. Evol. 2, 467–477 (2010)
29. Saks, M.E., Conery, J.S.: Anticodon-dependent conservation of bacterial tRNA gene sequences. RNA 13(5), 651–660 (2007)
30. Sankoff, D., Blanchette, M.: The Median Problem for Breakpoints in Comparative Genomics. In: Jiang, T., Lee, D.T. (eds.) COCOON 1997. LNCS, vol. 1276, pp. 251–263. Springer, Heidelberg (1997)
31. Tang, D.T., Glazov, E.A., McWilliam, S.M., Barris, W.C., Dalrymple, B.P.: Analysis of the complement and molecular evolution of tRNA genes in cow. BMC Genomics 10(188) (2009)
32. Tannier, E., Zheng, C., Sankoff, D.: Multichromosomal median and halving problems under different genomic distances. BMC Bioinformatics 10 (2009)
33. Tillier, E.R.M., Collins, R.A.: Genome rearrangement by replication-directed translocation. Nature Genetics 26 (2000)
34. Wapinski, I., Pfeffer, A., Friedman, N., Regev, A.: Natural history and evolutionary principles of gene duplication in fungi. Nature 449, 54–61 (2007)
35. Withers, M., Wernisch, L., Dos Reis, M.: Archaeology and evolution of transfer RNA genes in the *escherichia coli* genome. Bioinformatics 12, 933–942 (2006)

A   Proof of Theorem 1

Proof. Let A be a visible ancestor of X and Y and $(A, O_{A \to X}, O_{A \to Y})$ be a visible history of X and Y. Then construct a labeled alignment $\mathcal{A}$ of X and Y as follows:

1. *Initialization*: Define the two strings $X^- = Y^- = A$ on Σ^-, and define $\mathcal{A}$ as an alignment with all matches between X^- and Y^- (*i.e.* self-alignment of A).
2. Consider each operation of $O_{A \to X}$ in order.
   - If it is a duplication, then add the inserted string at the appropriate position in X^-, and add gaps ("$-$" characters) at the corresponding positions in Y^-. Label the inserted columns of $\mathcal{A}$ as a duplication in X, coming from the columns of the alignment representing the origin of the duplication.
   - If it is a loss, then replace the lost characters in X^- by gaps. Label the modified columns as a loss in X.
3. Consider each operation of $O_{A \to Y}$ and proceed in a symmetrical way.

As $(A, O_{A \to X}, O_{A \to Y})$ is a visible history of X and Y, by definition the origins and products of duplications remain unchanged by subsequent operations on each of $O_{A \to X}$ and $O_{A \to Y}$. Therefore, all intermediate labellings remain valid in the final alignment. Therefore, the constructive method described above leads to a labeled alignment of X and Y.

On the other hand, a substring $X_i \cdots X_j$ (resp. $Y_i \cdots Y_j$) that is labeled as a duplication in X (resp. Y) should not be present in A, as it is duplicated on the branch from A to X (resp. from A to Y). Also, a substring $X_i \cdots X_j$ (resp. $Y_i \cdots Y_j$) that is labeled as a loss in Y (resp. X) should be present in A, as it is lost on the branch from A to Y (resp. from A to X). This implies an obvious algorithm to reconstruct a unique ancestor.   □

B   Alignment Details

Ribosomal RNAs are grouped in bacteria into three families: the 16S, 5S and 23S rRNAs. As the major role of tRNAs is to serve as adapters between codons along the mRNA and the corresponding amino acids, we group them according to their anticodon. More precisely, the four-letter designation starts with one letter indicating functional class (either an IUPAC one-letter code for a charged amino acid, "X" for initiator or "J" for a special class of isoleucine tRNA) followed by an anticodon sequence in a DNA alphabet. The full alignment as given by our program can be retrieved at: http://lcbb4.epfl.ch/web/BacillusAlignements/.

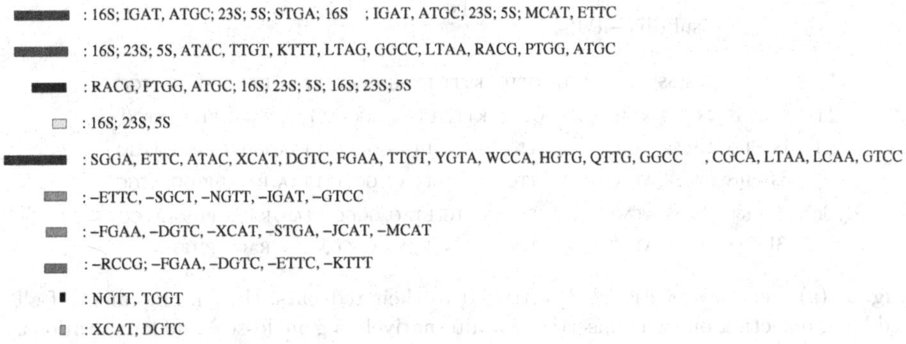

: 16S; IGAT, ATGC; 23S; 5S; STGA; 16S   ; IGAT, ATGC; 23S; 5S; MCAT, ETTC

: 16S; 23S; 5S, ATAC, TTGT, KTTT, LTAG, GGCC, LTAA, RACG, PTGG, ATGC

: RACG, PTGG, ATGC; 16S; 23S; 5S; 16S; 23S; 5S

: 16S; 23S, 5S

: SGGA, ETTC, ATAC, XCAT, DGTC, FGAA, TTGT, YGTA, WCCA, HGTG, QTTG, GGCC   , CGCA, LTAA, LCAA, GTCC

: –ETTC, –SGCT, –NGTT, –IGAT, –GTCC

: –FGAA, –DGTC, –XCAT, –STGA, –JCAT, –MCAT

: –RCCG; –FGAA, –DGTC, –ETTC, –KTTT

: NGTT, TGGT

: XCAT, DGTC

Fig. 4. The RNA gene clusters represented by each colored bar of Figure 3. Ribosomal RNAs are identified by their families' names: 16S, 23S, 5S. Each tRNA is identified by its anticodon preceded by the letter corresponding to the corresponding amino-acid. The symbol ',' (respectively ';') separates two genes that are inside (resp. not inside) the same operon.

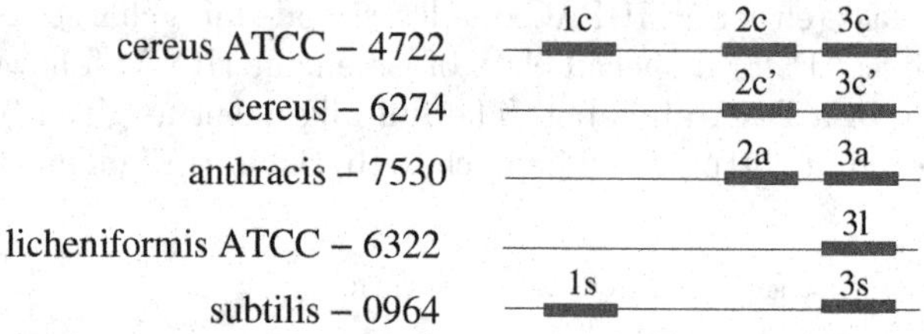

<pre>
                           1c   2c   3c
     cereus ATCC – 4722 ━━━━━■■■━━━━━━━■■━━━■■━━━━
                                       2c'   3c'
         cereus – 6274   ━━━━━━━━━━━━━━━■■━━━■■━━━━━━
                                          2a   3a
         anthracis – 7530   ━━━━━━━━━━━━━━━■■━━━■■━━━━
                                                 3l
  licheniformis ATCC – 6322 ━━━━━━━━━━━━━━━━━━━■■━━━━
                             1s   3s
         subtilis – 0964   ━━━━━■■■━━━━━━━━━━━━━■■━━━━
</pre>

2c, 2c' : 16S; 23S; 5S, ATAC, YGTA, QTTG, KTTT, LTAG, GGCC, LTAA, RACG, PTGG, ATGC

2a   : 16S; 23S; 5S, ATAC,   QTTG, KTTT, LTAG, GGCC, LTAA, RACG, PTGG, ATGC

1c, 1s : 16S; 23S; 5S, ATAC,   TTGT, KTTT, LTAG, GGCC, LTAA, RACG, PTGG, ATGC

   3s : 16S; 23S; 5S, ATAC,   TTGT, KTTT, LCAG, GGCC, LTAA, RACG, PTGG, ATGC

3c, 3c', 3a : 16S; 23S; 5S, ATAC,   TTGT, HGTG, LTAG, GGCC, LTAA, RACG, PTGG, ATGC

   3l : 16S; 23S; 5S, ATAC,   TCGT, KTTT, LCAG, GGCC,   RACG, PTGG, ATGC

Fig. 5. (a) Genomes of Figure 3 restricted to their red bars; (b) An alignment of all red bars, reflecting one gene insertion (or alternatively 5 gene losses) and substitutions, indicated by '*'

LoopWeaver – Loop Modeling
by the Weighted Scaling of Verified Proteins

Daniel Holtby[1], Shuai Cheng Li[2], and Ming Li[1]

[1] University of Waterloo
{djholtby,mli}@uwaterloo.ca
[2] City University of Hong Kong

Abstract. Modeling loops is a necessary step in protein structure determination even with experimental NMR data. It is well known to be difficult. Database techniques have the advantage of producing a higher proportion of predictions with sub-angstrom accuracy, when compared with *ab initio* techniques, but the disadvantage of also producing a higher proportion of clashing or highly inaccurate predictions. We introduce LoopWeaver, a database method that uses multidimensional scaling to achieve better clash-free placement of loops obtained from a database of protein structures. This allows us to maintain the above-mentioned advantage while avoiding the disadvantage. Test results show that we achieve significantly better results than all other methods, including Modeler, Loopy, SuperLooper, and Rapper before refinement. With refinement, our results (LoopWeaver and Loopy consensus) are better than ROSETTA, with 0.42Å RMSD on average for 206 length 6 loops, 0.64Å local RMSD for 168 length 7 loops, 0.81Å RMSD for 117 length 8 loops, and 0.98Å RMSD for length 9 loops, while ROSETTA has 0.55, 0.79, 1.16, 1.42, respectively, at the same average time limit (3 hours). When we allow ROSETTA run for over a week, it approaches, but does not surpass, our accuracy.

Keywords: molecular structural biology, loop modeling, loop prediction, database search.

1   Problem Definition

Loop modeling is a common and important problem in protein modeling. Loops are the least conserved portions of a protein structure, meaning they are the most likely parts to be different in proteins that are otherwise very similar.

When loop modeling, we will be given a *target protein structure* with a *gap*. This gap is a region of consecutive residues where the protein structure is missing atomic coordinates. The goal is to generate a realistic *loop* in order to fill in the gap and obtain a protein structure model with no break in the backbone. The edges of the gap are called *stems* and, where distinction is required, they are called the *C-stem* and *N-stem* after which atom is exposed to the gap being filled. The Euclidean distance between these two atoms is called the *span* of the gap, while the number of residues missing is the *length* of the gap.

B. Chor (Ed.): RECOMB 2012, LNBI 7262, pp. 113–126, 2012.
© Springer-Verlag Berlin Heidelberg 2012

Such gaps arise frequently in predictive protein models build using a template based approach, where a protein's structure is predicted using one or more *templates*, proteins with similar sequence and known structure. Any region that does not have at least one template match covering it must be modeled separately. Additionally, any indel events between the sequence and the template match will also result in gaps, since there will not be a one-to-one matching of residues between the two sequences surrounding the indel. Both problems are typically confined to the highly variable and flexible loop regions. Gaps can also occur in experimental protein models, especially NMR models. The flexible loop regions typically move around a great deal in solvent, making it difficult to get NMR data for these regions.

2   Related Work

There are many techniques used to model loops. Broadly, they can be grouped into two categories. The first we call *ab initio* or *statistical* methods. These methods work by sampling loops from a statistical model, and ensuring they fit properly into the gap. The most frequently cited example of this is the Mod-Loop package of MODELLER [7,6]. This program starts with a straight loop connecting the two end points, which is then adjusted and refined using several methods so that it takes on a more realistic shape while maintaining closure. Another technique is that used in the Loopy program [18], which generates loops by sampling torsion angle pairs, and then making minor changes that maintain the realistic nature of the random angles while working toward closure. The RAPPER tool [1] builds a fragment starting at one stem and working toward the other. Angles are sampled, and the fragment is extended with these sampled angles. If at any point a fragment is obtained that cannot connect to the far stem regardless of subsequent sampling, it is discarded.

The second category of loop modeling technique is the *knowledge based* or *database* method. These methods work by finding existing loops that can be placed into the gap. In 2003, Michalsky, Goede, and Preissner [14], presented the LIP database, and demonstrated that there are cases where a very close match from the database can be used to obtain a very accurate prediction. A more recent version of the LIP database is used by the SuperLooper server [8]. Another knowledge-based loop modeler is FREAD [5], which recently has been reevaluated with newer data [3] and improved methodology, giving improved accuracy at the cost of being less likely to produce a match. Both techniques are very similar. A database of known protein structures is searched for loops that will fit into the gap region well. Matches are found by similarity of the stems, to ensure that the loop will fit into the gap, and by sequence similarity, so that the match will be related to the loop being modeled. The matches are then placed into the gap, based only on the stems.

When compared to statistical methods, database methods have the advantage that, should a very similar loop exist in the database, the prediction based on that similar loop will be very accurate. As the protein databank continues to grow, database techniques become more accurate with no changes to the technique itself. On the other hand, if there is no match that fits into the gap, or all matches that do fit clash with the rest of the protein, then the database technique will be unable to produce a result. Even with low stem RMSD scores, there will often be unrealistic torsion angles at the edges of the loop, and even a few bad angles can cause serious problems with many energy potential functions.

Some tools, such as ROSETTA, when using the cyclic coordinate descent (CCD) algorithm [16], and FALC [10] are between the two, where loops are generated statistically, but based on sequence specific models derived from fragment matches, rather than generic models. Where a purely statistical tool would generate loop candidates by sampling from residue-specific ϕ/ψ tables, ROSETTA and FALC sample from position-specific ϕ/ψ tables that are built using fragment matches.

3   Method Overview

Our method, LoopWeaver is a database method. Matches are found in a database of known structures, using the RMSD of the stem regions. These matches are then placed into the gap and ranked. Our method for placing these loops into the gap is substantially different from that used in existing database methods such as SuperLooper and FREAD. Rather than using the RMSD of the stems to determine the placement, we formulate the problem as an instance of the Weighted Multi-Dimensional Scaling problem [11], and solve it using established heuristics. This improves the orientation by reducing the unreasonable angles at the edges of the loop, which means that energy functions are better able to rank our results. Additionally, this method fixes chain-breaks introduced if there are no matches with sufficiently low stem RMSD, and can be extended to fix most of the clashes that occur when placing the loop into the gap. A chain-break is when there is an unrealistic bond length where the loop connects with either stem (or internally within the loop). On a 2.2 GHz Opteron, it takes roughly 5 minutes to find and close 500 database matches for a length 10 loop. We use the DFIRE [23] energy potential function to rank the final loop candidates. As this is an all-atom potential function, the results must have accurate side-chains built. Side-chains are built using the TreePack tool [19,20]. This adds an average of 5 minutes to the running time. Side-chain packing involves all of the atoms surrounding the region being packed, so this step is also mostly independent of the length of the loop. For some tests, our best candidate is then refined using the KIC (KInematic Closure) refinement protocol included in ROSETTA 3.3.

Figure 1 gives example LoopWeaver results. In (a) we have a case where the database loop is nearly identical to the native loop, global RMSD 0.62Å. Meanwhile, both the ModLoop and Loopy candidates are pulled to the right, with global RMSD 4.37Å and 3.10Å respectively. In (b) we have a more difficult case. With no good match found in the database, selecting from the medium quality matches is difficult, and without scaling we would select the orange loop (4.06Å global RMSD). Multi-dimensional scaling improves the torsion angles and bond lengths where the loop connects to the anchors, allowing for better candidate ranking, and LoopWeaver selects the blue loop (1.47 Å).

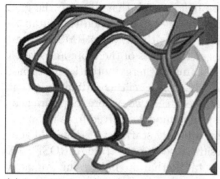 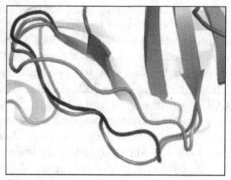

(a) LoopWeaver (blue) compared with ModLoop (orange), Loopy (magenta), and native structure (green). Protein T0513, loop length 10, loop B.

(b) LoopWeaver (blue) compared with unmodified database loop placement (orange) and native structure (green). Protein T0533, loop length 11, loop A.

Fig. 1. Example LoopWeaver results

4   Results

4.1   Test Sets

As LoopWeaver is a database driven tool, it cannot be benchmarked using the same target proteins as used in older loop modelling papers. Doing so either puts the database tool at substantial advantage by using a current database (because even if the exact matches are excluded, there still may be many similar matches in the database), or substantial disadvantage by using a database from the same time as the test set, which negates the advantage of rapidly expanding coverage in the PDB. Therefore, we tested our tool against others using more recent test sets. Specifically, we have selected all loop regions (identified by DSSP [9]) of length 6 through 11 from the x-ray targets presented at the CASP8 and CASP9 experiments, while excluding NMR models. The target proteins' native structures were obtained from Zhang Lab at the University of Michigan (http://zhanglab.ccmb.med.umich.edu/). To ensure fairness, our database consists of only protein structures released to the PDB prior to the start of the CASP8 experiments. The Method Details section contains the specifics of the database composition and selection.

Although the CASP targets mostly have full domain template matches available (since their purpose is to test template based protein modelling) they are selected by hand to be difficult, meaning that although there is a full domain template match, it is not identical to the actual tertiary structure. Since most of such deviations occur in the flexible loop regions, few of our loop targets can be modelled using the full domain template match.

4.2   Other Methods

We compare our results to those of several top loop modeling applications. First, we compare with the ModLoop program from the MODELLER package [6].

ModLoop was run with refinement set to "fast" and used to generate 50 loops. The candidates returned from MODELLER were then re-ranked according to their DFIRE [23] energy potential, which improves MODELLER's performance. Another statistical tool we tested against is RAPPER [1]. We generated 1000 candidates as described in their paper. As with MODELLER, we ranked the RAPPER results by the DFIRE energy potential, since this substantially improves their accuracy for all test sets. We also compare with the Loopy program, another *ab initio* method. Loopy was run with all parameters left to default, and with the number of initial models set to 2000 for loops shorter than length 10, and 4000 otherwise. These are the recommended number of candidates for loops of these lengths[18]. While RAPPER and MODELLER take an average of 3 hours (on a 2.2 GHz Opteron) per length 9 loop modeled, Loopy takes around 20 minutes. Finally, we test against version 3.3 of ROSETTA [12]. This version uses the KIC (KInematic Closure) algorithm to generate closed loops where all but 6 torsion angles can be exactly equal to the statistically sampled angle. ROSETTA is often excluded from benchmarks, as although it produces very accurate predictions, the Monte Carlo simulation required to do so is very time consuming. For example, while it may take 20 minutes for Loopy to generate a prediction for a given loop, ROSETTA can take upward of 5 days to do the same. If the number of candidates to generate is lowered drastically, ROSETTA can complete in time comparable with other tools, but is no longer accurate. However, the new KIC loop modeling algorithm, although slower than the older CCD (cyclic coordinate descent) algorithm, is accurate even with a greatly reduced running time, where CCD was not. ROSETTA tests used the parameters as described in their online guide [13], and generated 10 candidate structures. Additionally, we used the "-loops::fix_natsc" flag to prevent ROSETTA from refolding the native sidechains. By default ROSETTA will refold any native sidechains that lie within a certain distance (14 angstrom by default) of a portion of the loop that was remodeled. This may be more realistic for many loop modeling situations, but since other tools do not do this, it would result in ROSETTA solving a much harder problem, and being at a disadvantage.

We do not compare our method with FALC [10], a recent fragment assembly loop modeling server. As with the CCD closure protocol from ROSETTA, FALC is a statistical technique that samples angles from a position-specific ϕ/ψ distribution, built using fragment matches. We submitted the CASP8 portion of our length 10 test set to the FALC server, and observed an average score of 2.36 local RMSD, and 4.55 global RMSD. This is significantly larger than all other tools on this same subset, so we elected to cease submitting to their server. At the time of submission the FALC server is offline.

Our tables also do not compare our results with those of SuperLooper or FREAD as neither tool returns results for all loops, and it is meaningless to compare averages for different sets. For example, out of the 60 length 10 loops, FREAD (using the same database as LoopWeaver) returned matches for only 6 loops, and SuperLooper (using LIP from 2007) returned 45. For the six results returned by FREAD, the average score was 0.94Å local RMSD. Over the same six loops, LoopWeaver returns an average score of 0.44Å.

Finally, as the ROSETTA KIC paper [12] claims results to those of molecular mechanics refinement as used in PLOP[24], we have not examined molecular mechanics based solutions, or the LoopBuilder [15] protocol that uses PLOP for the refinement of Loopy generated loop candidates.

4.3   Scores

Table 1 shows the results of running our tool, as well as Loopy, MODELLER, and RAPPER, on the various test sets.

The "Loops" column indicates the number of targets in the given test set. Note that the longer gap, the fewer loops of that size there are to test against. There were too few loops of length 12 or longer to justify inclusion.

Included are both the local and global RMSD averages. The local RMSD captures the similarity of the overall shape of a loop, without being dependent on its orientation. (A small twist can have a very small effect on the local RMSD, but a huge effect on the global RMSD). The global RMSD captures both the shape and the orientation of the loop. In both instances all heavy backbone atoms are included in the calculation.

LoopWeaver performs quite well when compared with Loopy, MODELLER, and RAPPER.

ROSETTA results are not included in this first table. The reason for this exclusion is that ROSETTA includes a final refinement stage that the other tools lack. While all tools, ROSETTA included, start by generating loop candidates, closing those loop candidates if needed, and possibly making small adjustments to improve the energy of the candidates, ROSETTA then goes on to do a thorough refinement of its results using Monte Carlo simulation. Any comparison to ROSETTA should involve results that have undergone similar refinement efforts. In table 2 we present the LoopWeaver results after applying ROSETTA's KIC refinement tool, as well as the *ab initio* ROSETTA results. In both cases we generate a total of 10 candidates. When generating candidates for the refined LoopWeaver results, we generated all 10 refined loops based on the top LoopWeaver candidate, rather than generating 1 each for the top 10 candidates. LoopWeaver's refined results are better than both the unrefined results, and the

Table 1. Average RMSD scores for tested tools. The first value is the minimum (or local) RMSD, the second value is the unminimized (or global) RMSD.

Length	Loops	LoopWeaver	ModLoop	Loopy	Rapper
6	205	0.73/1.27	0.78/1.52	1.00/1.89	1.02/1.83
7	171	1.02/1.85	1.16/2.13	1.21/2.23	1.25/2.19
8	118	1.38/2.59	1.39/2.63	1.42/2.36	1.64/2.77
9	101	1.68/2.91	1.80/3.32	1.85/3.10	1.90/3.31
10	60	1.88/3.33	2.22/3.99	1.95/3.34	2.09/3.53
11	43	2.08/3.37	2.25/3.90	2.45/3.80	2.52/4.16

Table 2. Average RMSD scores with ROSETTA refinement. The first value is the minimum (or local) RMSD, the second value is the unminimized (or global) RMSD.

Length	Loops	LoopWeaver	ROSETTA
6	205	0.46/0.87	0.55/1.05
7	171	0.73/1.38	0.79/1.55
8	117	1.00/1.76	1.16/2.17
9	101	1.22/2.25	1.42/2.73
10	60	1.43/2.43	1.67/3.09
11	43	1.72/2.94	1.90/3.38

final ROSETTA results. The ROSETTA team recommends generating 1000 candidates for loop modeling of longer loops, and the extensive simulation involved is very CPU intensive. [13] Generating 1000 candidates for a single length 9 loop takes an average of 10 days on a 2.2 GHz Opteron, and almost all of this running time is used by the refinement stage. The ROSETTA team also notes that shorter loops *may* require fewer models. By generating 10 candidates, we get the running time average of 2.5 hours, comparable to the running time of MODELLER or RAPPER. Even with this restriction, the final ROSETTA results are substantially better than MODELLER and RAPPER, so this restriction is not unreasonable. Nevertheless, in the next section we also run ROSETTA for the recommended time and show that it does not catch up to our final results.

4.4 Consensus Scores

Database methods and *ab initio* methods are complementary. Because they work in very different ways, there are many cases where only one approach returns an accurate result. Hence it is desirable for us to combine Weaver and one of the *ab initio* methods.

In table 3 we present several consensus results. Consensus is done by employing the DFIRE energy potential function to select between the top results of the methods being combined. For all test sets the combination of Loopy and LoopWeaver yields an average score lower than either tool alone. For other combinations this is not always the case, and we often end up with a score that is somewhere between the two tools, rather than superior to both. This is because

Table 3. Average RMSD scores for consensus results. The first value is the minimum (or local) RMSD, the second value is the unminimized (or global) RMSD.

Length	Loops	LoopWeaver	Loopy + LoopWeaver	ModLoop + LoopWeaver	Loopy + ModLoop
6	205	0.71/1.36	0.69/1.30	0.77/1.48	0.80/1.54
7	171	1.02/1.85	0.96/1.76	1.01/1.84	1.11/2.06
8	118	1.38/2.59	1.21/2.19	1.34/2.53	1.32/2.51
9	101	1.68/2.91	1.61/2.72	1.56/2.90	1.73/3.19
10	60	1.88/3.33	1.81/3.11	2.03/3.59	2.14/3.81
11	43	2.08/3.37	1.96/3.18	2.03/3.53	2.16/3.65

Table 4. Average RMSD scores for refined consensus results. The first value is the minimum (or local) RMSD, the second value is the unminimized (or global) RMSD.

Length	Loops	LoopWeaver	Loopy + LoopWeaver	ROSETTA
6	205	0.46/0.87	0.42/0.79	0.55/1.05
7	171	0.73/1.38	0.64/1.15	0.79/1.55
8	118	1.00/1.76	0.81/1.37	1.16/2.17
9	101	1.22/2.25	0.98/1.76	1.42/2.73
10	60	1.43/2.43	1.12/1.92	1.67/3.09
11	43	1.72/2.94	1.43/2.42	1.90/3.38

of the method of selection. None of the methods used optimises its results against the DFIRE potential. So, although this function is accurate when selecting between candidates generated by the same method, it becomes less accurate when selecting between techniques. Because Loopy results have better DFIRE potential than MODELLER results with similar accuracy, they tend to be selected more often, even in cases where they are substantially worse.

We can improve our results by applying the ROSETTA KIC refinement step to our results, just as we have done before. In table 4 we show the results of applying KIC refinement to the Loopy and LoopWeaver consensus results. We selected these two tools for the refined consensus step because they had the best combined score prior to refinement. Additionally, running both tools takes a total of less than 30 minutes (average for the length 10 set on a 2.2 GHz Opteron), whereas RAPPER and MODELLER both take several hours to complete, on average. To obtain these results, we make two KIC refinement calls, one for each tool's candidate, with each generating 5 candidates so that we are still generating a total of 10, and are not doubling the refinement step's running time.

ROSETTA's KIC refinement makes substantial improvements to the average scores for all of our test sets, allowing us to attain much lower scores than ROSETTA alone, while on a similar timescale. These results even hold if we allow the ROSETTA *ab initio* execution substantially more time. We only have completed numbers for length 9, due to the large amount of CPU time required, but for length 9 and 1000 candidates generated, the average ROSETTA score is 1.01 for the local RMSD, and 1.82 for the global. So, if we allow ROSETTA 100 times the running time as our Loopy and LoopWeaver consensus method, ROSETTA approaches but does not overtake our results. Further increases to the number of ROSETTA candidates will result in only minor improvements, since almost all loops have converged by this point. It is also important to note that our own consensus results can also be improved by increasing the amount of refinement effort.

5   Method Details

5.1   Database Matches

We use a database comprising roughly 14,400 protein chains, selected using PISCES [17] with the cutoff values being 3.0 resolution, 90 percent identity, and 1.0 R value. Because our test sets include targets from CASP8, we also only allowed PISCES to select from proteins with a release date prior to the start of CASP8 (May of 2008). This database is used for all tests, even those involving targets from CASP9 rather than CASP8.

We examine the database of known protein structures, and look for appropriately spaced residues that are similar to the stems of the target gap. The metric used for determining the similarity between the stems is $RMSD_{stem}$, the RMSD for the C, $C\alpha$, and N atoms in each stem. Once we have obtained this large set of matches, we sort according to the stem RMSD, and take the top 500 matches. Where SuperLooper takes only matches that have a very low stem RMSD (otherwise the match will introduce a large chain break) our method of fitting the loop is able to resolve chain breaks, so we take the best matches regardless of quality.

5.2   Fitting the Loop

Once suitable loop candidates have been selected from the database, they must be placed into the gap in order to connect to the protein backbone correctly. Because we are selecting matches based on the stem RMSD, this can be done by using the superposition matrix used for computing the stem RMSD, and applying it to the rest of the match. If we restrict ourselves to matches with a very low stem RMSD, this will always fill the gap without having unreasonable bond lengths. If we are less strict with regards to the stem RMSD, then the loop will not necessarily fill the gap without having unrealistic atomic distances where the loop connects to the rest of the protein backbone.

One can view the placement of the loop into the gap as an attempt to satisfy two contradictory requirements. The first requirement is that the stems remain the same. The second requirement is that the loop we place should be the same shape as the database match, including the stem region of the database match. This requirement cannot be fully satisfied if we are not allowed to change the stems in the target protein. Using the superposition of the stems is not optimal. The inserted loop does not match the database loop because it has different stems, and although these differences are quite minor, they can have a large impact on the orientation of the loop. Our goal is then to satisfy these two sets of requirements in a more optimal way. This will hopefully not only improve the orientation of the loop, but also resolve any unrealistic bond lengths caused by a less restrictive RMSD cutoff value.

We have chosen to solve these requirements by formulating them as an instance of *weighted multi-dimensional scaling* (WMDS) as described in [11]. WMDS is a problem often used in statistics, as it can be used to turn high-dimensional data into 2 or 3 dimensional data suitable for graphing. It has also been used in MU-FOLD [22] as a method for assembling protein fragments. For a given dimension d and n points of data, we have D, a symmetric $n \times n$ matrix, as well as W, also a symmetric $n \times n$ matrix, and wish to find $X = \{x_1, x_2, \ldots, x_n\}$ where x_i is a coordinate in d-space, such that we minimize the *stress*, defined as $\sigma(X) = \sum_{0<i<j\leq n} W_{i,j}(\|x_i - x_j\| - D_{i,j})^2$. For our problem instances, $d = 3$ and n is the total number of heavy backbone atoms in the both the loop and the stems.

Formulating the Problem. The matrix D is defined as

$$d_{i,j} = \begin{cases} ||p_i - p_j|| & i,j \in \text{stem} \\ ||c_i - c_j|| & \text{otherwise} \end{cases} \tag{1}$$

Where $P = \{p_1, p_2, \ldots p_n\}$ is the set of atomic coordinates in the protein being modeled (1 being the first heavy atom in the C-terminus stem, and n being the last atom in the N-terminus stem), and $C = \{c_1, c_2, \ldots c_n\}$ is the set of atomic coordinates in the loop candidate, using the same numbering system. The atoms are listed in the order N Cα C O, so atom 0 will be the first N atom in the C-stem.

The reason we use the weighted version of this problem is that not all of the desired distances are equally important. Primarily, we do not want to make any changes to the stem atom, so the pairwise weights between two stem atoms should be very large. Beyond this, atoms that are closer together should be given higher weights. If two atoms pass within a few angstrom of each other within the loop, then they should remain close to this distance regardless of other changes, especially if they are adjacent atoms within the backbone. On the other hand, atoms that are far apart are free to move around a fair bit without changing the overall shape of the protein (if we ignore the effects of that movement on all of the other atoms, that is). The recommended weight for use with WMDS is either d^{-1} or d^{-2}, where d is the corresponding distance. A value of 2 is recommended [4] if you want more emphasis on close points than on distance points, which is the case here. We tested on a small number of length 8 loops, and tried exponents ranging from -1 to -3 in increments of 0.1. The best results were achieved at exactly -2. We define the matrix W as

$$w_{i,j} = w_{j,i} = \begin{cases} 10,000 & i,j \in \text{stem} \\ T(i \pmod 4), j - i) & j - i \le 4 \\ (\min\{d_{i,j}, r - \Phi d_{i,j}\})^{-2} & \text{otherwise} \end{cases} \tag{2}$$

where $0 \le i < j < n$, T is a 4×4 lookup table, r is the diameter of the database match (the largest pairwise distance between any two atoms in the loop), and Φ is the golden ratio conjugate $(2/(1 + \sqrt{5}))$.

The lookup table T is used to give very high weights to atoms in the same or adjacent residues. In particular, any atoms only one bond apart have very high weights. The weights for corresponding atoms in adjacent residues (Cα to Cα^{+1} for example) are also quite high so that torsion angles are not changed too much. T is reproduced here in table 5. For all atoms, the $j - i = 4$ cell will be the weight for the distance from that atom, to the corresponding atom in the next residue. For the $j - i = 1$ cell this will be the weight for the next atom in the protein backbone, with the exception of O. These values are the reciprocals of the variance of the corresponding distances in all loops of our protein structure database. The reasoning for this is that we want a move of one standard deviation to carry the same weight. Since the objective function involves squared distances, we use the variance of the distances rather than the standard deviations. This results in the Cα to Cα^{+1} weight being higher than

Table 5. Matrix T used for short-distance weights. Row corresponds to the type of the first atom. Column corresponds to how many atoms ahead the second atom is, in the same ordering. So, row 2 column 3 corresponds to the distance between a $C\alpha$ atom and the N atom in the next residue.

$$
\begin{array}{c}
N \\
C\alpha \\
C \\
O
\end{array}
\left(
\begin{array}{cccc}
32 & 5 & 6 & 32 \\
32 & 16 & 14 & 54 \\
32 & 40 & 13 & 24 \\
10 & 2 & 1 & 0.5
\end{array}
\right)
$$

the N to N^{+1} or C to C^{+1} for example, which is to be expected given that the omega torsion angle should change less than the ϕ and ψ angles. Since the higher weight will make this angle less likely to be changed, this is a desirable outcome.

The reason we have the golden ratio term in the last case of the piecewise equation, rather than simply using $d_{i,j}$ is that we want to preserve the overall shape of the loop as much as possible. Part of this is keeping the diameter of the loop roughly the same. The weight of a distance represents how much a pair of atoms are allowed to move relative to each other while still having on a minor effect on that pair's contribution to the stress function. The reason we have the weight equal to d^{-2} is that the farther the atoms are apart, the less it matters their exact distance apart. However, if they are near to the desired diameter of the loop, then if we wish to keep the diameter approximately the same, they should not be given as much freedom to move. So, we want the weights to start increasing again as d approaches r. The value Φ, which we use to cause the weights to increase after a certain point, was selected empirically. We selected target loops from the database, and for each target loop, found the matching loops with RMSD less than 1.5Å. That is, loops that are fairly similar in shape, though not identical. Then, we created bins with a width of 0.5Å, and for every pair of atoms within the loop, we populated the corresponding bin with the squared deviation of all corresponding pairs of atoms. So if a pair of atoms was 9.4Å apart in base loop, and we found a database match in which the same atoms were 10.1Å apart, we would add $(9.4 - 10.1)^2$ to the bin for the range 9.25 through 9.5 Å. For most loops tested, the standard deviation roughly follows the equation $(\min\{d_{i,j}, r - \Phi d_{i,j}\})^2$. Again, we set the weights to the reciprocal of this value, so that a difference of one standard deviation will contribute equally to the objective function regardless of where that distance occurs.

Solving the Problem. We use the *SMACOF* algorithm [11] for solving the WMDS problem. This algorithm works by minimizing a simple function that majorizes the stress function, yielding a fast, deterministic heuristic. This algorithm gives a good trade-off between speed and accuracy [2].

5.3   Clashes

One issue that the problem formulation does not address is the issue of collisions. As the algorithms for solving the WMDS problem rely on matrix functions being applied to the working set of atomic coordinates, no care is, or can be, paid to making sure the solution will not clash with the rest of the protein.

The solution we have adopted for resolving clashes is to keep track of all solutions which clash with the rest of the protein. Here, clash is defined as coming within 2.4Å, rather than a more complex method. After using SMACOF to close the database matches, we are left with a list of residues from the protein which clashed with one or more solution. For all of these residues, we evaluate the pairwise DFIRE energy potential between the clashing residues, and the loop candidate residues. We then keep only those clashing residues whose average contribution to the energy potential is positive over all of the loop candidates. After discarding these residues, we will be left with residues that clash with, or at least are unfavourably close to, many of the candidates, rather than residues that clashed with only a few database matches. The Cα atoms from the central residue of each remaining clashing residue (if any) are then included in the WMDS equations. As these atoms are from the input structure, they are fixed just like the stem atoms, and so treated the same. The desired distance d between clash atom a and loop atom b is computed by taking the average of $d_{a,b}$ for all loop candidates where the total DFIRE potential between a and the loop is negative (negative is good for energy potential functions), as long as the candidate did not clash with a. If the candidate clashed with another atom it is still included in the average, so that these distances can be computed even if there were no clash free candidates.

This approach is almost always able to resolve clashes. Even in instances where over 90% of the database matches clash after the first round of WMDS, we are able to resolve almost all clashes. In most cases this second round of WMDS results in less accurate predictions, but a less accurate candidate is more desirable than a clashing candidate.

5.4   Ranking and Selection

After we have obtained closed loops through our WMDS solution, we must determine which is the most likely loop to fill the target gap. We chose the DFIRE [23] energy function, as others have had good results using this function to rank loop candidates [21,5]. Because DFIRE is an all-atom model, we require the additional step of adding side-chain atoms to the protein backbone. We used the TreePack program [19,20] to do so.

6   Conclusions

This paper has made the following contributions: Firstly, by formulating the loop placement problem as an instance of the Weighted Multi-Dimensional Scaling problem, we have overcome the major weakness of the knowledge-based loop closure problem (chain-breaks and clashes). The resulting LoopWeaver system compares favorably with all *ab initio* methods, significantly changes the current status of database methods, and, when combined with an *ab initio* method such as Loopy, and KIC refinement, produces useful high quality loops (less than 1Å) on average for loops of length less than 9, and good loops for other lengths.

Availability

The set of test loops, the LoopWeaver predictions, and the ROSETTA-refined LoopWeaver predictions, are available for download from `http://www.cs.uwaterloo.ca/ djholtby/LoopWeaver/` We are currently working on an online version of LoopWeaver, which will be made available through the same URL.

Acknowledgements. We would like to thank Dong Xu for introducing us to the multi-dimensional scaling problem and its applications in protein structure prediction[22].

This work was made possible by the facilities of the Shared Hierarchical Academic Research Computing Network (SHARCNET:www.sharcnet.ca) and Compute/Calcul Canada.

Daniel Holtby's research is funded in part by an NSERC PGS D scholarship. Ming Li's research is supported in part by: NSERC Grant OGP0046506, Canada Research Chair program, an NSERC Collaborative Grant, OCRiT, Premier's Discovery Award, and the Killam Prize.

References

1. de Bakker, P.I.W., DePristo, M.A., Burke, D.F., Blundell, T.L.: Ab initio construction of polypeptide fragments: Accuracy of loop decoy discrimination by an all-atom statistical potential and the amber force field with the generalized born solvation model. Proteins: Structure, Function, and Bioinformatics 51, 21–40 (2003), `http://dx.doi.org/10.1002/prot.10235`
2. Basalaj, W.: Proximity visualization of abstract data. Technical report, University of Cambridge Computer Laboratory (2001), `http://pavis.org`
3. Choi, Y., Deane, C.M.: Fread revisited: Accurate loop structure prediction using a database search algorithm. Proteins: Structure, Function, and Bioinformatics 78(6), 1431–1440 (2010)
4. Cohen, J.D.: Drawing graphs to convey proximity: an incremental arrangement method. ACM Trans. Comput.-Hum. Interact. 4, 197–229 (1997), `http://doi.acm.org/10.1145/264645.264657`
5. Deane, C.M., Blundell, T.L.: Coda: A combined algorithm for predicting the structurally variable regions of protein models. Protein Science 10(3), 599–612 (2001)
6. Fiser, A., Do, R., Sali, A.: Modeling of loops in protein structures. Protein Science 9(9), 1753–1773 (2000)
7. Fiser, A., Sali, A.: Modeller: Generation and refinement of homology-based protein structure models. In: Charles, W., Carter, J., Sweet, R.M. (eds.) Macromolecular Crystallography, Part D. Methods in Enzymology, vol. 374, pp. 461–491. Academic Press (2003), `http://www.sciencedirect.com/science/article/B7CV2-4BT8FSC-R/2/efdaef0eaf06a1c59521dd71f7612a91`
8. Hildebrand, P.W., Goede, A., Bauer, R.A., Gruening, B., Ismer, J., Michalsky, E., Preissner, R.: SuperLoopera prediction server for the modeling of loops in globular and membrane proteins. Nucleic Acids Research 37(suppl. 2), W571–W574 (2009), `http://nar.oxfordjournals.org/content/37/suppl_2/W571.abstract`

9. Kabsch, W., Sander, C.: Dictionary of protein secondary structure: Pattern recognition of hydrogen-bonded and geometrical features. Biopolymers 22(12), 2577–2637 (1983), http://dx.doi.org/10.1002/bip.360221211

10. Lee, J., Lee, D., Park, H., Coutsias, E.A., Seok, C.: Protein loop modeling by using fragment assembly and analytical loop closure. Proteins: Structure, Function, and Bioinformatics 78(16), 3428–3436 (2010), http://dx.doi.org/10.1002/prot.22849

11. de Leeuw, J.: Applications of Convex Analysis to Multidimensional Scaling. In: Barra, J., Brodeau, F., Romier, G., van Cutsem, B. (eds.) Recent Developments in Statistics, pp. 133–146. North Holland Publishing Company (1977)

12. Mandell, D.J., Coutsias, E.A., Kortemme, T.: Sub-angstrom accuracy in protein loop reconstruction by robotics-inspired conformational sampling. Nat. Methods 6(8), 551–552 (2009), http://dx.doi.org/10.1038/nmeth0809-551

13. Mandell, D.J., Pache, R.A.: Rosetta projects: Documentation for kinematic loop modeling (October 2011), http://rosettacommons.org/manuals/archive/rosetta3.3_user_guide/app_kinematic_loopmodel.html

14. Michalsky, E., Goede, A., Preissner, R.: Loops In Proteins (LIP)-a comprehensive loop database for homology modelling. Protein Engineering 16(12), 979–985 (2003), http://peds.oxfordjournals.org/content/16/12/979.abstract

15. Soto, C.S.S., Fasnacht, M., Zhu, J., Forrest, L., Honig, B.: Loop modeling: sampling, filtering, and scoring. Proteins (August 2007), http://dx.doi.org/10.1002/prot.21612

16. Wang, C., Bradley, P., Baker, D.: Protein-protein docking with backbone flexibility. Journal of Molecular Biology 373(2), 503–519 (2007), http://www.sciencedirect.com/science/article/pii/S0022283607010030

17. Wang, G., Dunbrack, R.L.: PISCES: a protein sequence culling server. Bioinformatics 19(2), 1589–1591 (2003)

18. Xiang, Z., Soto, C.S., Honig, B.: Evaluating conformational free energies: The colony energy and its application to the problem of loop prediction. Proceedings of the National Academy of Sciences of the United States of America 99(17), 7432–7437 (2002), http://www.pnas.org/content/99/11/7432.abstract

19. Xu, J.: Rapid Protein Side-Chain Packing via Tree Decomposition. In: Miyano, S., Mesirov, J., Kasif, S., Istrail, S., Pevzner, P.A., Waterman, M. (eds.) RECOMB 2005. LNCS (LNBI), vol. 3500, pp. 423–439. Springer, Heidelberg (2005)

20. Xu, J., Berger, B.: Fast and accurate algorithms for protein side-chain packing. J. ACM 53, 533–557 (2006)

21. Zhang, C., Liu, S., Zhou, Y.: Accurate and efficient loop selections by the dfire-based all-atom statistical potential. Protein Science 13(2), 391–399 (2004), http://dx.doi.org/10.1110/ps.03411904

22. Zhang, J., Wang, Q., Barz, B., He, Z., Kosztin, I., Shang, Y., Xu, D.: Mufold: A new solution for protein 3d structure prediction. Proteins: Structure, Function, and Bioinformatics 78, 1137–1152 (2010)

23. Zhou, H., Zhou, Y.: Distance-scaled, finite ideal-gas reference state improves structure-derived potentials of mean force for structure selection and stability prediction. Protein Science 11, 2714–2726 (2002)

24. Zhu, K., Pincus, D.L., Zhao, S., Friesner, R.A.: Long loop prediction using the protein local optimization program. Proteins: Structure, Function, and Bioinformatics 65, 438–452 (2006)

A Robust Method for Transcript Quantification with RNA-seq Data

Yan Huang[1], Yin Hu[1], Corbin D. Jones[2], James N. MacLeod[3],
Derek Y. Chiang[4], Yufeng Liu[5], Jan F. Prins[6], and Jinze Liu[1],[*]

[1] Department of Computer Science
[2] Department of Biology, University of North Carolina at Chapel Hill
[3] Department of Veterinary Science, University of Kentucky
[4] Department of Genetics
[5] Department of Statistics and Operations Research
[6] Department of Computer Science, University of North Carolina at Chapel Hill
{yan,yin,liuj}@netlab.uky.edu, {cdjones,yfliu}@email.unc.edu,
jnmacleod@uky.edu, chiang@med.unc.edu, prins@cs.unc.edu

Abstract. The advent of high throughput RNA-seq technology allows deep sampling of the transcriptome, making it possible to characterize both the diversity and the abundance of transcript isoforms. Accurate abundance estimation or *transcript quantification* of isoforms is critical for downstream differential analysis (e.g. healthy vs. diseased cells), but remains a challenging problem for several reasons. First, while various types of algorithms have been developed for abundance estimation, short reads often do not uniquely identify the transcript isoforms from which they were sampled. As a result, the quantification problem may not be identifiable, i.e. lacks a unique transcript solution even if the read maps uniquely to the reference genome. In this paper, we develop a general linear model for transcript quantification that leverages reads spanning multiple splice junctions to ameliorate identifiability. Second, RNA-seq reads sampled from the transcriptome exhibit unknown position-specific and sequence-specific biases. We extend our method to simultaneously learn bias parameters during transcript quantification to improve accuracy. Third, transcript quantification is often provided with a candidate set of isoforms, not all of which are likely to be significantly expressed in a given tissue type or condition. By resolving the linear system with LASSO our approach can infer an accurate set of dominantly expressed transcripts while existing methods tend to assign positive expression to every candidate isoform. Using simulated RNA-seq datasets, our method demonstrated better quantification accuracy than existing methods. The application of our method on real data experimentally demonstrated that transcript quantification is effective for differential analysis of transcriptomes.

Keywords: Transcript quantification, Transcriptome, RNA-seq.

---

[*] Corresponding author.

B. Chor (Ed.): RECOMB 2012, LNBI 7262, pp. 127–147, 2012.

1 Introduction

Recent studies have estimated that as many as 95% of all multi-exon genes are alternatively spliced, resulting in more than one transcript per gene [23, 33]. *Transcript quantification* determines the steady state levels of alternative transcripts within a sample, enabling the detection of differences in the expression of alternative transcripts under different conditions. Its application in detecting biomarkers between diseased and normal tissues can greatly impact biomedical research.

High-throughput sequencing technology ·(e.g. RNA-seq with Illumina, ABI Solid, etc.) provides deep sampling of the mRNA transcriptome. It allows the parallel sequencing of large number of mRNA molecules, generating tens of millions of short reads with lengthes up to 100bp at one end or both ends of mRNA fragments. Recent studies using RNA-seq have significantly expanded our knowledge on both the variety and the abundance of alternative splicing events [7, 36].

However, transcript quantification remains a challenging problem. First, it is commonly observed that "the more the isoforms, the harder to predict" [19]. Intuitively, transcript isoforms from the same gene often overlap significantly and a short read may be mapped to more than one transcript isoform. Determining the expression of individual transcripts from short read alignment, therefore, can lead to an *unidentifiable* model, where no unique solution exists. Secondly, transcript quantification often takes the candidate set of transcript isoforms, either from annotation databases such as Ensembl [2] and Refseq [3], or inferred from the splice graph using programs like Scripture [10], IsoInfer [8], IsoLasso [19], or Cufflinks [31]. It is biologically unlikely to expect all candidate transcripts for a given gene to be significantly expressed concurrently in a cell. However, existing analytical approaches tend to assign positive expression values to every candidate transcript provided, thereby creating a situation in which large errors in abundance estimation can be computationally introduced for transcript isoforms that may, in reality, barely be expressed. An improved transcript quantification method, therefore, would determine or logically infer the subset of expressed transcript isoforms. Finally, various sampling biases have been observed regularly in RNA-seq datasets as a result of library preparation protocols. These biases typically include position-specific bias [6, 17, 25, 37] such as 3' bias and transcription start and end biases, and sequence-specific bias [18, 25, 32], where the read sampling in the transcriptome favors certain subsequences. How to compensate for these biases during transcript quantification is an open problem.

Transcript isoforms can differ not only in exons alternatively included or excluded but also in where two or more exons are connected together. In RNA-seq data, this information typically is implied by the spliced reads, i.e., the reads that cross one or more splice junctions. We have developed a general linear model for transcript quantification that leverages discriminative features in spliced reads to ameliorate the issue of identifiability and to simultaneously correct the sampling bias. Our contribution in this paper is three-fold: (1) We explicitly identify *MultiSplice*, a novel structural feature consisting of a contiguous set of exons that are expected to be spanned by the RNA-seq reads or transcript fragments of a given length. The MultiSplice, which includes single splice junctions as a special case, is used in two ways: its presence in the sample will infer the host transcript while its absence may reject it. MultiSplices are more powerful than single exons in disambiguating transcript isoforms, making more transcript quantification problems identifiable with long or paired-end reads; (2) We set up a linear system which minimizes the summed mean squared errors between the expected expression and the observed expression across all structure features along a gene while taking into account various bias effects; (3) We develop an iterative minimization algorithm in combination with LASSO [30] to resolve the aforementioned linear system in order to achieve the most accurate set of dominantly expressed transcripts while simultaneously correcting biases.

We have demonstrated the efficacy of our methods on both simulated RNA-seq datasets and real RNA-seq data: (1) We conducted the first study to investigate the question: what is the maximum read length needed in order to disambiguate all possible transcript isoforms in transcriptomes from different species; (2) We compared the proposed method with several state-of-the-art methods including Cufflinks, the Poisson model, and the ExonOnly model. Our results using simulated data from the human mRNA transcriptome demonstrated superior performance of the proposed method in most cases. When applied to 8 RNA-seq datasets from two breast cancer cell lines (MCF-7 and SUM-102), the quantification obtained from MultiSplice demonstrated good consistency within technical replicates from each transcriptome-wide assessment and substantial differences between the two biological groups (cell lines) in a small percentage of genes.

2   Related Work

Various transcript quantification algorithms have been published recently. These methods can be divided mainly into two categories: read-centric and exon-centric. The representative methods using read-centric approaches include but are not limited to Cufflinks [31], IsoEM [22], and RSEM [17]. The central idea with read-centric approaches is to assign probability for each fragment to one transcript by maximizing the joint likelihood of read alignments based on the distribution of transcript fragments, and thereby estimating the transcript expression. When it is impossible to precisely allocate a fragment to a unique transcript, Cufflinks, for example, simply disregards or randomly assigns the read, causing information loss or inaccurate quantification. The second strategy, called exon-centric, considers the read abundance on an exonic segment as the cumulative abundance of all transcript isoforms. Methods in this category represent the transcript as a combination of exons and aim at estimating individual transcript abundance from the observed read counts or read coverage at each exon. The representative models in this category include the Poisson model [13, 24, 29] and linear regression approaches, such as rQuant [6], IsoLasso [19] and SLIDE [20].

Transcript abundance estimations can be unidentifiable, where no unique quantification exists. Both exon-centric and read-centric models may suffer from this problem. The paper by Lacroix et al. is one of the theoretical studies that have considered the identifiability problem of transcript quantification [16].

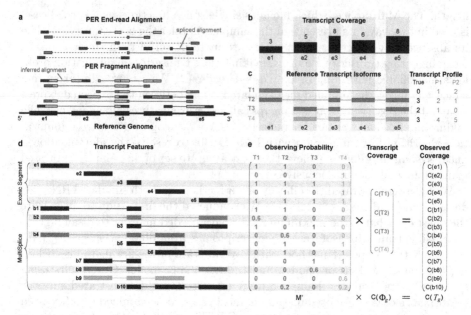

Fig. 1. Overview of the MultiSplice model. **a.** Sequenced RNA-seq short-reads are first mapped to the reference genome using an RNA-seq read aligner such as MapSplice [34]. In the presence of paired-end reads, MapPER [12] can be applied to find *PER fragment alignments* for the entire transcript fragment based on the distribution of insert size. **b.** Observed coverage on each exonic segment. **c.** Four transcripts originate from the alternative start and exon skipping events. Provided with these transcripts, abundance estimates would be unidentifiable for methods that only use coverage on exonic segments. Both transcript profiles P_1 and P_2, for instance, can explain the observed read coverage on each exon, but deviate from the true transcript expression profile. **d.** MultiSplices that connect multiple exonic segments in a transcript. **e.** A linear model can be set up where the expected coverage on every exonic or MultiSplice feature approximates its observed coverage. The transcript expression is solved as the one that minimizes the sum of squared relative error.

3   Method

In this section, we propose a method designated *MultiSplice*, for mRNA isoform quantification. We first define the observed features used in the Multi-Splice model and the statistics collected. Then, we derive a general linear model to relate transcript level estimates to the observed expression on every feature.

Preliminaries. For a gene g, we use $\mathcal{E}_g$ to denote the set of exonic segments [13, 19] in g, which are disjoint genomic intervals on the genome that can be included in a transcript in its entirety. We use $\mathcal{T}_g$ to denote the set of mRNA isoforms transcribed from g. These mRNAs can be a set of annotated transcripts retrieved from a database such as Ensembl [2] or Refseq [3]. A transcript $t \in \mathcal{T}_g$ is defined by a sequence of exon segments, $t = e_1^t e_2^t \cdots e_{n_t}^t$, where $e \in \mathcal{E}_g$ and n_t denotes the number of exonic segments in the transcript t. The length of each exonic segment

e is defined as the number of nucleotides in the exonic segment, denoted as $l(e)$. Hence, the length for every transcript is $l(t) = \sum_{i=1}^{n_t} l(e_i^t)$.

3.1   MultiSplice

In a typical RNA-seq dataset, a significant percentage of the read alignments are spliced alignments that connect more than one exon. With paired-end reads, the transcript fragment where its two ends are sampled can be inferred based on the distribution of the insert size [25]. Transcript fragments are typically between 200bp and 300bp, making them more likely to cross multiple exons, indicating these exons are present together in one transcript. This information can be crucial in distinguishing alternative transcript isoforms. However, they are often ignored in current computational approaches.

In this subsection, we consider a sequence of adjacent exons in an mRNA transcript covered by transcript fragments. These structural features are the basis of *MultiSplice*. For generality, we assume that the RNA-seq reads are sampled from transcript fragments whose lengths follow a given distribution F_{fr} with probability density function f_{fr}. For example, the fragment length distribution F_{fr} is often modeled as a normal distribution with mean and variance learned from the genomic alignment of the RNA-seq reads. We also assume the maximum fragment length is l_{fr}.

Definition 1. *Let* $b = e_i^t e_{i+1}^t \cdots e_{i+n_b}^t$ *be a substring of a transcript sequence* $t = e_1^t e_2^t \cdots e_{n_t}^t$, $n_b \geq 1$ *and* $i + n_b \leq n_t$. *Then* b *is a MultiSplice in* t *if and only if*

$$\sum_{q=1}^{n_b-1} l(e_{i+q}) \leq l_{fr} - 2. \tag{1}$$

The condition in Equation 1 guarantees that a MultiSplice b connects $n_b + 1$ adjacent exons with at least 1 base landed on the left most exon e_i^t and the right most exon $e_{i+n_b}^t$. We use $\mathcal{B}_G$ to denote the set of all MultiSplices in gene G. From the definition, the set of MultiSplices vary according to the fragment length l_{fr}. The longer the fragments, the more MultiSplices are expected to function as structural features, and the higher power in disentangling highly similar alternative isoforms.

In Figure 1, for example, assume the maximum fragment length is $l_{fr} = 300bp$ with the expected fragment length of $250bp$ and the exonic segments of this gene have lengths of $l(e_1) = 200bp, l(e_2) = 200bp, l(e_3) = 100bp, l(e_4) = 200bp, l(e_5) = 200bp$. In reference transcript $T_1 = e_1 e_3 e_5$, $b_2 = e_1 e_3 e_5$ is a substring of T_1, and we have $l(e_3) = 100bp < 300bp = l_{fr}$ which allows a fragment to cover b_2. Therefore, b_2 is a MultiSplice feature of the gene. Combining MultiSplices from all the reference transcripts, b_1, b_3, b_5, b_6, and b_7 are MultiSplices consisting of a single splice junction, b_2, b_4, b_8, b_9, and b_{10} are MultiSplices consisting of two splice junctions.

3.2 Expected Coverage and Observed Coverage

Given the gene g and a transcript $t \in \mathcal{T}_g$, let c_i be the number of transcript fragments covering the ith nucleotide of t. We define the coverage on t as the averaged number of transcripts covering each base in the transcript, $C(t) = \frac{1}{l(t)} \sum_{i=1}^{l(t)} c_i$. Then $C(t)$ is an estimator for the quantity of t in the sample, which provides a direct measure for the expression level of t. In our model, $C(t)$ is the unknown variable. The feature space that can be observed from the given RNA-seq sample is the union of all exonic segments and MultiSplices of the gene, $\Phi_g = \mathcal{E}_g \cup \mathcal{B}_g$. We aim at resolving the transcript expressions that minimize the difference between the observed expression and the expected expression of every feature.

For every $\phi \in \Phi_g$ and every transcript $t \in \mathcal{T}_g$, the expected coverage of feature ϕ from t can be expressed as a function of the transcript quantity $C(t)$, i.e., $E[C(\phi|t)] = m(\phi, t)C(t)$, where $m(\phi, t)$ contains the probability of observing ϕ in t assuming uniform sampling. Next, we define the *expected* coverage on exonic segments and MultiSplice respectively.

For an exonic segment e in t, assuming N_t fragments were sampled from t, the number of fragments falling in e then follows a binomial distribution with parameters N_t and $p_{e|t}$, $N_{e|t} \sim Bin(N_t, p_{e|t})$. When N_t is sufficiently large, the binomial distribution is well approximated using a normal distribution with mean $N_t p_{e|t}$ and variance $N_t p_{e|t}(1 - p_{e|t})$, $N_{e|t} \dot\sim N(N_t p_{e|t}, N_t p_{e|t}(1 - p_{e|t}))$. In expectation, $\frac{N_{e|t} l_{fr}}{l(e)}$ calculates the fragment coverage on e contributed by t, $C_{e|t}$, and $\frac{N_t l_{fr}}{l(t)}$ calculates the transcript coverage of t, C_t. Therefore, we have $\frac{N_{e|t} l_{fr}}{l_e} \dot\sim N(\frac{N_t p_{e|t} l_{fr}}{l_e}, \frac{r^2}{l_e^2} N_t p_{e|t}(1 - p_{e|t}))$ or $C_{e|t} \dot\sim N(C_t, \frac{r(l_t - l_e)C_t}{l_t l_e})$, and hence the expectation of observed coverage on e contributed by t equals the coverage of t, with $m(e, t) = 1$.

For a MultiSplice $b = e_i^t e_{i+1}^t \cdots e_{i+n_b}^t$, we are interested in the number of fragments containing it. Should a transcript fragment f_t cover b, f_t must start no later than the 3' end boundary of the leftmost exonic segment e_i^t and have at least 1 base landed on the rightmost exonic segment $e_{i+n_b}^t$. Therefore, there exists a window $w(b)$ before the 3' end of e_i^t with length $l(w(b)) = l(f_t) - \sum_{q=1}^{n_b-1} l(e_{i+q}) - 1$, where b can be covered by the transcript fragment f_t. The probability that f_t covers b is hence $p_{b|t,l(f_t)} = \frac{l(f_t) - \sum_{q=1}^{n_b-1} l(e_{i+q}) - 1}{l(t)}$. Because $l(f_t)$ follows the fragment length distribution F, the expectation of $p_{b|t,l(f_t)}$ is then $p_{b|t} = E[p_{b|t,l(f_t)}] = \int p_{b|t,x} \cdot f(x)\, dx$, for x is the domain where the density function f is defined, resulting in $p_{b|t} = \frac{E[l(f_t)] - \sum_{q=1}^{n_b-1} l(e_{i+q}) - 1}{l(t)}$. Accordingly, the expected coverage of MultiSplice b gained from t is $E[C(b|t)] = m(b, t)C(t)$ if $m(b, t) = \frac{p_{b|t} l(t)}{E[l(f_t)]}$. In Figure 1, $m(b_2, T_1) = \frac{E[l(f_t)] - l(e_3) - 1}{E[l(f_t)]} = \frac{250 - 100 - 1}{250} = 0.6$.

In summary, the probability that a MultiSplice feature ϕ contained in a transcript fragment f_t uniformly sampled from transcript t is:

$$m(\phi, t) = \begin{cases} 1 & \text{if } \phi \subset t \text{ and } \phi \in \mathcal{E}_G \\ \frac{E[l(f_t)] - \sum_{q=1}^{n_b-1} l(e_{i+q}) - 1}{E[l(f_t)]} & \text{if } \phi \subset t \text{ and } \phi \in \mathcal{B}_G \\ 0 & \text{if } \phi \not\subset t. \end{cases} \qquad (2)$$

with $\phi \subset t$ standing for that ϕ is in t.

The *observed* coverage on an exonic segment $e \in \mathcal{E}_G$ as $C(e) = \frac{1}{l(e)} \sum_{i=1}^{l(e)} c_i$, where c_i is the number of reads covering the ith nucleotide in e. The read coverage $C(e)$ provides an estimator for the number of transcript copies that flow through the exonic segment e assuming uniform sampling. For a MultiSplice $b \in \mathcal{B}_G$, we use $C(b)$ to denote the read coverage on b defined as the number of transcript fragments that include b.

3.3 A Generalized Linear Model for Transcript Quantification

We construct a matrix $\mathbf{M}' \in \Re^{|\Phi_G| \times |\mathcal{T}_G|}$ to represent the structure of the transcripts, whose entry on the row of ϕ and the column of t corresponds to the probability of observing feature ϕ from transcript t, $\mathbf{M}'(\phi, t) = m(\phi, t)$. The linear model is set up for every feature $\phi \in \Phi_G$ by equating the observed coverage on ϕ to the expected coverage from all transcripts:

$$C(\phi) = \sum_{t \in \mathcal{T}_G} \mathbf{M}'(\phi, t) C(t) + \epsilon_\phi, \text{for any } \phi \in \Phi_G. \qquad (3)$$

Here $C(t) \geq 0$ for every $t \in \mathcal{T}_G$, ϵ_ϕ is the error term for feature ϕ in transcript t.

Lemma 1. *The MultiSplice model for transcript quantification is identifiable if the rank of M' is no less than the number of transcripts $|\mathcal{T}_G|$.*

Lemma 1 directly follows the the Rouché-Capelli theorem [11].

4 Bias Correction

Under uniform sampling, the sampling probability is the same at every nucleotide of a transcript. The observed coverage on ϕ is unbiased for the expected coverage on t. In this case, $\sigma(\phi, t)$ is set to 1 for all transcripts and features. However, sampling bias is often introduced in RNA-seq sample preparation protocols and has been demonstrated to have significant effects in RNA-seq analysis [14, 35]. Therefore, we discuss in the following subsections how MultiSplice corrects various sampling bias via learning of the bias coefficients and simultaneously solves the linear model for transcript coverage $C(t)$ of every transcript t.

Figure 2(a-e) shows how various types of sampling bias alter the sampling probability and hence the coverage. Two types of sampling bias are commonly observed in RNA-seq data, namely, the position-specific bias and the sequence-specific bias [6, 4, 21, 27]. In our model, sampling bias may affect the sampling

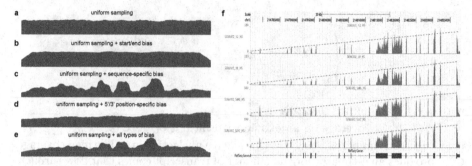

Fig. 2. Sampling bias present in the RNA-seq data. **a.** RNA-seq read coverage under uniform sampling. **b.** RNA-seq read coverage under uniform sampling with transcript start/end bias. **c.** RNA-seq read coverage under uniform sampling with sequence-specific bias. **d.** RNA-seq read coverage under uniform sampling with 5'/3' position-specific bias. **e.** RNA-seq read coverage under uniform sampling with all aforementioned types of bias. **f.** Sampling bias on gene CENPF in the breast cancer dataset used in Section 6. Please note that the second peak in the coverage plot is not an exon in CENPF. The observed coverage on each exon decreases almost linearly from the 3' end to the 5' end. The coverage also drops at the bases near the end of the gene. The non-uniformity in the two middle large exons is likely to be due to the sequence-specific sampling bias.

probability of exonic segments and MultiSplices. Therefore, we calculate the bias coefficient $\sigma(\phi, t)$ for every feature $\phi \in \Phi_G$ and every transcript t so that $E[C(\phi|t)] = \sigma(\phi, t)m(\phi, t)C(t)$. Next, we introduce each independent bias individually.

Sequence-Specific Bias. The sequence-specific bias refers to the perturbation of sampling probability related to certain sequences at the beginning or end of transcript fragments [4, 18]. The characteristic of this type of bias in the given RNA-seq sample can be learned in advance by examining the relationship between GC content and the observed coverage on single-isoform genes. To derive the sequence-specific bias at an arbitrary exonic position, we look into 8bp upstream to the 5' start to 11bp downstream according to [25]. A Markov chain is constructed to model the effect on the sampling probability at the position from the sequence of surrounding nucleotides. Then we use an approach based on the probabilistic suffix tree [5] to learn the sequence-specific bias coefficient $\alpha(t, i)$ for ith nucleotide in transcript t.

Transcript Start/End Bias. Sampling near transcript start site or transcript end site is often insufficient. The read coverage in these regions is typically lower than expected because the positions where a sampled read can cover are restricted by the transcript boundaries. The bias coefficient for start/end bias at the ith nucleotide in transcript t is written as:

$$\beta(t, i) = \begin{cases} i/E[l(fr)] & \text{if } i < E[l(fr)] \\ 1 & \text{if } E[l(fr)] \le i \le l(t) - E[l(fr)] \\ (l(t) - i)/E[l(fr)] & \text{if } i > l(t) - E[l(fr)]. \end{cases}$$

5'/3' Position-Specific Bias. Position-specific bias refers to the alteration on sampling probability according to position in the transcript. For example, nucleotides to the 3' end of the transcript have higher probability to be sampled in Figure 2(d). Here we model the position-specific bias coefficient as a linear function, $\gamma(t, i) = \gamma_1^t \cdot i + \gamma_0^t$. The intercept γ_0^t gives the bias coefficient at the 5' transcript start site. The slope γ_1^t measures the extent of the bias: a positive γ_1^t indicates that 3' transcript end site has higher sampling probability than the start site; a zero γ_1^t indicates no positional bias in the transcript t.

Combined Bias Model. Assuming the above three types of bias have independent effect on read sampling, we derive the bias coefficient at ith nucleotide in transcript t as $\sigma(t, i) = \alpha(t, i) \cdot \beta(t, i) \cdot \gamma(t, i)$. The bias coefficient of an exonic segment $e \in \mathcal{E}_g$ is then the averaged bias coefficient on all positions in the exonic segment e, and the bias coefficient of a MultiSplice $b \in \mathcal{B}_g$ is the averaged bias coefficient on all positions in its sampling window $w(b)$. In summary, the bias coefficient for a MultiSplice feature $\phi \in \Phi_g$ in transcript t is

$$
\sigma(\phi, t) = \begin{cases} \frac{\sum_{i \in \phi} \sigma(t,i)}{l(\phi)} & \text{if } \phi \subset t \text{ and } \phi \in \mathcal{E}_g \\ \frac{\sum_{i \in w\phi} \sigma(t,i)}{E[l(w(\phi))]} & \text{if } \phi \subset t \text{ and } \phi \in \mathcal{B}_g \\ 0 & \text{if } \phi \not\subset t. \end{cases} \tag{4}
$$

5 Solving the Generalized Linear Models with Bias Correction

Conventionally, we are interested in the set of transcript expressions that minimize the sum of squared errors, the absolute residuals between the expected coverage and the observed coverage. This solution is relatively sensitive to unexpected sampling noise which often occurs in real RNA-seq samples and may lead to a highly unstable extrapolation when the expression of the alternative splicing events discriminating the transcripts is notably lower than the average level of gene expression. Therefore, we define the sum of squared relative errors (SSRE), which measures the relative residual regarding the ratio of the expected coverage against the observed coverage.

$$
SSRE = \sum_{f \in \mathcal{F}_G} \left(\frac{\sum_{t \in \mathcal{T}_G} \sigma(f,t) \mathbf{M}'(f,t) C(t)}{C(f)} - 1 \right)^2. \tag{5}
$$

Bias Parameter Estimates. Among all the bias parameters, the sequence-specific bias is learned in advance while the start and end bias is a function of transcript fragment length. The only bias parameters unknown related to the 3' bias defined by the intercept γ_0^t and slope γ_1^t for every transcript $t \in \mathcal{T}_g$. Therefore, we use an iterative-minimization strategy and search for a set of bias coefficients γ_0^t's and γ_1^t's that better fit the RNA-seq sample than the uniform sampling model. We start with the transcript coverage $C(t)$'s that are solved from the uniform sampling model (with $\gamma_0^t = 1$ and $\gamma_1^t = 0$ as initial condition).

Analogous to the hill climbing algorithm [26], we then iteratively probe a locally optimal set of transcript coverage together with the bias coefficients around the uniform solution through minimizing the SSRE. In each iteration, a candidate solution is obtained through sequentially setting the partial derivatives to 0 with respect to every unknown parameter γ_0^t, γ_1^t, $C(t)$, and for every transcript $t \in \mathcal{T}_G$. If the candidate solution results in a smaller SSRE, the candidate solution is taken and the iteration continues. For details of the step to estimate the bias parameters, please refer to the Appendix section.

Solving the Linear Model with LASSO Regularization. Lastly, we solve for the level of individual transcript expression with additional regularization, based on the bias coefficients from the previous step. One common problem in transcript quantification is that the set of expressed transcripts are not known a priori. Hence it becomes crucially important to identify the set of truly expressed transcripts provided in a candidate set. Therefore, we further apply the L1 regularization (known as LASSO) for its proven effectiveness in irrelevance-removal and solve for the set of transcript expression $C(\mathcal{T}_G)$ that minimizes the following loss function

$$L = \text{SSRE} + \text{L1 penalty} = \sum_{\phi \in \Phi_G} \left(\frac{\sum_{t \in \mathcal{T}_G} \sigma(\phi, t) \mathbf{M}'(\phi, t) C(t)}{C(\phi)} - 1 \right)^2 + \lambda \|C(\mathcal{T}_G)\|_1, \quad (6)$$

where $\lambda \geq 0$ denotes the weight of the L1 shrinkage and $C(t) \geq 0$ for every $t \in \mathcal{T}_G$.

6   Experimental Results

To evaluate the performance of the MultiSplice model, we compared it with three other approaches. The *ExonOnly* model, where only exonic segments are used to represent transcript composition as proposed in SLIDE [20], was implemented using a linear regression approach with LASSO. The ExonOnly model provided the baseline comparison for MultiSplice. The *Poisson* model, which was originally proposed by [24], was implemented in C since it is not publicly available. Cufflinks [31] is a representative of read-centric model. Cufflinks 1.1.0 was downloaded from its website in September, 2011.

These algorithms were run on both simulated datasets and real datasets. Reads were first mapped by MapSplice 1.15.1 [34] to the reference genome. If the read was paired-end, MapPER [12] was applied to infer the alignment of the entire transcript fragment.

6.1   Transcriptome Identifiability with Increasing Read Length

We first study how the increase in read length may alleviate the lack of identifiability issues in transcript quantification using MultiSplice. We downloaded UCSC gene models in human (track UCSC genes:GRCh37/hg19), mouse (track UCSC Genes:NCBI37/mm9), worm (track WormBase Genes:WS190/ce6) and fly (track FlyBase Genes:BDGP R5/dm3). We computed the feature matrix used in MultiSplice given variable read length and determined its rank. The transcript isoforms of a gene are identifiable if the rank of the feature matrix is no less than the number of transcripts. Figure 3 plots the additional number of genes that become identifiable as the read length increases from 50bp assuming single-end read RNA-seq data. For all four species, as the read length increases, MultiSplice is capable of resolving the transcript quantification issues of more genes. With 500bp reads, about 98% genes in both human and mouse become identifiable. Surprisingly, for worm and fly, 500bp reads do not gain significant improvement over 50bp reads. This is mostly due to the fact that the exon lengths of fly and worm are comparably much longer [9] than human and mouse, making it difficult for reads of moderate size to take effect. With current short read technology where read length is typically 100bp or less, paired-end reads with the size of transcript fragments around 500bp may be the most economical and effective for transcription quantification for genes with identifiability issues. This is under the assumption that it is possible to infer the transcript fragment from paired-end reads based on the tightly controlled distribution of insert-size.

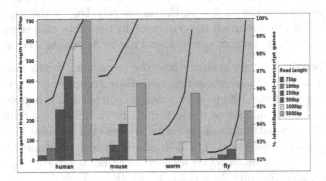

Fig. 3. Changes in mRNA identifiability as a function of transcript fragment/read length. Starting from levels achieved with 50bp single-end reads, the left side of the y-axis shows the additional number of genes that become identifiable using MultiSplice as the read length increases. The y-axis on the right side shows the total percentage of genes for which mRNA transcript structures are resolved. The UCSC annotated transcript sets of four species: human, mouse, fly and worm were used for this analysis.

6.2 Simulated Human RNA-seq Experiment

Data Simulation. Due to the lack of the ground truth within real datasets, simulated data has become an important resource for the evaluation of transcript quantification algorithms [6, 17, 22]. We developed an in-house simulator to generate RNA-seq datasets of a given sampling depth using human hg19 Refseq annotation. The simulation process consists of three steps: (1) randomly assign relative proportions to all the transcripts within a gene and set this as the true profile; (2) calculate the number of reads to be sampled from each transcript; (3) sample transcript fragments of a given length along the transcripts according to the per base coefficient $\sigma(t, i) = \frac{ki\alpha(t,i)\beta(t,i)}{l(t)} + 1$ for the ith base on transcript t, where $\alpha(t, i)$ and $\beta(t, i)$ are the sequence-specific bias and the transcript start/end bias as defined in Section 4 and k is the slope of the position-specific bias. Paired-end reads will be generating by taking the two ends of the transcript fragment. Please note the sequence bias per base has been learned from a real dataset, a technical replicate of MCF-7 data that will be introduced in the next section.

Accuracy Measurement. Due to inconsistencies in the normalization scheme used by different software, the estimated abundance may not be comparable among different approaches. Hence, we computed relative proportions of transcript isoforms for each method. The similarity between the estimated result and the ground truth is measured by both Pearson correlation and Euclidean distance. Let X denote the vector of real isoform proportions of a gene and $\hat{X}$ denote the estimated proportions. The formula of the correlation is: $r(X, \hat{X}) = cov(X, \hat{X})/(\sigma_X \cdot \sigma_{\hat{X}})$. A value close to 1 means that our estimation is highly accurate and vice versa. Below, we adopt a boxplot to illustrate the performance of each method. The box is constructed by the 1st quartile, the median, and the 3rd quartile. The ends of the upper and lower whisker are given by the 3rd quartile $+1.5 \times IQR$ (inner quartile range) and 1st quartile $-1.5 \times IQR$, respectively. Due to the space limit, we present the result of correlation measurement in the main manuscript. Results measured by Euclidian distance can be found in the Appendix section.

Sampling Depth. Next we evaluate how the sequencing depth may affect the accuracy of transcript abundance estimation. Four groups of 2x50bp paired-end synthetic data (insert size 100bp) were generated on the whole human transcriptome with increasing number of reads: 6 million, 12 million, 18 million and 24 million. 14530 genes with multiple isoforms are selected for analysis. The genes were divided into three subsets: (1) 13576 genes to which identifiability holds for all methods. (2) 455 genes to which identifiability holds for MultiSplice. (3)499 genes to which identifiability does not hold for all methods.

For each subplot in Figure 4(a, b, c), the estimation accuracy for all methods generally improves as more reads are sampled. Cufflinks seems to be affected most by the sampling depth. For the genes whose identifiability conditions are satisfied for all methods, the correlation between the estimated transcript proportion is highly similar with the ground truth, with an average correlation close to 0.9 for all methods. In the second category, when the genes are still identifiable with MultiSplice, the estimation accuracy of MultiSplice remains high, with an average correlation above 0.6 while others slip below 0.5. For the category when identifiability is not satisfied for all methods, the estimation accuracy is degraded even more. However, MultiSplice still consistently gives better estimation results indicating that the inclusion of MultiSplice features make transcript quantification more stable than other methods. Cufflinks demonstrated the worst performance in this category, mainly because the unidentifiability conditions make it difficult to assign these reads to a transcript. Instead, it throws out most of

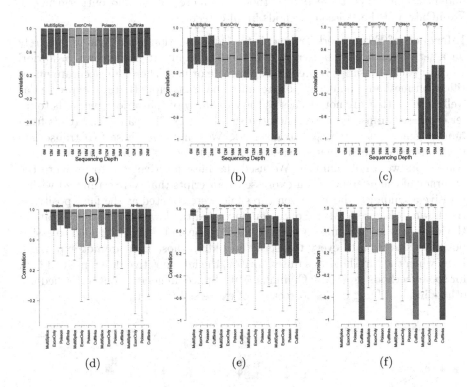

Fig. 4. a-c. Boxplots of the correlation between estimated transcript proportions and the ground truth under varying number of sampled reads: 6M, 12M, 18M and 24M over a total of 14530 human genes with more than one isoforms. (a),(b) and (c) correspond to the gene set that is identifiable with basic exon structure, identifiable with additional MultiSplice features, and unidentifiable, respectively. d-f: Boxplots of the correlation between estimated transcript proportions and the ground truth under four circumstances: uniform sampling, sampling with positional bias only, with sequence bias only and with all bias. (d),(e) and (f) correspond to the gene set that is identifiable with basic exon structure, identifiable with additional MultiSplice features, and unidentifiable, respectively.

multi-mapped reads. Apparently, increasing sampling depth cannot alleviate the issue of unidentifiability.

Bias Correction. To study the effect of the bias correction, we have simulated data with uniform sampling, sampling with only positional bias, sampling with only sequence bias, and sampling with the combined positional and sequence bias. Here, we set the slope of the position-specific bias k to 2 with 24 million 2x50bp paired-end reads sampled from the whole transcriptome for each case. All the approaches achieve the best results when the sampling process is uniform. As positional or sequence bias is introduced, their performance tapers down. The presence of both positional and sequence biases has the largest impact in all methods. Meanwhile, because MultiSplice and Cufflinks correct both sequence

and positional bias, these two methods are more robust and outperform the
ExonOnly and the Poisson methods in all categories.

Inference of Expressed Transcripts. Quantification of mRNAs usually rely
on a set of candidate transcript structures as input. It is unknown in apriori
whether each transcript is present in a sample or not. Therefore, accurate quantification methods should be able to infer the transcripts that are expressed as
well as those that are not. To assess the capability of the various methods to infer expressed transcripts, we generated simulated 2x50bp paired-end reads from
human genes with at least 3 transcripts. We randomly chose two transcripts
from one gene and simulated reads only from these transcripts. The remaining
transcripts were not sampled. We used the false positive rate to measure the
accuracy of the inference. Non-expressed transcripts that were estimated with a
positive abundance above a given threshold were counted as the false positives.
As shown in Figure 5, MultiSplice demonstrated the lowest false positive rate in
the identification of dominant transcripts. Poisson and Cufflinks tended to assign
positive expression to every transcript including those that are not expressed.
Even when the threshold was raised to 10%, the false positive rate remained high
for some methods especially Cufflinks. MultiSplice, in general, outperformed the
others in identifying of the correct set of expressed transcripts.

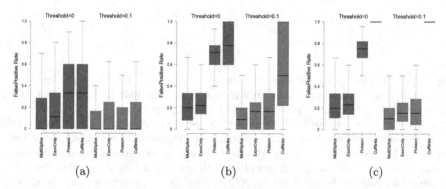

Fig. 5. Comparison of false positive rates in the inference of the expressed transcripts.
Thresholds represent the minimum fraction of a transcript that is considered expressed.
(a),(b) and (c) correspond to the gene set that is identifiable with the basic exon structure, identifiable with additional MultiSplice features, and unidentifiable, respectively.

6.3   Real Human RNA-seq Experiment

We attempted to use RNA-seq data generated from the samples in the Microarray Quality Control (MAQC) Project [15] with TaqMan qRT-PCR measurements of the abundance for approximate 1000 genes. Our primary interest is in
disambiguating multiple isoforms using MultiSplice features. However, most of
these genes express only a single isoform. Therefore, we applied the set of transcript quantification methods to a dataset that was originally used by Singh et
al. to study differential transcription [28]. In this study, two groups of RNA-seq

datasets were generated from SUM-102 and MCF-7, two breast cancer cell lines. Each group contains 4 samples as technical replicates. The RNA-seq data were generated from Illumina HISEQ2000. Each sample had 80 million 100bp single-end reads. About 60 million reads can be aligned to the reference genome by MapSplice. The Refseq human annotated transcripts were fed into each software for transcript quantification.

Since ground truth expression profiles do not exist for the real datasets, we investigated whether the different methods provided a consistent estimation within samples of technical replicates which only vary by random sampling. In contrast, a significant number of genes between MCF-7 and SUM-102 were expected to be differentially expressed [28]. To evaluate this, we computed Jensen–Shannon divergence (JSD), used in Cuffdiff [1] to measure the dissimilarity between two samples and calculated the *within-group* and *between-group* differences. As detailed in Figure 6(a), both MultiSplice and Cufflinks had smaller average within-group difference than the average between-group difference while the other two methods do not show clear difference. MultiSplice demonstrated higher between-group difference than Cufflinks, but also had relatively higher within-group differences as well. Most of these, however, were well below a JSD of 0.2 and considered to be insignificant. A closer look at a number of cases showed that occasionally MultiSplice and Cufflinks may overestimate or underestimate the between-group difference respectively. Figure 6(b) (The complete figure with 8 samples can be found in the Appendix Figure 8(a)) shows a gene where Cufflinks underestimated the difference between the two groups. The second isoform of the gene AIM1 has a unique first exon (chr6:106989461-106989496). Clear difference in the read coverage on this exon can be observed between the two groups, indicating strong differential levels of expression, i.e., the second isoform is barely

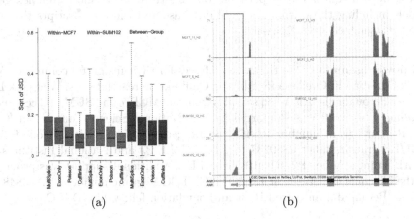

(a)   (b)

Fig. 6. a. Boxplots of the within-MCF-7, within-SUM-102, and between-group square root of JSD of all genes for all methods. **b.** A case where Cufflinks underestimated the difference between the two groups. The second isoform of Gene AIM1 has a unique first exon, whose read coverage differs significantly between the two groups. A detailed plot with all 8 samples can be found in the Appendix Figure 8(a).

expressed in MCF-7 while almost comparable to the first isoform in SUM-102 cells. The between group square root of JSD is 0.21 by Cufflinks, much lower than 0.50 by MultiSplice.

The exon-skipping event found in gene CD46 is also differentially expressed (Figure 8(b), Appendix). The estimation of transcript quantification with MultiSplice was consistent with the observation in the qRT-PCR data showing that steady state levels of transcripts with the skipped exon were present in amounts more than two fold higher expression in SUM-102 than in MCF-7 cells.

7   Conclusion

In this paper, we propose a generalized linear system for the accurate quantification of alternative transcript isoforms with RNA-seq data. We introduce a set of new structural features, namely MultiSplice, to ameliorate the issue of *identifiability*. With MultiSplice features, 98% of Refseq transcript models in human and mouse become identifiable with 500bp reads (or paired-end reads with 500bp transcript fragments), an 8% increase from 50bp. Therefore, longer reads or paired-end reads with longer insert-sizes rather than further increases in sequencing depths can be crucial for the accurate quantification of mRNA isoforms with complex alternative transcription, even though a majority of the genes have relatively simple transcript variants. The results also demonstrate the robustness of the MultiSplice method under various sampling biases, consistently outperforming three other methods: Cufflinks, Poisson and ExonOnly. The application of our approach to real RNA-seq datasets for transcriptional profiling successfully identified a number of isoforms whose proportion changes differed significantly between two distinct breast cancer cell lines. In the near future, we will continue to experiment our algorithms with more complex gene models including those from Ensembl database and those transcripts that are directly assembled from RNA-seq.

Acknowledgements. We thank the referees for their insightful comments. We thank Christian F. Orellana for running Cufflinks on real datasets. We thank Dr. Charles Perou for the RNA-seq samples from the MCF-7 and SUM-102 cell lines. *Funding*: This work was supported by The National Science Foundation [CAREER award grant number 1054631 to J.L.]; the National Science Foundation [ABI/EF grant number 0850237 to J.L. and J.F.P.] and National Institutes of Health [grant number P20RR016481 to J.L.]. Additional support was provided by the National Institutes of Health: NCI TCGA [grant number CA143848 to Charles Perou] and an Alfred P. Sloan Foundation fellowship [D.Y.C.].

References

[1] Cufflinks, http://cufflinks.cbcb.umd.edu
[2] Ensembl Genome Browser, http://useast.ensembl.org/index.html
[3] NCBI Reference Sequence (RefSeq), http://www.ncbi.nlm.nih.gov/RefSeq

[4] Roberts, A., Trapnell, C., Donaghey, J., Rinn, J., Pachter, L.: Improving rna-seq expression estimates by correcting for fragment bias. Genome Biology 12(3), R22 (2011)

[5] Bejerano, G.: Algorithms for variable length markov chain modeling. Bioinformatics 20, 788–789 (2004)

[6] Bohnert, R., Gunnar, R.: rquant.web: a tool for rna-seq-based transcript quantitation. Nucleic Acids Research 38(suppl. 2), W348–W351 (2010)

[7] Brosseau, J.-P., Lucier, J.-F., Lapointe, E., Durand, M., Gendron, D., Gervais-Bird, J., Tremblay, K., Perreault, J.-P., Elela, S.A.: High-throughput quantification of splicing isoforms. RNA Society 16, 442–449 (2010)

[8] Feng, J., Li, W., Jiang, T.: Inference of Isoforms from Short Sequence Reads. In: Berger, B. (ed.) RECOMB 2010. LNCS, vol. 6044, pp. 138–157. Springer, Heidelberg (2010)

[9] Fox-Walsh, K.L., Dou, Y., Lam, B.J., Hung, S.-P., Baldi, P.F., Herte, K.J.: The architecture of pre-mrnas affects mechanisms of splice-site pairing. Proc. Natl. Acad. Sci. 102(45), 16176–16181 (2005)

[10] Guttman, M., Garber, M., Levin, J.Z., Donaghey, J., Robinson, J., Adiconis, X., Fan, L., Koziol, M.J., Gnirke, A., Nusbaum, C., Rinn, J.L., Lander, E.S., Regev, A.: Ab initio reconstruction of cell type-specific transcriptomes in mouse reveals the conserved multi-exonic structure of lincrnas. Nature Biotechnology 28, 503–510 (2010)

[11] Horn, R.A., Johnson, C.R.: Matrix analysis. Cambridge University Press (1990)

[12] Hu, Y., Wang, K., He, X., Chiang, D.Y., Prins, J.F., Liu, J.: A probabilistic framework for aligning paired-end rna-seq data. Bioinformatics 26, 1950–1957 (2010)

[13] Jiang, H., Wong, W.H.: Statistical inferences for isoform expression in rna-seq. Bioinformatics 25, 1026–1032 (2009)

[14] Kozarewa, I., Ning, Z., Quail, M.A., Sanders, M.J., Berriman, M., Turner, D.J.: Amplification-free illumina sequencing-library preparation facilitates improved mapping and assembly of (g+c)-biased genomes. Nuc. 6, 291–295 (2009)

[15] Shi, L., Reid, L.H., Jones, W.D., et al.: The microarray quality control (maqc) project shows inter- and intraplatform reproducibility of gene expression measurements. Nature Biotechnology 24(9), 1151–1161 (2006)

[16] Lacroix, V., Sammeth, M., Guigo, R., Bergeron, A.: Exact Transcriptome Reconstruction from Short Sequence Reads. In: Crandall, K.A., Lagergren, J. (eds.) WABI 2008. LNCS (LNBI), vol. 5251, pp. 50–63. Springer, Heidelberg (2008)

[17] Li, B., Ruotti, V., Stewart, R.M., Thomson, J.A., Dewey, C.N.: Rna-seq gene expression estimation with read mapping uncertainty. Bioinformatics 26 (4), 493–500 (2010)

[18] Li, J., Jiang, H., Wong, W.H.: Modeling non-uniformity in short-read rates in rna-seq data. Genome Biology 11 (2010)

[19] Li, W., Feng, J., Jiang, T.: IsoLasso: A LASSO Regression Approach to RNA-Seq Based Transcriptome Assembly. In: Bafna, V., Sahinalp, S.C. (eds.) RECOMB 2011. LNCS, vol. 6577, pp. 168–188. Springer, Heidelberg (2011)

[20] Lia, J.J., Jiangb, C.-R., Browna, J.B., Huanga, H., Bickela, P.J.: Sparse linear modeling of next-generation mrna sequencing (rna-seq) data for isoform discovery and abundance estimation. PNAS (2011)

[21] Olejniczak, M., Galka, P., Krzyzosiak, W.J.: Sequence-non-specific effects of rna interference triggers and microrna regulators. Nucl. Acids Res. 38(1), 1–16 (2010)

[22] Nicolae, M., Mangul, S., Mandoiu, I.I., Zelikovsky, A.: Estimation of alternative splicing isoform frequencies from rna-seq data. Algorithms for Molecular Biology 6, 9 (2011)

[23] Pan, Q., Shai, O., Lee, L.J., Frey, B.J., Blencowe, B.J.: Deep surveying of alternative splicing complexity in the human transcriptome by high-throughput sequencing. Nature Genetics 40, 1413–1415 (2008)

[24] Richard, H., Schulz, M.H., Sultan, M., Nrnberger, A., Schrinner, S., Balzereit, D.,
Dagand, E., Rasche, A., Lehrach, H., Vingron, M., Haas, S.A., Yaspo, M.-L.: Pre-
diction of alternative isoforms from exon expression levels in rna-seq experiments.
Nucleic Acids Research 38, e112 (2010)

[25] Roberts, A., Trapnell, C., Donaghey, J., Rinn, J.L., Pachter, L.: Improving rna-
seq expression estimates by correcting for fragment bias. Genome Biology 12, R22
(2011)

[26] Russell, S., Norvig, P.: Artificial intelligence: A modern approach, R22 (2003)

[27] Srivastava, S., Chen, L.: A two-parameter generalized poisson model to improve
the analysis of rna-seq data. Nucleic Acids Research, 1–15 (2010)

[28] Singh, D., Orellana, C.F., Hu, Y., Jones, C.D., Liu, Y., Chiang, D.Y., Liu, J., Prins,
J.F.: Fdm: A graph-based statistical method to detect differential transcription
using rna-seq data. Bioinformatics (2011)

[29] Srivastava, S., Chen, L.: A two-parameter generalized poisson model to improve
the analysis of rna-seq data. Nucleic Acids Research 38, e112 (2010)

[30] Tibshirani, R.: Regression shrinkage and selection via the lasso. Journal of Royal
Statistical Society Series B. 58, 267–288 (1996)

[31] Trapnell, C., Williams, B.A., Pertea, G., Mortazavi, A., Kwan, G., van Baren,
M.J., Salzberg, S.L., Wold, B.J., Pachter, L.: Transcript assembly and quantification by rna-seq reveals unannotated transcripts and isoform switching during cell
differentiation. Nature Biotechnology 28, 511–515 (2010)

[32] Turro, E., Su, S.-Y., Gonçalves, Â., Coin, L.J.M., Richardson, S., Lewin, A.: Haplotype and isoform specific expression estimation using multi-mapping rna-seq
reads. Genome Biology 12, R13 (2011)

[33] Wang, E.T., Sandberg, R., Luo, S., Khrebtukova, I., Zhang, L., Mayr, C.,
Kingsmore, S.F., Schroth, G.P., Burge, C.B.: Alternative isoform regulation in
human tissue transcriptomes. Nature 456, 470–476 (2008)

[34] Wang, K., Singh, D., Zeng, Z., Huang, Y., Coleman, S., Savich, G.L., He,
X., Mieczkowski, P., Grimm, S.A., Perou, C.M., MacLeod, J.N., Chiang, D.Y.,
Prins, J.F., Liu, J.: Mapsplice: Accurate mapping of rna-seq reads for splice junc-
tion discovery. Nucleic Acid Research 38(18), 178 (2010)

[35] Wang, Z., Gerstein, M., Snyder, M.: Rna-seq: a revolutionary tool for transcriptomics. Nature Reviews Genetics 10, 57–63 (2009)

[36] Wu, J., Akerman, M., Sun, S., Richard McCombie, W., Krainer, A.R., Zhang,
M.Q.: Splicetrap: a method to quantify alternative splicing under single cellular
conditions. Bioinformatics (2011)

[37] Wu, Z., Wang, X., Zhang, X.: Using non-uniform read distribution models to
improve isoform expression inference in rna-seq. Bioinformatics 27, 502–508 (2011)

Appendix

Iterative-Minimization Algorithm

In Section 5, we use an iterative-minimization strategy to search for a set of bias
coefficients γ_0^t's and γ_1^t's for every transcript $t \in \mathcal{T}_g$ that better fit the RNA-seq
sample than the uniform sampling model. We initiate the iterations with the

transcript coverage $C(t)$'s solved from the uniform sampling model and the bias coefficients $\gamma_0^t = 1$ and $\gamma_1^t = 0$. In each iteration, for transcript t we set:

1. $\frac{\partial SSRE}{\partial C(t)} = 0$; 2. $\frac{\partial SSRE}{\partial \gamma_1^t} = 0$; 3. $\frac{\partial SSRE}{\partial \gamma_0^t} = 0$.

$$\frac{\partial SSRE}{\partial C(t)} = 0$$

$$\Rightarrow \sum_{\phi \in \Phi_g} 2(C(\phi) - \sum_{s \in T_g} \sigma(\phi, s)\mathbf{M}'(\phi, s)C(s)) \cdot \sigma(\phi, t)\mathbf{M}'(\phi, t) = 0$$

$$\Rightarrow \sum_{s \in T_g} C(s)(\sum_{\phi \in \Phi_g} \sigma(\phi, s)\mathbf{M}'(\phi, s)\sigma(\phi, t)\mathbf{M}'(\phi, t)) = \sum_{\phi \in \Phi_g} C(\phi)\sigma(\phi, t)\mathbf{M}'(\phi, t)$$

$$\Rightarrow C(t) = \frac{\sum_{\phi \in \Phi_g} C(\phi)\sigma(\phi, t)\mathbf{M}'(\phi, t) - \sum_{s \in T_g, s \neq t} C(s)(\sum_{\phi \in \Phi_g} \sigma(\phi, s)\mathbf{M}'(\phi, s)\sigma(\phi, t)\mathbf{M}'(\phi, t))}{\sum_{\phi \in \Phi_g} \sigma(\phi, t)\mathbf{M}'(\phi, t)\sigma(\phi, t)\mathbf{M}'(\phi, t)}.$$

$\sigma(\phi, t)$ is the only function related to γ_1^t and γ_0^t.

$$\frac{\partial SSRE}{\partial \gamma_1^t} = 0$$

$$\Rightarrow \sum_{\phi \in \Phi_g} 2(C(\phi) - \sum_{s \in T_g} \sigma(\phi, s)\mathbf{M}'(\phi, s)C(s)) \cdot \frac{\partial \sigma(\phi, t)}{\partial \gamma_1^t}\mathbf{M}'(\phi, t)C(t) = 0$$

$$\Rightarrow C(t) \sum_{\phi \in \Phi_g} \sigma(\phi, t)\mathbf{M}'(\phi, t)\frac{\partial \sigma(\phi, t)}{\partial \gamma_1^t}\mathbf{M}'(\phi, t)$$

$$= \sum_{\phi \in \Phi_g} C(\phi)\frac{\partial \sigma(\phi, t)}{\partial \gamma_1^t}\mathbf{M}'(\phi, t) - \sum_{s \in T_g, s \neq t} C(s)(\sum_{\phi \in \Phi_g} \sigma(\phi, s)\mathbf{M}'(\phi, s)\frac{\partial \sigma(\phi, t)}{\partial \gamma_1^t}\mathbf{M}'(\phi, t)).$$

Similarly,

$$\frac{\partial SSRE}{\partial \gamma_0^t} = 0$$

$$\Rightarrow \sum_{\phi \in \Phi_g} 2(C(\phi) - \sum_{s \in T_g} \sigma(\phi, s)\mathbf{M}'(\phi, s)C(s)) \cdot \frac{\partial \sigma(\phi, t)}{\partial \gamma_0^t}\mathbf{M}'(\phi, t)C(t) = 0$$

$$\Rightarrow C(t) \sum_{\phi \in \Phi_g} \sigma(\phi, t)\mathbf{M}'(\phi, t)\frac{\partial \sigma(\phi, t)}{\partial \gamma_0^t}\mathbf{M}'(\phi, t)$$

$$= \sum_{\phi \in \Phi_g} C(\phi)\frac{\partial \sigma(\phi, t)}{\partial \gamma_0^t}\mathbf{M}'(\phi, t) - \sum_{s \in T_g, s \neq t} C(s)(\sum_{\phi \in \Phi_g} \sigma(\phi, s)\mathbf{M}'(\phi, s)\frac{\partial \sigma(\phi, t)}{\partial \gamma_0^t}\mathbf{M}'(\phi, t)).$$

Because $\sigma(\phi, t)$ is a linear combination of γ_1^t and γ_0^t, and hence $\sum_{\phi \in \Phi_g} \sigma(\phi, t)\mathbf{M}'\phi, t$ is also the linear combination of γ_1^t and γ_0^t. Then we can directly calculate $\frac{\partial \sigma(\phi,t)}{\partial \gamma_1^t}$ and $\frac{\partial \sigma(\phi,t)}{\partial \gamma_0^t}$.

Supplementary Figures

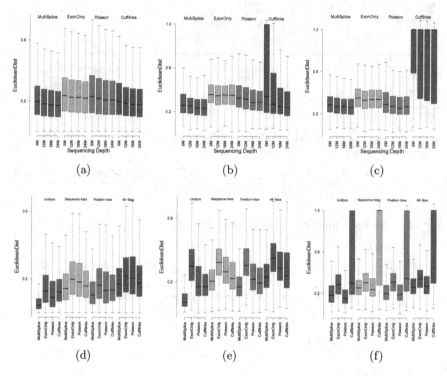

Fig. 7. a-c. Boxplots of the Euclidean distance between estimated transcript proportions and the ground truth under varying number of sampled reads: 6M, 12M, 18M and 24M over a total of 14530 human genes with more than one isoforms. (a),(b) and (c) correspond to the gene set that is identifiable with basic exon structure, identifiable with additional MultiSplice features, and unidentifiable, respectively. d-f: Boxplots of the Euclidean distance between estimated transcript proportions and the ground truth under four circumstances: uniform sampling, sampling with positional bias only, with sequence bias only and with all bias. (d),(e) and (f) correspond to the gene set that is identifiable with basic exon structure, identifiable with additional MultiSplice features, and unidentifiable, respectively.

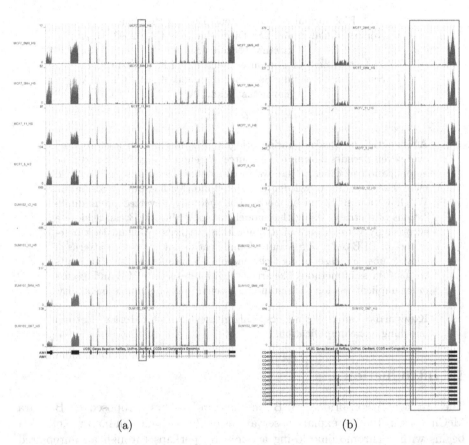

(a)   (b)

Fig. 8. a. The coverage plot of Gene AIM1 in all 8 breast cancer cell line samples. Please note the first exon of the second isoform is barely expressed MCF-7 but its expression significantly increased in the SUM-102 samples. b. The coverage plot of Gene CD46. The exon-skipping event on the 13th exon has been confirmed by qRT-PCR.

Modeling the Breakage-Fusion-Bridge Mechanism: Combinatorics and Cancer Genomics

Marcus Kinsella[1] and Vineet Bafna[2]

[1] Bioinformatics and Systems Biology Program,
University of California, San Diego, La Jolla CA 92093, USA
mckinsel@ucsd.edu
[2] Department of Computer Science and Engineering,
University of California, San Diego, La Jolla CA 92093, USA
vbafna@cs.ucsd.edu

Abstract. The breakage-fusion-bridge (BFB) mechanism was proposed over seven decades ago and is a source of genomic variability and gene amplification in cancer. Here we formally model and analyze the BFB mechanism, to our knowledge this first time this has been undertaken. We show that BFB can be modeled as successive inverted prefix duplications of a string. Using this model, we show that BFB can achieve a surprisingly broad range of amplification patterns. We find that a sequence of BFB operations can be found that nearly fits most patterns of copy number increases along a chromosome. We conclude that this limits the usefulness of methods like array CGH for detecting BFB and discuss other implications for understanding mechanisms of genomic instability.

Keywords: cancer genomics, algorithms, combinatorial pattern matching, gene amplification.

1 Introduction

The breakage-fusion-bridge (BFB) mechanism was first proposed by Barbara McClintock in 1938 to explain observations of chromosomes in maize [5,6]. BFB begins with a chromosome losing a telomere, perhaps through an unrepaired DNA break or through telomere shortening. As the chromosome replicates, the broken ends of each of its sister chromatids fuse together. During anaphase, as the centromeres of the chromosome migrate to opposite ends of the cell, the fused chromatids are torn apart. Each daughter cell receives a chromosome missing a telomere, and the cycle can begin again (Figure 1A).

When the fused chromosome is torn apart during anaphase, it likely does not tear exactly in the middle of the two centromeres. As a result, one daughter cell receives a chromosome with a terminal inverted duplication while the other receives a chromosome with a terminal deletion. After many BFB cycles, repeated inverted duplications can result in a dramatic increase in the copy number of segments of the unstable chromosome (Figure 1B).

B. Chor (Ed.): RECOMB 2012, LNBI 7262, pp. 148–162, 2012.

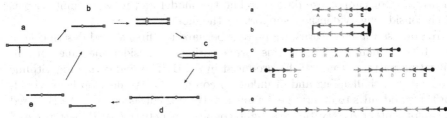

(A) The BFB mechanism is a multiple step process. First, a chromosome loses a telomere (a). Then, the telomere-lacking chromosome (b) replicates. The sister chromatids lacking telomeres fuse together (c). During anaphase, the centromeres separate, forming a dicentric chromosome (d). As the centromeres migrate to opposite ends of the cell, the chromosome is torn apart, and each daughter cell gets a chromosome lacking a telomere (e).

(B) After telomere loss (a), sister chromatid fusion (b), and centromere separation (c), the dicentric chromosome may not break in the center of the two centromeres (d). In this case, the break was between the green and cyan segments, so one daughter cell will have a a deletion of those segments while the other will have an inverted duplication. Multiple rounds of inversion and duplication can lead to amplification of chromosomal segments (e).

Fig. 1. The Breakage Fusion Bridge mechanism

BFB's ability to amplify chromosomal segments suggests a role for the mechanism in cancer. Gene amplification is common in tumors. A recent review identified 77 genes whose amplification is implicated in cancer development [7]. Multiple lines of evidence indicate that BFB may be responsible for much of this amplification. Telomere dysfunction and crisis is associated with tumorigenesis [1]. Such dysfunction is consistent with the initiation and continuation of BFB cycles. Some tumors also display the cytogenetic hallmarks of BFB: chromosomes that stretch across spindle poles during anaphase, dicentric chromosomes, and homogeneously staining regions. Thus, an improved understanding of BFB may shed light on how genomic instability leads to tumor formation and progression.

Observing BFB through classical cytogenetic techniques can be difficult and only provides coarse detail. Recent studies have begun to apply modern methods to the problem of detecting BFB and elucidating its role in generating genomic aberrations. Kitada and Yamasaki performed FISH and array CGH on a lung cancer cell line and showed that the pattern of amplification and rearrangement they observed was consistent with BFB [4]. Later, Bignell *et al.* sequenced breakpoints in a breast cancer cell line and confirmed that the copy count and breakpoint patterns on 17q were consistent with a BFB model, purportedly the first demonstration of a sequence-level hallmark of BFB in human cancer [2].

These studies, among others, illustrate the promise of new methods for gaining a more complete understanding of BFB. However, making observations that are consistent with a model does not allow one to conclude that the model is correct. Moreover, without a precise definition of the model under consideration, an investigator cannot use data to refine the model and may succumb to bias when considering evidentiary support for the model.

To address these concerns, we present perhaps the first formal description of the BFB mechanism. We use this formalization to consider the range of amplification patterns that can be produced by BFB. The underlying algorithmic problems are challenging and of unknown complexity. We develop heuristic algorithms and rules to determine if a given pattern is consistent with BFB, based on, among other observations, a tight connection between BFB patterns and trees with certain symmetries. The methods make the problems tractable for practical instances.

Using these methods, we obtain a result useful for experimental interpretation: BFB-associated amplification patterns are common. In fact, if some experimental uncertainty is allowed, a majority of possible patterns are consistent with BFB. This suggests that observing that an amplification pattern could have been produced by BFB is not conclusive evidence that BFB produced the amplification. Conversely, some cells have amplification patterns that are very unlikely to have been produced by BFB.

2   Formalizing the BFB Schedule

Our first task is to describe a model of the breakage-fusion-bridge mechanism that is both consistent with the biological features of the mechanism and is amenable to computational techniques. We begin by considering a chromosome arm that has lost its telomere. Suppose we label potentially unequally sized intervals along this chromosome arm A,B,C,... from the telomeric end to the centromere. Then, we trace the fate of these intervals through a BFB cycle.

For illustration, suppose that the chromosome arm, after loss of the telomere, is composed of five intervals, ABCDE. Then, the chromosome is analogous to Figure 1B(a) where A,B,C,D,E correspond to the magenta, cyan, green, red, and blue segments, respectively. After replication, fusion, and centromere separation, the whole arm has been duplicated to form a palindrome, EDCBAABCDE, as in Figure 1B(c). This palindromic stretch of DNA is then torn apart. Unless it breaks in the center, between the two A segments, one daughter cell will have a deletion and the other will have an inverted duplication. In Figure 1B(d), one daughter cell gets CDE while the other gets BAABCDE. Thus, BFB cycles can be thought of as operations on a string of chromosomal segments. The only significant restriction this creates is that breaks must occur at segment boundaries, but this is not a loss of generality since the segment boundaries can be defined freely. Moreover, BFB breakpoints may be reused in subsequent BFB cycles [8], so it is likely useful to keep track of candidate breakpoints.

We are now ready to describe our model. Let Σ ($|\Sigma| = k$) be an alphabet where each symbol corresponds to a chromosomal segment, and the ordering of Σ corresponds to the ordering of the segments from telomere to centromere. Let x^t denote the string representing chromosomal segments after t BFB operations. Before any BFB operation, the string is just the initial segments of the chromosome in order. Therefore, x^0 consists of all k lexicographically ordered characters from Σ. Let x^{-1} be the reverse of string x, pref(x) be a prefix of the string x, and suff(x) be a suffix of the string x.

$$x^t = \begin{cases} \text{pref}(x^{(t-1)})^{-1}x^{(t-1)} & \text{if inverted duplication} \\ \text{suff}(x^{(t-1)}) & \text{if deletion} \end{cases} \tag{1}$$

Define a *BFB-schedule* as a specific sequence of BFB operations, that is, inverted prefix duplications and prefix deletions. Define x as a *BFB(x^0)-string* if it can be generated from x^0 through a series of BFB operations. For ease of notation, x^0 is implied, and we refer to x being a BFB-string. The following simple lemmas establish basic properties and ensure that we do not need to worry about deletions. Proofs will be presented in an accompanying journal publication.

Lemma 1. *Let x be a BFB-string. If x^0 is a suffix of x, then x can be obtained from x^0 using only inverted prefix duplications.*

Lemma 2. Suffix Lemma: *Any suffix of a BFB-string x that includes x^0 is itself a BFB-string.*

3 Algorithms for BFB

We begin with a simple problem: *Given string x of length n, determine if x is a BFB-string.* This can be solved in $O(n)$ time using the following algorithm:

Algorithm *CheckBFB(x)*
Input: String x ($|x| = n$) containing suffix x^0 ($|x^0| = k$)
Output: True if x is a BFB-string
1. (* Find the longest even palindrome beginning at each character in x *)
2. **for** ($1 \leq i < n - k$)
3.   P[i] =max{$\ell \mid x[i \ldots i + 2\ell]$ is palindromic})
4. i = 1
5. **while** ($i < n - k$)
6.   **if** ($P[i] = 0$) **return false**
7.   i = i+P[i]
8. **return true**

Theorem 1. *Algorithm CheckBFB checks if x is a BFB-string in $O(n)$ time.*

Proof. (Sketch) A string is a BFB-string iff it can be formed by an inverted prefix duplication from another BFB-string or it is the original string, x^0. An inverted prefix duplication forms an even palindrome at the beginning of a string. CheckBFB finds such a palindrome and then, in effect, recurses on the string that ends at that palindrome's center. Lemma 2 guarantees that that string must be a BFB-string if x is a BFB-string. If a string does not begin with a palindrome, and it is not x^0, then it is not a BFB-string.

A linear time algorithm for finding maximum palindrome sizes will be described elsewhere. The remainder of CheckBFB visits each character at most once, so CheckBFB is in $O(n)$.

For illustration, consider the palindrome array P_1 for the string $x_1 =$ BAABC-CBAAAABC. Starting with $i = 1$, we can advance i as $1 \rightarrow 3 \rightarrow 6 \rightarrow 10 \rightarrow 11 \rightarrow$ True. For $x_2 =$BBCCCCBBBAABC, we advance i as $1 \rightarrow 5 \rightarrow 6 \rightarrow$ False.

i:	1	2	3	4	5	6	7	8	9	10	11	12	3
x_1	B	A	A	B	C	C	B	A	A	A	A	B	C
$P_1[i]$	2	4	3	2	1	4	3	2	1	1	0	0	0
x_2	B	B	C	C	C	C	B	B	B	A	A	B	C
$P_2[i]$	4	3	2	1	1	0	1	1	2	1	0	0	0

Thus, if we are given the full ordering of segments of a chromosome arm, we can determine if BFB could have produced that ordering. However, such complete information is often not available from current technologies. For example, an array CGH or sequencing experiment may only give the count of each segment, not the order. So, we would like to determine if a pattern of *copy counts* of chromosomal segments could have been produced by BFB. Formally, we define a sequence of positive integer copy counts from telomere to centromere along a chromosome arm as a *count-vector* ($\boldsymbol{n} = [n_1, n_2, ..., n_k]$). We say $\boldsymbol{n}$ admits a *BFB-schedule* if there is some BFB-string x whose character counts equal $\boldsymbol{n}$. For example, the count vector $[6, 3, 5]$ admits the BFB schedule

$$\underline{ABC} \rightarrow \underline{C}BAABC \rightarrow \underline{C}CBAABC \rightarrow \underline{AA}BCCCBAABC \rightarrow AAAABCCCBAABC$$

In this case, x^4 has 6, 3, and 5 of characters A, B, and C. We now define our problem: *The BFB-count-vector problem: Given a count vector $\boldsymbol{n}$, does $\boldsymbol{n}$ admit a BFB schedule?*

BFB-Pivot Algorithm: Each inverted prefix duplication reverses the order in which characters in the BFB-string appear. Consider the BFB schedule $\underline{ABC} \rightarrow \underline{CBAABC} \rightarrow BCCBAABC$. From left to right, the characters appear in proper lexicographical order, then reversed, then proper again. And each character is preceded by either itself or the next character higher or lower depending on whether the string is increasing or decreasing. With this knowledge, we can create a simple algorithm that adds single characters to the beginning of a candidate BFB-string until either a string with character counts satisfying $\boldsymbol{n}$ is found or one is shown not to exist (Figure 2A).

Lemma 3. *BFB-Pivot($\boldsymbol{n}, x$) returns True if and only if x can be extended via BFB operations to a string with counts satisfying $\boldsymbol{n}$.*

Proof. (Sketch) As we note above, segments of a BFB-string oscillate between "increasing" and "decreasing". So, when a character is prepended to an existing candidate BFB-string, it must either follow the existing trend, or it can be the beginning of a new inverted prefix duplication. In the latter case, another instance of the current first character of the string will be prepended.

BFB-Pivot attempts to prepend both eligible characters to the existing string until either an acceptable string is found or the current string fails because it is not a BFB-string or the count of a character exceeds the count in the count vector. By checking all such candidate BFB-strings, BFB-Pivot is guaranteed to find a BFB-string satisfying $\boldsymbol{n}$ if such a string exists.

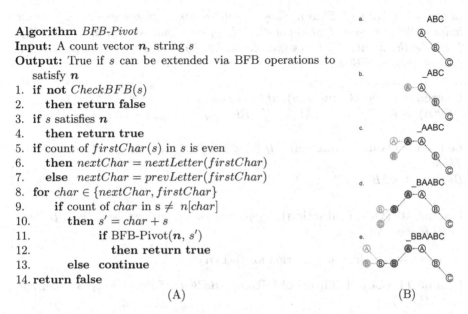

Algorithm *BFB-Pivot*
Input: A count vector n, string s
Output: True if s can be extended via BFB operations to
   satisfy n
1. **if not** $CheckBFB(s)$
2.   **then return false**
3. **if** s satisfies n
4.   **then return true**
5. **if** count of $firstChar(s)$ in s is even
6.   **then** $nextChar = nextLetter(firstChar)$
7.   **else** $nextChar = prevLetter(firstChar)$
8. **for** $char \in \{nextChar, firstChar\}$
9.   **if** count of $char$ in s $\neq n[char]$
10.   **then** $s' = char + s$
11.   **if** BFB-Pivot(n, s')
12.   **then return true**
13.   **else continue**
14. **return false**

(A)   (B)

Fig. 2. The BFB-Pivot algorithm and an illustration of BFB-Pivot searching for candidate BFB strings. nextLetter and prevLetter return the character immediately lower or higher lexicographically, and firstChar returns the leftmost character of a string.

It is useful to consider a graphical representation of BFB-Pivot, shown in Figure 2B. As nodes of the same character are added, the color of the added node oscillates. For each leftmost node, two possible edges are considered, a left edge ($\leftarrow$) and either an up ($\nwarrow$) or down ($\swarrow$) edge, depending on the color of the node. Therefore, the worst case complexity of BFB-Pivot is $O(2^n)$ where $n = \sum n_i$. Note that the input size of the problem is only $O(k \log n)$, so the output of a consistent BFB string as a certificate of correctness is already exponential in the input size, and BFB-Pivot's running time is doubly exponential. It is possible that some combinatorial rules completely define the set of count-vectors that admit a BFB schedule; we present seven conditions below.

Rules for BFB: Consider a count vector $n = [n_1, n_2, \ldots, n_k]$. We use $BFB(n)$ to denote that n admits a BFB-sequence.

Lemma 4 (Subsequence Rule). *Let n' be a subsequence of n. $BFB(n) \Rightarrow BFB(n')$.*

For example, if [6,4,6] admits a BFB schedule, then so must [6,4], [6,6], [4,6], etc.

Lemma 5 (Rule of One). *If $\exists\, i < j \leq k$ such that $n_i = 1$, $n_j > 1$, then $\neg BFB(n)$.*

Lemma 6 (Odd-Even Rule). *If $\exists\, i < j \leq k$ such that n_i is odd and n_j is even, then $\neg BFB(n)$.*

Lemma 7 (Rule of Four). *Suppose, all counts in $\boldsymbol{n}$ are even. Let i be the index of the first count that is not divisible by four, that is $4 \nmid n_i$ and $\forall\, j < i\; 4|j$. Let f_ℓ be the number of times that divisibility by four changes in $n_i, \ldots, n_\ell$. If $f_\ell > \frac{n_\ell}{2}$ then $\neg BFB(\boldsymbol{n})$.*

Lemma 8 (Two Reduction). *If $n_i = 2$ for some i. Then,*
$$BFB(\boldsymbol{n}) \Leftrightarrow BFB([\tfrac{1}{2}n_1, \ldots, \tfrac{1}{2}n_{i-1}]) \bigwedge BFB([n_k - 1, n_{k-1} - 1, \ldots, n_i - 1])$$

Lemma 9 (Four Reduction). *If $\exists\, i < j < \ell$ s.t. $n_i = 4$, $n_\ell = 4$, and $n_j > 4$. Then*
$$BFB(\boldsymbol{n}) \Rightarrow BFB(\tfrac{n_1}{2}, \tfrac{n_2}{2}, \ldots, \tfrac{n_i}{2}).$$

Lemma 10 (Odd Reduction). *Suppose all counts in $\boldsymbol{n}$ are odd. Then, $BFB(\boldsymbol{n})$ $\Rightarrow BFB(reverse(\boldsymbol{n} - 1))$.*

We also have a sufficient condition for $BFB(\boldsymbol{n})$.

Lemma 11 (Count Threshold Rule). *$BFB(\boldsymbol{n})$ if for all i n_i is even, and $n_i > 2(i - 1)$.*

Once the rules have been applied, we apply BFB-Pivot, or a second heuristic that we call the *BFB-Tree algorithm*.

BFB-trees. Consider a count-vector $\boldsymbol{n}$ that admits a BFB schedule and the graph representation of the resulting BFB-string, as in Figure 2B. Denote the layers of the graph corresponding to characters as $\sigma_1, \sigma_2, \ldots, \sigma_k$. Denote a *return-block* as a substring of a BFB-string that begins and ends at the same layer and only visits higher layers in between. A count vector $\boldsymbol{n}$ that admits a BFB schedule and has a final count of 2 yields a return-block; we denote this a *RB-BFB-schedule*. Return-blocks are related to BFB-strings by a single BFB-operation, and we can focus on them w.l.o.g. Formally,

Lemma 12. *The count-vector $\boldsymbol{n} = [n_1, n_2, \ldots, n_k]$ admits a BFB schedule iff $[2n_1, 2n_2, \ldots, 2n_k, n_{k+1} = 2]$ admits an RB-BFB-schedule.*

Return-blocks have a recursive structure that allows us to represent them as trees (Figure 3b,c). Consider a BFB-string for a count-vector $\boldsymbol{n} = [n_1, n_2, \ldots, n_k]$, converted into an RB-BFB-string for $[2n_1, 2n_2, \ldots, 2n_k, n_{k+1} = 2]$ (Fig. 3b). Create a root r of the tree (with label corresponding to σ_{k+1}). The path through the BFB graph starts at σ_{k+1}, traverses other return-blocks at level k, and finally returns to σ_{k+1}. Each return-block at level k is the root of a subtree with r as its parent.

We use this idea to define rooted *labeled* trees. Each node is labeled so all nodes an identical distance from the root have the same label. For a node v with $2\ell + 1$ (odd) children, number the child nodes as $v_{-\ell}, \ldots v_0, \ldots, v_\ell$. For a node with 2ℓ children the child nodes are labeled $v_{-\ell}, \ldots, v_{-1}, v_1, \ldots, v_\ell$; there is no v_0 node. $T(v)$ denotes the subtree rooted at v. Define a *labeled-traversal* of $T(r)$,

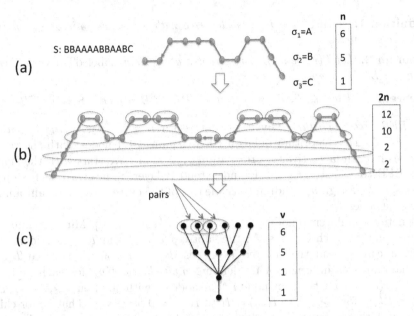

Fig. 3. BFB-tree generated from an RB-BFB-schedule. (a) A graph for BBAAAAB-BAABC that supports $[6, 5, 1]$. (b) A single BFB operation begets an RB-BFB graph for $[12, 10, 2, 2]$. Dotted ellipses denote the nodes of a BFB-tree. (c) A BFB-tree for the BFB-string, with 3 levels. The single node at level 3 has 5 children, ordered as $\{-2, -1, 0, 1, 2\}$. Pairs are illustrated at levels 1, and 2.

as the string obtained by traversing the labels in an ordered, depth-first-search as given below:

Algorithm *LabelTraverse(T(r))*
Input: A labeled tree rooted at r
Output: The string given by a labeled traversal
1. Let $\sigma = \text{Label}(r)$
2. Let $S_1 = \epsilon$
3. **for** (each child r_j ordered from least to max)
4. $S_1 = S_1 \cdot LabelTraverse(T(r_j))$
5. **return** $\sigma S_1 \sigma$

A labeled tree is *mirror-symmetric* if for all nodes v, $\text{LabelTraverse}(T(v)) = \text{LabelTraverse}(T(v))^{-1}$. In other words, 'rotation' at v results in the same tree. The labeled traversal of a tree T visits each node exactly twice, outputting its label each time. Define a partial order on the nodes of T in which $u < v$ if the last appearance of u precedes the first appearance of v in the labeled traversal. For any $u < v$, let $T_<(u, v)$ denote the subtree of T containing the LCA of u, v and all nodes w s.t. $u < w < v$. We call (u, v), with $u < v$, a *label-pair* if u and v have the same label σ, and no node in $T_<(u, v)$ has label σ. A labeled tree is *pair-symmetric* if for all label-pairs (u, v), $T_<(u, v)$ is mirror-symmetric. Finally, we say a labeled tree has *long-ends* if starting from the root, and following the least numbered node at each step, we can reach each layer $k, k - 1, \ldots, 2, 1$.

Definition 1. *A BFB-tree is a labeled tree with long-ends, mirror-symmetry, and pair-symmetry.*

Theorem 2. *Let $T(r)$ be a BFB-tree rooted at r. Then* LabeledTraversal$(T(r))$ *is an RB-BFB-sequence.*

Theorem 3. *The tree T_S derived from an RB-BFB-sequence S is a BFB-tree.*

The BFB-Tree Algorithm: Given a count vector $\boldsymbol{n} = [n_1, n_2, \ldots, n_k]$ $(\sum_i n_i = n)$, the BFB-Tree algorithm builds a BFB-tree on $n + 1$ nodes, with the count of nodes in each layer given by $\boldsymbol{\nu} = [n_1, n_2, \ldots, n_k, 1]$. We start with the single node BFB-tree T at level $k + 1$, and extend it layer by layer. In each step j, $1 \leq j \leq k$, we assign n_j children to the leaves of the current tree T, maintaining BFB-properties.

Denote the children of node v by the set $C_v = \{v_{-\ell}, \ldots, v_\ell\}$. Mirror-symmetry ensures that for each $0 \leq i \leq \ell$, the subtrees $T(v_{-i})$ and $T(v_i)$ are identical. Thus, it can be said the subtree $T(v_{-i})$ is *dependent* on the subtree $T(v_i)$. We maintain this information by defining *multiplicity*, $I(v)$, for each node as follows: $I(r) = 1$ for the root node r. For nodes v with children $v_{-\ell}, \ldots, v_\ell$, set $I(v_j) = 2I(v)$ for $\ell \geq j > 0$, $I(v_0) = I(v)$, $I(v_j) = 0$ otherwise. That is, for child nodes v_j with a corresponding dependent child v_{-j} we assign a multiplicity of double the parent's multiplicity. For nodes with an odd number of children, there will be a child v_0 without a corresponding dependant child node. This node is assigned the same multiplicity as the parent node. Finally, for each dependent node pair, one is assigned a multiplicity of zero since, by mirror-symmetry, it is completely defined by its dependent whose multiplicity has doubled. Valid assignments at level j must then satisfy the diophantine equations $\sum_v I_v = n_{j+1}$, and $\sum_v I(v)|C_v| = n_j$.

Algorithm *BFB-Tree(T, j, I)*
1. **if** $j = 0$, return CheckBFB(LabeledTraversal(T))
2. **for** each assignment C_v s.t. $\sum_v I(v)|C_v| = n_j$
3.   Extend T according to the assignment
4.   Adjust I
5.   **if** (Construct_BFB_tree$(T, j - 1, I)$) **return true**
6. **return false**

Note that once a node is dependent $(I(v) = 0)$, its descendants remain dependent. Further, the multiplicity of the node is a power of 2, and is doubled each time an independent node is a non-central child of its parent. As the multiplicities increase, the number of valid assignments decreases quickly, improving the running time in practice. Further analysis of the algorithm will be presented elsewhere.

4   Results

Performance on Realistic Data-Sets. Using practical sized examples, we investigated the performance of the 3 approaches to checking BFB: rules, BFB-Pivot, and BFB-Tree. We applied the BFB rules to all $\sum_{i=1}^{5} 20^i = 3,368,420$ count vectors with $k \leq 5$ and each $n_i \leq 20$. Remarkably, the rules were able to resolve 3,034,440, or 90%, of the count vectors. As BFB-Pivot and BFB-Tree are both exponential time procedures, this check helped speed up the entire study.

Nearly all of the count vectors that the rules could not resolve admitted a BFB schedule (Table 1), so BFB-Pivot and BFB-Tree could usually halt once an acceptable BFB string was found rather than exhausting all possible paths or trees.

Table 1. Method and result for the count vectors used to analyze algorithm speed

	Rules	Tree/Pivot	Total
Admits BFB	170,576	333,840	504,416 (15.0%)
Not BFB	2,863,864	140	2,864,004 (85.0%)
Total	3,034,440	333,980	3,368,420
Fraction	(90.1%)	(9.9%)	

We ran BFB-Pivot and BFB-Tree on each of the 333,980 count vectors that could not be resolved by rules. The running times plotted against n for each algorithm is shown in Figure 4. As expected, both algorithms' worst-case running times grew exponentially with n. However, the worst case running times for BFB-Tree were orders of magnitude lower than for BFB-Pivot. For example, the longest running count vector for BFB-Tree was [12,3,19,19,19] which took 10 seconds to complete. The longest running count vector for BFB-Pivot was [20,20,18,20,6] which took 23,790 seconds to complete.

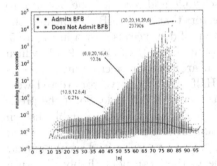 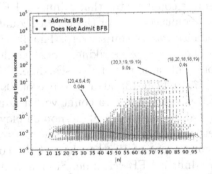

Fig. 4. Pivot (left) and tree (right) algorithm running time. Each point refers to the size and running time of a specific example. The line represents median running time. Note that since the rules excluded nearly all count vectors that did not admit a BFB schedule, almost all points are blue.

Consequences for Experimental Interpretation. Consider experimental data, such as array CGH or read mapping depth of coverage, that reveal the copy counts of segments of a chromosome arm, or in the terminology of this paper, a count vector. If this count vector admits a BFB schedule, one might infer from this observation that BFB occurred.

On its face, this is a reasonable conclusion. Only 15% of the count vectors with $k \leq 5$ admit a BFB schedule. If we expand that analysis to the 64,000,000 count vectors with $k = 6$ and $n_i \leq 20$, only 7.3% admit a BFB schedule. So, it is plausible that observing a count vector that admits a BFB schedule is much more likely if BFB did in fact occur. However, using our model of BFB, we show that this inference is usually incorrect.

Note first that experimentally-derived count vectors are often imprecise. Experiment typically provides a small range of values for each element in the count vector rather than a single definite value. This is a result of the imprecision of the experimental method as well as potentially high levels of structural variation or aneuploidy in the genome being studied. Therefore, the observation made about the data may not be that a particular count vector admits a BFB schedule but that there is a count vector that admits a BFB schedule "nearby", that is, within experimental precision of the observed count vector.

Further, the uncertainty in a count tends to increase with the magnitude of the count. It is easier to distinguish between copy counts 1 vs. 2 than between 32 vs. 33. For simplicity, we assume that the relationship between uncertainty and magnitude is linear and use the Canberra distance [3] to compare count vectors:

$$d(x,y) = \sum_i \frac{|x_i - y_i|}{|x_i| + |y_i|} \qquad (2)$$

For each of the 64,000,000 count vectors with 6 segments, we searched for the nearest count vector that admits a BFB schedule and recorded the distance to that count vector. For the 7.2% of count vectors that admit a BFB schedule, the distance was, of course, zero. The distribution of distances is shown in Figure 5.

The results are striking. Consider the count vector $[14, 7, 18, 16, 9, 12]$, which does not admit a BFB schedule. The nearest count vector that does is $[14, 7, 19, 17, 9, 13]$, at a distance of .097. Half of all count vectors tested were at least this close to a count vector admitting a BFB schedule. So, if the precision of the experimental method employed is such that it can not reliably discern between a copy count of 12 and a copy count of 13 or a copy count of 18 and a copy count of 19, then half of the count vectors we examined would appear to admit a BFB schedule. Similarly, if the method can not reliably discern a copy count of 9 and and 8 or 7 and 8, then 70% of count vectors will appear to admit a BFB schedule. Figure 5A has distances and and example vector pairs for additional percentiles.

Thus, even with small amounts experimental uncertainty, a majority of count vectors admit a BFB schedule, so mechanisms other than BFB are likely to produce count vectors that look like they were created by BFB. Therefore, finding a

Distance (%ile)	Count Vector	Nearest Admitting BFB
.097 (50)	[14,7,18,16,9,12]	[14,7,19,17,9,13]
.129 (60)	[7,13,6,4,19,12]	[8,14,6,4,20,12]
.192 (70)	[9,7,7,8,2,14]	[8,8,8,8,2,14]
.362 (80)	[16,17,14,18,1,19]	[16,18,14,18,2,19]
.458 (90)	[20,1,7,3,10,6]	[20,2,7,3,11,7]
.566 (95)	[15,8,8,1,15,2]	[16,8,8,2,15,3]
.889 (99)	[9,1,5,9,1,15]	[10,2,5,9,3,15]

(A)

(B)

Fig. 5. A. Percentage of count vectors at least as close to a count vector admitting a BFB schedule as the shown count vector pair. B. Distribution of distances to nearest count vector admitting a BFB schedule.

count vector consistent with BFB should only slightly increase one's belief that BFB occurred.

On the other hand, observing a count-vector that is *distant* from *any* BFB admitting count vector, provides strong evidence that BFB was *not* the cause of the amplification. The likelihood of observing a count vector that is distant from a count vector that admits a BFB schedule is low in any case, and is surely even lower if the chromosome arm underwent BFB.

Application to Experimental Data. We will now apply these insights to previously reported chromosomal amplification patterns and propose interpretations of the evidence as to whether BFB was the source of the observed amplifications.

First, we examine the amplification in the small cell lung carcinoma cell line NCI-H2171 published by Zhao et al. [9]. They used quantitative SNP arrays to find a count vector of $[1, 2, 1, 2, 8, 1, 3, 1, 3]$ along the long arm of chromosome 11. This count vector does not admit a BFB schedule. The nearest count vector that does is $[2, 2, 2, 2, 8, 3, 3, 3, 3]$. For this to have been the true count vector, multiple segments with a true count of two or three would have to have been called as one, which is unlikely. Moreover, because some of the copy counts are one, it is not possible that there was, for example, one intact chromosome along with another that underwent BFB. So, this amplification pattern is evidence against the hypothesis that BFB was the source of the amplification in this chromosome.

Second, we examine the amplification in the non-small cell lung cancer cell line PTX250 published by Kitada and Yamasaki [4]. They used aCGH and FISH and showed that a BFB schedule producing the count vector $[34, 6, 4, 3, 3]$ fit their copy count observations along an arm of chromosome 7 and thus concluded that BFB cycles produced this amplification pattern. However by our reasoning above, we know that finding a count vector that admits a BFB schedule is not conclusive evidence that BFB has occurred. If the measurement of copy counts is at all imprecise, then a large proportion of all count vectors, perhaps even a majority, will admit a BFB schedule. And, there does appear to be some uncertainty in the measurement of the $[34, 6, 4, 3, 3]$ count vector. Neither aCGH nor FISH could reliably determine that the copy count of the first segment was 34 rather than some nearby value. So, the observation that the count vector along chromosome 7 admits a BFB schedule supports the hypothesis that BFB occurred, it certainly does not resolve the question definitively.

Kitada and Yamasaki also used another type of evidence that has not been discussed in this paper to argue that BFB produced their observed amplifications. Using FISH, they show that the relative positions of the amplified segments of chromosome 7 are also consistent with BFB. That is, if each of the original five segments were labeled A,B,C,D,E from telomere to centromere, they observe a chromosome with segments

CBAAAAAAAABAAAAAAAAAAAAAAAABAAAAAAAABCDEEDCBAABCDE

which is a string that can be produced by BFB and therefore it is more likely that BFB produced the amplifications.

This string, in fact, could not have been produced by BFB. But, by using the methods developed in this paper, we can show that there are 25 string with character counts [34, 6, 4, 3, 3] that can be produced by BFB (Table A1). One of these strings may also fit the observed positions, but this would pose a question similar to the question we have investigated in this paper: how should this observation affect our credence that BFB occurred? The answer is not obvious, and we suspect that techniques like those used here could point toward a solution.

5 Discussion

We present perhaps the first formalization of the BFB mechanism. Our main result is that BFB can produce a surprisingly broad range of patterns of copy number increases along a chromosome arm. Indeed, for most contiguous patterns of amplification, it is possible to find a BFB schedule that yields either the given pattern or one that is very similar. As a result, one must be cautious when interpreting copy count data as evidence for BFB. The presence of amplification at all, or the presence of a terminal deletion from a chromosome arm can both suggest the occurrence of BFB, though they may not distinguish well between BFB and other amplification hypotheses. However, unless the counts are known precisely, a specific pattern of copy counts along the chromosome arm cannot offer compelling support for BFB.

This study suggests several avenues for future work. There are other types of evidence that can be deployed to argue that BFB has occurred. FISH can reveal, to some extent, the arrangement of segments along a chromosome arm. Next-generation sequencing can reveal the copy counts of breakpoints between rearranged chromosomal segments and may allow for a fuller characterization of a chromosome arm. Methods similar to those presented in this paper may prove useful for evaluating and interpreting these different types of evidence. Modeling may also be helpful for evaluating proposed refinements to models of BFB.

Finally, we have outlined two algorithms for determining whether a count vector admits a BFB schedule, but neither is polynomial in reasonable measures of the input size. This problem, along with related problems, is interesting as computational problems *per se*. We hope in the future to have either faster solutions to these problems or proofs of hardness.

References

1. Artandi, S.E., DePinho, R.A.: Telomeres and telomerase in cancer. Carcinogenesis 31, 9–18 (2010)
2. Bignell, G.R., Santarius, T., Pole, J.C., Butler, A.P., Perry, J., Pleasance, E., Greenman, C., Menzies, A., Taylor, S., Edkins, S., Campbell, P., Quail, M., Plumb, B., Matthews, L., McLay, K., Edwards, P.A., Rogers, J., Wooster, R., Futreal, P.A., Stratton, M.R.: Architectures of somatic genomic rearrangement in human cancer amplicons at sequence-level resolution. Genome Res. 17, 1296–1303 (2007)
3. Deza, M.M., Deza, E.: Encyclopedia of Distances. Springer (2009)
4. Kitada, K., Yamasaki, T.: The complicated copy number alterations in chromosome 7 of a lung cancer cell line is explained by a model based on repeated breakage-fusion-bridge cycles. Cancer Genet. Cytogenet. 185, 11–19 (2008)
5. McClintock, B.: The Production of Homozygous Deficient Tissues with Mutant Characteristics by Means of the Aberrant Mitotic Behavior of Ring-Shaped Chromosomes. Genetics 23, 315–376 (1938)
6. McClintock, B.: The Stability of Broken Ends of Chromosomes in Zea Mays. Genetics 26, 234–282 (1941)
7. Santarius, T., Shipley, J., Brewer, D., Stratton, M.R., Cooper, C.S.: A census of amplified and overexpressed human cancer genes. Nat. Rev. Cancer 10, 59–64 (2010)
8. Shimizu, N., Shingaki, K., Kaneko-Sasaguri, Y., Hashizume, T., Kanda, T.: When, where and how the bridge breaks: anaphase bridge breakage plays a crucial role in gene amplification and HSR generation. Exp. Cell Res. 302, 233–243 (2005)
9. Zhao, X., Li, C., Paez, J.G., Chin, K., Janne, P.A., Chen, T.H., Girard, L., Minna, J., Christiani, D., Leo, C., Gray, J.W., Sellers, W.R., Meyerson, M.: An integrated view of copy number and allelic alterations in the cancer genome using single nucleotide polymorphism arrays. Cancer Res. 64, 3060–3071 (2004)

A   Tables

Table A1 BFB strings with character counts 34,6,4,3,3

```
CBBBBCDEEDCBAAAAAAAAAAAAAAAAAAAAAAAAAAAAAAAAAABCDE
CBAABBAAABCDEEDCBAAAAAAAAAAAAAAAAAAAAAAAAAAAAABCDE
CBAAAABBAAAABCDEEDCBAAAAAAAAAAAAAAAAAAAAAAAAAABCDE
CBAAAAAABBAAAAAABCDEEDCBAAAAAAAAAAAAAAAAAAAAAABCDE
CBAAAAAAAABBAAAAAAAABCDEEDCBAAAAAAAAAAAAAAAAAABCDE
CBAABCDEEDCBAAAAAAAAAAAAAAAABBAAAAAAAAAAAAAAAABCDE
CCDEEDCBAAAAAAAAAAAAAAAABBAABBAAAAAAAAAAAAAAAABCDE
CBAAAAAABCDEEDCBAAAAAAAAAAAAABBAAAAAAAAAAAAAAABCDE
CCDEEDCBAAAAAAAAAAAAAABBAAAAAABBAAAAAAAAAAAAAABCDE
CBAAAAAAAAAABBAAAAAAAAAABCDEEDCBAAAAAAAAAAAAAABCDE
CBAAAAAAAAAABCDEEDCBAAAAAAAAAAABBAAAAAAAAAAAAABCDE
```

CCDEEDCBAAAAAAAAAAAABBAAAAAAAAAAAABBAAAAAAAAAAAABCDE
CCDEEDCBAAAAAAAAAABBAAAAAAAAAAAAAAAABBAAAAAAAAAABCDE
CBAAAAAAAAAAAAAAAABCDEEDCBAAAAAAAAAABBAAAAAAAAAABCDE
CBAAAAAAAAAAAABBAAAAAAAAAAAAAABCDEEDCBAAAAAAAAAABCDE
CCDEEDCBAAAAAAAABBAAAAAAAAAAAAAAAAAAAABBAAAAAAAABCDE
CBAAAAAAAAAAAAAAAAAAAABCDEEDCBAAAAAAAABBAAAAAAAABCDE
CCDEEDCBAAAAAABBAAAAAAAAAAAAAAAAAAAAAAAABBAAAAAABCDE
CBAAAAAAAAAAAAAAAAAAAAAAAABCDEEDCBAAAAAABBAAAAAABCDE
CBAAAAAAAAAAAAAABBAAAAAAAAAAAAAAAABCDEEDCBAAAAAABCDE
CCDEEDCBAAAABBAAAAAAAAAAAAAAAAAAAAAAAAAAAABBAAAABCDE
CBAAAAAAAAAAAAAAAAAAAAAAAAAAAABCDEEDCBAAAABBAAAABCDE
CCDEEDCBAABBAAAAAAAAAAAAAAAAAAAAAAAAAAAAAAAABBAABCDE
CBAAAAAAAAAAAAAAAAAAAAAAAAAAAAAAAABCDEEDCBAABBAABCDE
CBAAAAAAAAAAAAAAAABBAAAAAAAAAAAAAAAAAABCDEEDCBAABCDE

TrueSight: Self-training Algorithm for Splice Junction Detection Using RNA-seq

Yang Li[1,2], Hong-Mei Li[2,3], Paul Burns[4], Mark Borodovsky[4,5], Gene E. Robinson[2,3,*], and Jian Ma[1,2,*]

[1] Department of Bioengineering
[2] Institute for Genomic Biology
[3] Department of Entomology,
University of Illinois at Urbana-Champaign
[4] Wallace H. Coulter Department of Biomedical Engineering
[5] School of Computational Science & Engineering,
Georgia Institute of Technology
{jianma,generobi}@illinois.edu

1 Introduction

RNA-seq has proven to be a powerful technique for transcriptome profiling based on next-generation sequencing (NGS) technologies. However, due to the limited read length of NGS data, it is extremely challenging to accurately map RNA-seq reads to splice junctions, which is critically important for the analysis of alternative splicing and isoform construction. Several tools have been developed to find splice junctions by RNA-seq *de novo*, without the aid of gene annotations [1-3]. However, the sensitivity and specificity of these tools need to be improved. In this paper, we describe a novel method, called TrueSight, that combines information from (i) RNA-seq read mapping quality and (ii) coding potential from the reference genome sequences into a unified model that utilizes semi-supervised learning to precisely identify splice junctions.

2 Methods

TrueSight considers RNA-seq gapped alignment under a semi-supervised learning framework. In its initial alignment step, TrueSight searches all possible gapped alignments for each RNA-seq read which cannot be fully aligned to the reference genome, and carefully *labels* high confidence positive/negative junctions, based on several stringent empirical criteria (semi-supervised) on all junctions inferred by initial gapped alignments. TrueSight extracts nine features (including genome sequence coding potential, splicing signals, and seven features from RNA-seq mapping) for each junction and learns a logistic regression model using these features on the *labeled* positive/negative junctions. TrueSight then iteratively updates the logistic regression model using a classification EM algorithm [4] on those junctions not initially *labeled* and finally assigns a reliability score for each of them. Since it is quite common to see multiple gapped alignments for a particular read, TrueSight only retains the junction with the highest score for each read.

---

* Corresponding authors.

B. Chor (Ed.): RECOMB 2012, LNBI 7262, pp. 163–164, 2012.
© Springer-Verlag Berlin Heidelberg 2012

3   Results

We evaluated the performance of TrueSight in comparison with TopHat [1] and MapSplice [2], using both real and simulated human RNA-seq data. Overall, TrueSight achieved the best sensitivity and specificity out of all three programs, especially on low coverage junctions (Figure 1). We also applied TrueSight to discover novel splice forms in honeybee transcriptomes that cannot be detected by other methods and experimentally validated some of them.

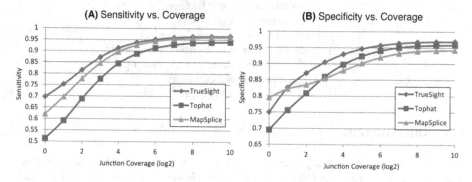

Fig. 1. SN/SP vs. Coverage on simulated paired-end (100bp) reads

4   Conclusions

We propose a new RNA-seq gapped alignment algorithm, incorporating both DNA sequence and RNA-seq features into a logistic regression model which assigns reliability scores to splice junctions inferred by RNA-seq reads. Tested on both real and simulated datesets, TrueSight shows better sensitivity and specificity, suggesting that our self-learning algorithm is superior to existing methods in *de novo* splice junction detection using RNA-seq. We believe this new tool will be highly useful to comprehensively study splice variants based on RNA-seq.

References

1. Trapnell, C., Pachter, L., Salzberg, S.L.: TopHat: discovering splice junctions with RNA-Seq. Bioinformatics 25(9), 1105–1111 (2009)
2. Wang, K., et al.: MapSplice: accurate mapping of RNA-seq reads for splice junction discovery. Nucleic Acids Res. 38(18), e178 (2010)
3. Au, K.F., et al.: Detection of splice junctions from paired-end RNA-seq data by SpliceMap. Nucleic Acids Res. 38(14), 4570–4578 (2010)
4. Celeux, G., Govaert, G.: A classification EM algorithm for clustering and two stochastic versions. Computational Statistics & Data Analysis 14(3), 315–332 (1992)

Synthetic Sequence Design
for Signal Location Search

Yaw-Ling Lin[1], Charles Ward[2], and Steven Skiena[2]

[1] Department of Computer Science and Information Engineering,
Providence University, 200 Chung Chi Road, Shalu, Taichung, Taiwan 433
`yllin@pu.edu.tw`
[2] Department of Computer Science,
Stony Brook University, Stony Brook, NY 11794-4400
`{charles,skiena}@cs.sunysb.edu`

Abstract. We present a new approach to identify the locations of critical DNA or RNA sequence signals which couples large-scale synthesis with sophisticated designs employing combinatorial group testing and balanced Gray codes. Experiments in polio and adenovirus demonstrate the efficiency and generality of this procedure. In this paper, we give a new class of consecutive positive group testing designs, which offer a better tradeoff of cost, resolution, and robustness than previous designs for signal search.

Let n denote the number of distinct regions in a sequence, and d the maximum number of consecutive positives regions which can occur. We propose a design which improves on the consecutive-positive group testing designs of Colbourn. Our design completely identifies the boundaries of the positive region using t tests, where $t \approx \log_2(1.27n/d) + 0.5 \log_2(\log_2(1.5n/d)) + d$.

Keywords: Combinatorial group testing, non-adaptive group testing, Gray codes, synthetic biology.

1 Introduction

The exciting new field of *synthetic biology* is emerging with the goal of designing novel organisms at the genetic level. While DNA sequencing technology can be thought of as reading DNA molecules, DNA *synthesis* is the inverse operation, where one can take any desired DNA sequence and construct DNA molecules with that exact sequence. Commercial vendors such as GeneArt and Blue Heron today charge under 40 cents per base, or only a few thousand dollars to synthesize virus-length sequences, and prices are rapidly dropping [4,9].

The advent of cheap synthesis clearly will have many exciting new applications throughout the life sciences, and the ability to design sequences to specification leads to a variety of new computational problems. We believe that imaginative sequence design will change the way molecular biology is done – and require tools from algorithm design and discrete mathematics not in the vocabulary of molecular biologists [18]. Computer Science-oriented research is necessary to

B. Chor (Ed.): RECOMB 2012, LNBI 7262, pp. 165–179, 2012.

frame the questions and develop the experimental methods employing large-scale synthesis, not merely to analyze the resulting data.

The identification of critical DNA or RNA sequence signals such as binding domains, secondary structures, splice junctions, slippage sites, and mRNAi targets, is essential to understanding the biology of any system, but efficient assays to perform genome-wide scans for critical signals have been elusive. Bioinformatic methods are useful, but not perfect at predicting them, and so laboratory experiments are used to prove the significance of signals in coding sequences. Site-directed mutagenesis, another widely used method, allows only small changes, and is a slow, tedious process, impractical for blindly searching even relatively small regions.

In this paper, we present a new, experimentally validated approach to identify the locations of critical signals coupling large-scale synthesis with sophisticated sequence designs based-on combinatorial group testing. We believe that our synthesis-based genome scanning procedure will become a standard technique to understand viral biology. Certainly such genome scans will be an essential part of any effort to apply the synthetic attenuated virus engineering (SAVE) technology for designing vaccines [8,20] beyond well-studied model organisms.

Our methodology is very attractive in practice. Indeed, we have already applied this methodology in experiments in two distinct systems: polio and adenovirus, as described in Section 2. The contributions of this paper are:

- *A Group-Testing Approach for Biological Signal Location: proposal and validation* – Here we describe our novel algorithmic design/synthesis approach to biological signal detection and describe results from the first two rounds of laboratory experiments validating the procedure.[1] Although the basic ideas underlying our approach (non-adaptive group testing and balanced Gray codes) are well known to the algorithmic/discrete mathematics communities, our biological application of them is both novel and significant, serving as an intriguing demonstration of the power of combinatorial methods.

- *Group Testing for Expensive Pools* – Each test in our search strategy requires *de novo* synthesis of a gene variant, at a cost of a few thousand dollars per test. Even optimistic reductions in the cost of large-scale synthesis will not soon make search procedures involving more than two dozen or so tests practical. Theoretical results in combinatorial group testing usually depend heavily on asymptotic analysis, which are difficult to interpret for small instances. Notably, one recent paper defines "practical cases" as $n \leq 10^{10}$ [14].

  We review the literature for relevant group testing designs, with an eye towards realistic values of n. For example, the best non-adaptive design for two defects (Lindstrom ([13], p. 70)) for $n = 16$ proves no better than brute-force tests for each single region. This result, and the need to detect biological signals across

---

[1] The signal detection through synthesis approach for *in vivo* systems which we present here is indeed novel, although consecutive positive group testing has been applied to the quite different problem DNA clone map screening (see [1,3,7]). The designs we propose are also applicable to clone map screening.

region boundaries, motivates consideration of the special class of *consecutive positive* designs [7].

- *Improved Designs for Consecutive Positive Group Testing* – Let n denote the number of distinct regions in a sequence, and let there be a set of at most d consecutive regions which are all positive. We propose a design which identifies the boundaries of the positive region using t tests, where

$$t \approx \log_2(1.27n/d) + 0.5\log_2(\log_2(1.5n/d)) + d.$$

In contrast, Colbourn's consecutive positive design [7] uses $t = \lceil \log_2(n/(d-1)) \rceil + 2d + 1$ tests. As a practical example, Colbourn's construction uses 10 tests to detect up to 3 consecutive positives for 16 regions. By contrast, our design (denoted $M_3(7,3)$) can detect up to 3 consecutive positives for 105 regions, offering six times the resolution with the same number of tests.

Our improved designs require parameter optimization to customize them for a particular experimental setting, namely finding the appropriate tradeoff, for a given number of tests, between the number of distinct regions n and the maximum detectable signal size m. We show how to optimize this tradeoff, and to aid the biological community in using our experimental approach, we have made available software to construct sequences adhering to these designs at *http://www.algorithm.cs.sunysb.edu/signalSearch*.

- *Middle-Levels Conjecture Equivalence* – The well-known middle levels (or revolving door) conjecture [26] states that the middle levels M_{2k+1} of the Hasse diagram of the partial-order defined by inclusion on all subsets of $\{1, \ldots, 2k+1\}$ is Hamiltonian. As a by-product of our design construction, we show that the more general *adjacent level* conjecture is equivalent to the middle levels conjecture.

Our paper is organized as follows. Section 2 presents our basic group testing signal-search procedure, discussing its application in laboratory experiments. Section 3 surveys related results in the group testing literature, notably the notion of *consecutive positive* tests. Section 4 relates consecutive positive designs to the middle levels conjecture, and its equivalences to the adjacent level conjecture. Finally, Section 5 presents our improved designs for biological signal search.

2   Signal Location Search Experience

We have developed a novel signal search procedure employing large-scale DNA synthesis and algorithmic ideas from combinatorial group testing and Gray codes. Further, we have employed our synthesis-based sequence scanning approach to study two distinct biological systems, Poliovirus and Adenovirus, in collaboration with two different laboratories. In the subsections below, we briefly describe our method and experiences with these biological experiments. Finally, we summarize the criteria demanded by sequence search design, which will drive and motivate the more sophisticated designs that follow.

2.1 Biological Signal Search

The advent of large-scale synthesis opens new doors for rapid signal detection. Suppose we replace a wildtype gene coding sequence (W) with a different but synonymous encoding (D) for that region. If the phenotype changes (e.g., the organism dies), it implies that there must be a critical signal at some location within that region. To pinpoint the position of this signal, we can construct a sequence that is half wildtype (W) and half synonymous encoding (D). The phenotype of the resulting construct identifies which half the signal must lie in. By applying this type of binary search process, either through additional synthesis or subcloning, the signal can be located in $\lceil \log_2 n \rceil$ sequential rounds of experiment. Unfortunately, each such round takes on the order of weeks to complete, so it remains a very slow and laborious process.

Instead, we propose the following procedure: To narrow down the location of a single signal to, say, (1/16)th of the original sequence in a single round of synthesis, we can synthesize four different designs slicing each gene into $2^4 = 16$ equal-sized regions interleaving the wildtype (W) and synonymous encoding (D). As shown in Figure 1(left), each of the 16 "columns" contains a unique signature of W and D. For example, if sequences 3 and 4 exhibit the phenotype while genes 1 and 2 behave as wildtype, then the critical signal should occur in the fourth region from the left, because that is the only region represented by wildtype in genes 1 and 2 but not in 3 and 4.

One concern with this approach is that the critical signal may overlap the boundary between a pair of regions. This first designs assumes that the n regions are each sufficiently large with respect to the size of the defect that it is unlikely that this will happen. However, this assumption is not robust and motivates the balanced Gray code design [2] of Figure 1(right), which we employed in our experiments. *Gray codes* [15,17,24,23,30] are subset orderings where neighboring subsets differ in exactly one element, thus reducing the number of transitions. The four sequences have the same power to distinguish the 16 regions as those Figure 1(left), but they are more interpretable when a signal crosses region boundaries, restricting the possible locations of the signal to a contiguous region. As discussed later, this problem motivates the consideration of consecutive-positive group testing models.

```
1. WWWWWWWWDDDDDDDD        1. WWWWWWWWDDDDDDDD
2. WWWWDDDDWWWWDDDD        2. WWWWDDDDDDDDWWWW
3. WWDDWWDDWWDDWWDD        3. WWDDDDWWWWDDDDWW
4. WDWDWDWDWDWDWDWD        4. WDDWWDDWWDDWWDDW
```

Fig. 1. (left) Designs of four synthetic genes interleaving wildtype (W) and synonymous designs (D) to identify the position of a single essential signal in wildtype, based on combinatorial group testing. (right) Improved designs using balanced Gray codes to minimize the number of W/D transitions.

2.2   Practical Experience

Adenovirus provides an attractive vector in human gene therapy research, particularly when coupled with the site-specific integration ability of Adeno-associated virus (AAV). Bahou's group, at the Stony Brook Medical Center [28], hoped that by inserting into Adenovirus the proteins necessary for AAV's integration into chromosome 19, site-specific integration with an arbitrary gene payload could be achieved. Unfortunately, AAV's Rep78 protein has an inhibitory effect on replication. Synthetic scrambled designs indicated an apparent cis-acting element in the Rep78 ORF sequence, apparently governing the toxicity of Rep78 when incorporated into Adenovirus.

In order to localize this toxic signal, we employed our signal search procedure. Four sequences (this time omitting the two homogenous columns $WWWW$ and $DDDD$ to ensure experimental controls) were synthesized and grown in culture, with favorable results. In particular, we pinpointed the signal to a particular 130-base region out of an 1.8kb gene. Subcloning later verified the apparent location of the signal predicted by the experiment.

A similar experiment in poliovirus is on-going, the details of which will be reported in [29].

2.3   Lessons Learned: Design Criteria for Sequence Signal Search

Our experiences with polio and adenovirus inform our basic design problem: we seek to construct t tests (design sequences) capable of pinpointing the location of a signal of length at most m bases as tightly as possible in a region of length g. Typically, g is on the order of 1-2,000 bases, while m depends upon the expected type of signal. For restriction sites and RNAi signals, m might be reasonably set around 20, but secondary structures result in larger signals, such as the $61 - nt$ Cre signal in Poliovirus.

Given a particular g, m, and t, we aim to partition the region into n segments and construct t tests to determine which segment contains the critical signal. In our previous designs, $n = 16$ and $t = 4$. Our experience reveals several issues which motivate the development of improved group testing designs for signal search:

- *Multiple Signals and Region Boundaries* – The combinatorial search procedure above fails when two or more unknown signals contribute to phenotype. The combinatorial group testing literature reviewed in Section 3 extensively analyzes the general case of d arbitrary positives, but such designs require prohibitively many tests even for the case of $d = 2$. However, we find a much more compelling cost/performance tradeoff in designs identifying at most d *consecutive* positives, which works to address the previously observed difficulty of a signal spanning a region boundary.

  Our previous designs assumed that the n regions were each sufficiently large with respect to m that it is unlikely that a signal overlapped a boundary between regions. But for many realistic values of n and m, the signal may potentially span multiple regions, as in Figure 2. Employing our simple previous

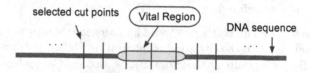

Fig. 2. A vital region in the DNA induces consecutive positives in a group testing for the sequence

designs in such a case would produce uninterpretable results. However, signals which overlap boundaries can be modeled as *consecutive positives*, and unlike the case of arbitrary positives, the problem of detecting consecutive positives is feasible for practical signal search instances. Our improved designs in Section 5 optimize the tradeoffs between region size and maximum consecutive positives.

- *Experimental Robustness* – Molecular biology is a messy business, and many things can go wrong in the course of synthesizing a molecular sequence, cloning this into a backbone, transfecting onto cells, and observing the presence/absence of phenotype. Designs should incorporate some level of redundancy or error-correction to increase the robustness of this experiment. For example, our adenovirus designs dropped the two regions corresponding to pure wildtype or synthetic design (columns 1 and 16 of Figure 1(left)). This ensured that in any valid experimental outcome at least one design exhibits phenotype while another avoids it, thus serving as an internal check against systematic error in the experiment.

  The improved designs we propose in Section 5 contain several features to improve robustness. In particular, exactly $\lfloor (t-1)/2 \rfloor + \ell$ tests will exhibit the phenotype if the signal spans ℓ (consecutive) segments in the probe region.

3   Group Testing Designs

The classical and well-studied group testing problem [12] is to find a set of at most d positive elements in a set of n elements through the use of group tests. A group test, or *pool*, is a subset of elements which tests whether the subset contains at least one positive. Group testing may be *non-adaptive*, in which all pools must be declared before any results are obtained, or *adaptive*, in which pools may be defined sequentially in *phases* (groups of one or more tests). For the purposes of the signal search problem, the delay between phases is quite long, and we therefore prefer designs which are non-adaptive, or else utilize as few stages as possible.

By the information-theoretic lower bound, $\lceil \log_2(\binom{n}{d}) \rceil \sim d \log_2(n/d)$ tests are necessary for combinatorial group testing [12]. Performing independent binary searches for each element yields a trivial upper bound of $n \lg n$ tests for arbitrary d, but better results are possible with pooling. Below, we summarize results of relevance to our problem.

Arbitrary Positives: For fully adaptive strategies, there are many results for the minimum number of tests $(M(d,n))$ if the number d of positives is known. The upper bound $M(d,n) \leq \lceil \log_2 \binom{n}{d} \rceil + d - 1$, for example, due to Hwang [12]. When the number d of positives is unknown, a group testing strategy using $O(d \log n)$ queries is "competitive", and the hidden constant factor is the competitive ratio. The best competitive ratio is presently 1.5 for sequential tests [25].

Group testing strategies with minimal adaptivity are preferable for applications where the tests are time-consuming, such as our signal search problem. Such strategies work in a few stages, where queries in a stage are prepared prior to the stage and then asked in parallel. For 1-stage group testing, or *non-adaptive group testing*, at least $\Omega((d^2/\log d) \log n)$ queries are needed [5], and $O(d^2 \log n)$ pools are sufficient [6] in the case of a given known d.

When the signal length d is not given, it can be estimated. Damaschke [11] showed that $\Omega(\log n)$ queries are needed for population size n using randomized non-adaptive group tests, to estimate within a constant factor c with a prescribed probability ϵ of underestimating d. Damaschke [10] proposes a Las Vegas strategy that uses $O(d \log n)$ pools in 3 stages and succeeds with any prescribed constant probability, arbitrarily close to 1.

Disjunct and separable matrices play an important role in non-adaptive testing. A $\overline{d}$-separable matrix has the property that the union of each set of d columns is unique. A $\overline{d}$-disjunct matrix has the stronger property these unions do not contain any other columns from the matrix. Therefore, a $\overline{d}$-separable or a $\overline{d}$-disjunct $t \times n$ matrix enables determination of up to d positive items from among n or fewer items using t tests.

For the case of $d = 2$, Kautz and Singleton [16] construct a 2-disjunct matrix with $t = 3^{q+1}$ and $n = 3^{(2^q)}$, for any positive integer q. Macula and Reuter [19] describe a $\overline{2}$-separable matrix with $t = (q^2 + 3q)/2$ and $n = 2^q - 1$, for any positive integer q. Lindstrom ([13], p. 70) gives a $\overline{2}$-separable matrix with $t = 4q$ and $n = 2^q$, for any positive integer q.

For $d = 3$, Du and Hwang [12] (p. 159) describe the construction of a $\overline{3}$-separable matrix with $t = 4\binom{3q}{2} = 18q^2 - 6q$ and $n = 2^q - 1$, for any positive integer q. Eppstein et al. [14] constructs a 2-separable matrix that has a time-optimal analysis algorithm can be constructed with $t = (q^2 + 5q)/2$ and $n = 3^q$, for any positive integer q; they also constructs a 3-separable matrix with $t = 2q^2 - 2q$ and $n = 2^q$, for any positive integer q. In practical cases, when $n \leq 10^{10}$, Eppstein et al.'s construction usually uses fewer tests for $d = 2$ or 3, comparing to these mentioned algorithms.

Consecutive Positives: Unfortunately, it is clear that non-adaptive testing for arbitrary positives is expensive for realistic values of our signal search problem, even for small values of d. For example, Kautz and Singleton's construction for $\overline{2}$-separable construction yields 20 tests for 31 regions. Du and Hwang's $\overline{3}$-separable construction yields 60 tests for 64 regions!

More feasible constructions are possible, however, when we restrict the d positives to be consecutive. The standard binary test design with t tests, denoted by B_t, can cover as much as $n = 2^t$ regions. Extending this, Colbourn [7] constructs an $B_t^+ = (t+3) \times 2^t$-matrix by adding three suitable rows. From this matrix he derives a group test with $m = \lceil \log_2 n \rceil + 3$ pools which are able to identify up to 2-consecutive positives. Müller and Jimbo [21] improve this result and construct 2-consecutive positive detectable matrices H_t of size $(t+1) \times 2^t$ $(t > 2)$ by recursive construction of 2-consecutive positive detectable matrices from H_t to H_{t+2}. The basic construction theme of H_{t+2} is by concatenating 4 retro-reverse copies of H_t's with a 2-bit gray code. While it is impossible to use 3 tests to detect up to 2-consecutive positives in 4 regions, they construct the following two base case matrices:

$$H_3 = \begin{pmatrix} 1\,0\,1\,0\,0\,0\,1\,0 \\ 1\,0\,0\,1\,0\,1\,0\,0 \\ 0\,1\,0\,0\,0\,1\,0\,1 \\ 0\,1\,1\,0\,1\,0\,0\,0 \end{pmatrix}, \quad H_4 = \begin{pmatrix} 0\,0\,1\,0\,0\,0\,0\,0\,1\,0\,1\,0\,1\,0\,1\,1 \\ 0\,1\,0\,0\,0\,1\,1\,0\,1\,0\,0\,0\,1\,1\,0\,0 \\ 1\,0\,1\,0\,0\,0\,1\,1\,0\,1\,0\,1\,0\,0\,0\,0 \\ 0\,0\,0\,1\,0\,1\,0\,1\,0\,0\,0\,1\,1\,0\,0\,1 \\ 1\,0\,1\,0\,1\,0\,0\,1\,0\,0\,0\,0\,0\,1\,1\,0 \end{pmatrix}$$

Thus from these matrices they obtain a group test with $m = \lceil \log_2 n \rceil + 1$ pools, which is optimal to identify up to 2-consecutive positives for $n \geq 5$.

In view of applications, however, the maximal 2-consecutive positive detectable matrices are not robust to errors. There is no room for error checking or repair, and moreover, the number of possible positive outcomes in the valid test result is unbounded, making it difficult to discriminate a valid test result from systematic error.

In order to detect any *false positives* or *false negatives* on the test result. Müller and Jimbo [22] investigate cyclic sequences which contain elements of all k-subsets such that the unions of any two consecutive k-subsets of order is distinct and with the same parity. They derive group testing procedures which detect up to one error for 2-consecutive positives, while sacrificing a moderate amount of resolution. Unfortunately, maximal cyclic sequences of this form have not been found for even some relatively small problem sizes (e.g., $k = 4, t = 10$).

Proposed Testing Approach: We propose an approach wherein each column has $k = \lceil (t-1)/2 \rceil$ positives and each adjacent column union has $k+1$ positives. This design enjoys the nice property of being able to detect many single errors on the region, as well systematic (consistent) multiple errors. Moreover, there are only two classes of single test errors which are not detected: (1) those which confuse a single region positive with a 2-consecutive positive which includes that region, and (2) those which confuse a 2-consecutive positive with a single region positive contained within. Thus, both types of confusion do not generate greatly misleading results.

Consider the following test design:

$$M(5,2) = \begin{pmatrix} 1\,1\,1\,0\,0\,0\,0\,1\,0\,0 \\ 1\,0\,0\,0\,0\,1\,0\,0\,1\,1 \\ 0\,0\,1\,1\,1\,0\,0\,0\,0\,1 \\ 0\,1\,0\,0\,1\,1\,1\,0\,0\,0 \\ 0\,0\,0\,1\,0\,0\,1\,1\,1\,0 \end{pmatrix} \tag{1}$$

It gives a $(t = 5, n = 10)$ test that distinguishes any single region or 2-consecutive positives over the sequence. In general, it is possible to construct a *cyclic 2-consecutive positive detectable matrix* $M(t,k) = t \times \binom{t}{k}$-matrix such that its column is a k-set (out of t elements) such that each two adjacent k-sets has distinct unions which are $(k+1)$-sets. Note that $n = \binom{t}{k}$ is maximized when $k = \lfloor(t-1)/2\rfloor$. Note that the vertical binary code design with 3 extra boundary detective tests, namely B_3^+, gives a $(t = 7, n = 10)$ design to achieve the same effect.

Given $t \geq 5$, it is possible to design a $t \times n$ matrix which enables determination of 1 positive or 2 consecutive positives from among $n = \binom{t}{\lfloor(t-1)/2\rfloor}$ or fewer items using t tests. It is not hard to verify that given t tests, the number of covered regions, n is approximately $0.798 \cdot 2^t / \sqrt{t}$. Given the number regions n, the number of necessary tests is therefore $\leq \lceil \log_2 n + 0.55 \log_2 \log_2 n + 0.95 \rceil$.

For all cases practical for our problem, we have $t \leq \lceil \log_2 n \rceil + 3$. In other words, compared to standard binary test design, which correctly detects only a single region defect and provides no error detection, this approach uses no more than 3 additional tests while identifying 1 or 2-consecutive positives and providing the aforementioned error detection properties.

4   Maximal Cycles Crossing Adjacent Levels of the Hasse Diagram

The construction of our 2-consecutive group test design is related to the *the middle levels (Revolving Door) conjecture* [26]. It is conjectured that there is always a Hamilton cycle in the middle levels, M_{2k+1}, of the Hasse diagram of (the partially-ordered set of subsets of a $2k+1$-element set ordered by inclusion). See, for example, the simple case of Figure 3. Currently, the conjecture is known to be true for $k = 19$ (as of 3/17/2011 [27]); note that $\binom{39}{19} = 68,923,264,410$. The middle level problem can be consider as a special case of the more generalized *adjacent level problem*, where it is conjectured that there always exists a long simple cycle in any two adjacent levels of the Hasse diagram that spans at least one level of the the graph.

Here we show that the adjacent level graph conjecture is equivalent to the middle level conjecture. Given the Hasse diagram of B_n being the undirected version of the partially-ordered set of subsets of n element set ordered by inclusion. Each vertex v of B_n represent a subset of $\{1..n\}$, denoted by S_v. S_v is called a k-set if its cardinality is k, or $|S_v| = k$. Let $M(n, i)$ denote the subgraph induced by vertices of i-set and $(i+1)$-set in B_n. Note that $M(n, i)$ is a bipartite graph, where all edges link between a i-set, u, and $(i+1)$-set, v, where $uv \in E[M(n,i)]$ iff $S_u \subset S_v$. The middle level conjecture is equivalent to saying that $M(2k+1, k)$ is Hamiltonian for $k \geq 3$.

Theorem 1. *Let $\phi(n,k)$ denote the proposition that "there exists a simple cycle in $M(n,k)$ spanning vertices of level k." It follows that $\phi(n, k-1)$ and $\phi(n,k)$ implies that $\phi(n+1, k)$.*

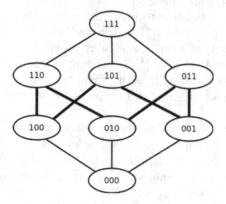

$$
\begin{array}{r|cc|cccc}
1 & 0 & 1 & 1 & 0 & 0 & 0 \\
2 & 1 \dashrightarrow 1 & & 1 & 1 \dashrightarrow v & 1 \\
3 & 1 & 1 & 1 & 1 & v & 1 \\
\cdots & \cdots \dashrightarrow \cdots & & \cdots \dashrightarrow \cdots \\
k & 1 & 1 & 1 & 1 & v & 1 \\
k+1 & 1 \dashrightarrow 0 & & 0 & 0 \dashrightarrow v & 1 \\
k+2 & 0 & 0 & 0 & 0 & 0 & 0 \\
\cdots & \cdots \; \pi_2 \; \cdots & & \cdots \; \pi_1 \; \cdots \\
n & 0 & 0 & 0 & 0 & 0 & 0 \\
n+1 & 0 \; \cdots \; 0 & & 1 & 1 \; \cdots \; 1 & 1 \\
\end{array}
$$

Fig. 3.   The Hasse diagram of B_3 showing a Hamilton cycle through the middle two levels

Fig. 4.   Constructing $\phi(n+1,k)$ from $\phi(n,k-1)$ and $\phi(n,k)$, as described in Theorem 1

Proof.   Here we show how to join two cycles in $M(n, k-1)$ and $M(n, k)$ into one larger cycle in $M(n+1, k)$ spanning vertices of level k. Without loss of generality, we can assume that $n \geq 2k+1$; otherwise, the theorem is trivially proven. It follows that $\binom{n}{k} > \binom{n}{k-1}$; that is, the number of k-set is strictly larger than the number of $(k-1)$-set of $[1..n]$.

Let C_1 be the cycle in $M(n, k-1)$ given by the hypothesis. Without loss of generality, we assume that the k-set $[1..k]$ does *not* appear in C_1 (otherwise, simply relabel vertices). Furthermore, we assume that the $(k-1)$-set vertex $[2..k]$ is connected to the k-set vertex $[2..k+1]$ in $M(n, k-1)$ (again, otherwise relabel). It follows that C_1 can be denoted by $([2..k], \pi_1, v, [2..k+1])$. Here $[2..k] \neq v \subset [2..k+1]$ is a $(k-1)$-set, and π_1 is a path of length $2\binom{n}{k-1} - 3$. Let C_2 be the cycle in $M(n, k)$; with a similar argument, we assume that $[1..k]$ is connected to $[1..k+1]$, then connected to $[2..k+1]$. That is, C_2 can be denoted by $([2..k+1], \pi_2, [1..k], [1..k+1])$. Note that π_2 is a long path of length $2\binom{n}{k} - 3$.

Now observe the cycle $C = ([2..k+1], \pi_2, [1..k], [1..k, n+1], [2..k, n+1], \pi_1 \cup [n+1], v \cup [n+1], [2..k+1, n+1])$. Here the term $\pi_1 \cup [n+1]$ denotes a new path built from π_1 by adding one extra number $[n+1]$ into the original vertices of π_1. This is illustrated in Figure 4. It is easily verified that the length of C is exactly $2 \cdot [\binom{n}{k-1} + \binom{n}{k}] = 2\binom{n+1}{k}$. Furthermore, is readily seen that C exhausts vertices of k-set in $M(n+1, k)$. It follows that C is the long cycle of $M(n+1, k)$ spanning vertices of level k. $\square$

Theorem 2. *The Middle Level Conjecture is true if and only if the adjacent level conjecture is true.*

Proof.   Sketch only, due to space; see Appendix for full details. The middle level is a specific adjacent level, therefore the first direction is trivial. To prove the reverse direction, induction using Theorem 1 suffices. $\square$

5   Consecutive Positives Detectable Matrices

Critical signals can cross multiple regions under test, as shown in Figure 2, which motivates the need for a general consecutive positives group testing design. Consider the following test design:

$$
M_3(5,2) = \left(\begin{array}{c}
1\,1\,1\,1\,1\,1\,1\,1\,1 \cdots 1\,1\,1\,0\,0\,0\,0\,0\,0 \\
1\,1\,1\,0\,0\,0\,0\,0\,0 \cdots 0\,0\,0\,1\,1\,1\,1\,1\,1 \\
0\,0\,0\,0\,0\,0\,1\,1\,1 \cdots 0\,0\,0\,0\,0\,0\,1\,1\,1 \\
0\,0\,0\,1\,1\,1\,0\,0\,0 \cdots 0\,0\,0\,0\,0\,0\,0\,0\,0 \\
0\,0\,0\,0\,0\,0\,0\,0\,0 \cdots 1\,1\,1\,1\,1\,1\,0\,0\,0 \\
\hline
1\,0\,0\,0\,1\,0\,0\,0\,1 \cdots 0\,1\,0\,0\,0\,1\,1\,0\,0 \\
0\,1\,0\,0\,0\,1\,1\,0\,0 \cdots 0\,0\,1\,1\,0\,0\,0\,1\,0 \\
0\,0\,1\,1\,0\,0\,0\,1\,0 \cdots 1\,0\,0\,0\,1\,0\,0\,0\,1
\end{array}\right) \tag{2}
$$

It gives a $(t = 8, n = 30)$ test that distinguishes any up to three consecutive positives over the sequence. That is, by adding in three extra check pools, the design triples the original 2-consecutive positive regions, and detect up to 3 consecutive positives. In other words, we have a design using t tests that detects up to three consecutive positives over $n \sim 2^{t-4}$ regions.

In general, it is possible to construct a d-consecutive positive detectable matrix $(d \geq 3\ M_d(r, k) = (t = r + d) \times d\binom{r}{k})$-matrix such that its column consist of d duplicated k-sets (out of r elements) such that each two adjacent distinct k-sets has distinct unions. Furthermore, the union of two adjacent distinct columns (sets) are exactly a $(k+1)$-set.

Here we describe the details of the construction for $M_d(r, k)$. Note that Hamiltonian cycles of the middle level graph of B_{2k+1}, denoted by $M(2k+1, k)$, have been verified to exist for $k \leq 19$ [27,26]. Furthermore, we use the constructive method shown in the proof of Theorem 1 to obtain the cycle in $M(2k + 2, k)$ spanning the k-set. Denote the k-set-spanning cycle found in $M(t, k)$ to be H_t (here $k = \lfloor (t-1)/2 \rfloor$.) Note that H_t is a sequence of alternative k-set and $(k+1)$-set out of $[1..t]$ with size exactly $\binom{t}{k}$ on each set. Given H_t, let $A(H_t)$ denote the 2-consecutive detectable boolean matrix construct from H_t. That is, $A(H_t)$ is a $t \times \binom{t}{\lfloor (t-1)/2 \rfloor}$ boolean matrix, where the i-th column of $A(H_t)$ is the (vertical) boolean string of the i-th k-set in H_t. As an example, $A(H_5)$, a 5×10 boolean matrix is shown in (1).

Given an arbitrary $d \geq 3$, a d-consecutive positive detectable matrix $M_d(t, k)$, which is a $(t+d) \times d\binom{t}{k}$ matrix with $k = \lfloor (t-1)/2 \rfloor$, can be constructed according to the corresponding $A(H_t)$. In particular, each column of $M_d(t, k)$ is a $(t + d)$-length vertical boolean vector; furthermore, the first t entries of the i-th column of $M_d(t, k)$ is equal to the $(1 + \lfloor (i - 1)/d \rfloor)$-th column of $A(H_t)$. Let e_j denote a length-d boolean vector such that every entry of e_j equal to 0 except its j-th entry equal to 1. That is, $e_j[j] = 1$, while $e_j[i] = 0$ for all $i \neq j$. Let $M_d(t, k)[i]$ denote the i-th column of the matrix of $M_d(t, k)$ and for the similar notation as of $A(H_t)[j]$, we have the following:

$$
M_d(t, k)[i] = A(H_t)[1 + \lfloor (i-1)/d \rfloor] \circ e_{\psi(i)}, \quad \text{for } 1 \leq i \leq d\binom{t}{k} \tag{3}
$$

with the function $\psi(i) = 1 + (i - 1 - \lfloor (i-1)/d \rfloor \bmod d)$. As an example, see Eq. (2) for the construction for $M_3(5,2)$.

Theorem 3. *The non-adaptive test corresponding to the matrix $M_d(t, \lfloor (t - 1)/2 \rfloor)$ in (3) correctly identifies up to d consecutive positives over $d \cdot \binom{t}{\lfloor (t-1)/2 \rfloor}$ items, using $t + d$ pools. Furthermore, there are ℓ (consecutive) positive items in the probe region if and only if there are exactly $\lfloor (t-1)/2 \rfloor + \ell$ observed "1"s in the production string.*

Proof. Sketch only, due to space; see Appendix for full details. Each column from the original 2-consecutive design is duplicated d times, therefore the first t pools localize the defect to an area of either d or $2d$ regions. The remaining d columns suffice to determine the boundaries of the defective region. □

Similar to our original design, since $\binom{r}{\lfloor (r-1)/2 \rfloor} \sim 0.798 \cdot 2^r / \sqrt{r}$, and $r = t - d$, the maximum covered item number $n = 0.798d \cdot 2^{t-d} / \sqrt{t-d}$. It thus follows that $t \approx \log_2(1.27n/d) + 0.5 \log_2(\log_2(1.5n/d)) + d$. In contrast, the original Colbourn's method [7] constructs a $(r + 2d + 1) \times (d - 1)2^r$-matrix; that is, $t = \log_2(n/(d-1)) + 2d + 1$. In particular, for $r = 3, d = 3$, the Colbourn's construction leads to a 10×16 matrix, using 10 pools to detect up to 3 consecutive positives for 16 regions. Note that with 10 pools, $M_3(7,3)$ can detect up to 3 consecutive positives for 105 regions.

5.1　Optimizing Experimental Design

Let g denote the length of the genome being inspected, m be some conservative maximum length of the signal being sought, and let t denote the number of tests (pools) to be used. Finally, let n denote the number of regions in the resulting designs, and $b = g/n$ denote the block size of these regions. As m grows with respect to b, the number of consecutive positives our design must handle (d) also grows (it is required that $(d - 1)b = (d - 1)\lceil g/n \rceil \geq m - 1$). Given t, we should choose a value of d such that

$$n(t,d) = \binom{t-d}{\lfloor (t-d-1)/2 \rfloor} d \sim 0.798d \cdot 2^{t-d} / \sqrt{t-d}$$

One-Round Design: In the non-adaptive case, our goal is achieved by maximizing the number of inspected items, $n(t,d)$ that are *permissible*. That is, consider the permissible designs $D = D(g,t,m) = \{d \geq 2 | (d-1)\lceil g/n(t,d) \rceil \geq m - 1\}$. It follows that $n^* = \max_{d \in D(g,t,m)} \{n(t,d)\}$. The corresponding block resolution $b = b(g,t,m) = \lceil g/n^* \rceil$ is therefore minimized.

Two-Round Design: We also consider splitting our t tests across a 2-stage adaptive design. The idea to locate a rough assessment of a consecutive segment within the genome in the first round. If the first round uses t_1 tests, then we can use the remaining $t_2 = t - t_1$ tests to further refine the smaller segment identified by the first round. Note that, after identifying $b_1 = b(g, t_1, m)$, the second round examination faces a shorter genome of length: $g' = 2 \cdot b_1$. That is, both the left and right block need to be further examined to further reduce the underlying genome. That is, $b_2(g, t, m) = \min_{2 \le t^* \le t-2}\{b(\, 2b(g, t^*, m), t-t^*, m\,)\}$. The values of t_1 and t_2 which give the finest resulting resolution are chosen.

Comparison: It is clear that the 2-stage design is sometimes, though not always, advantageous. Consider, for example, $g = 10,000, m = 1,000$ and $t = 20$. The one-run approach finds the largest permissible n when $d = 16, n = \binom{4}{2} \cdot 16 = 96$. The resolution of this non-adaptive approach will be $b_1 = \lceil g/n \rceil = 105$. Note that the 2-stage approach obtains the first stage parameters $t_1 = 12, d_1 = 8, n_1 = 48, b_1 = 209$. While the second stage now has $g_2 = 2b_1 = 418, m = 1,000, t_2 = 8$; it will come down to $n_2 = 8, d_2 = 8$. Thus, the final resolution will be $b_2 = \lceil 418/8 \rceil = 53$ base pairs, which is superior to the non-adaptive approach.

Figure 5 shows the resolution achieved by the 1 and 2 stage designs, for a genome of size $10,000$, and various values of m and t.

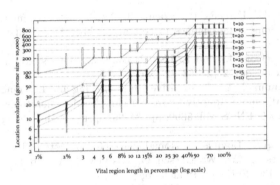

Located resolution: ℓ		
$g = 10,000$		
m	t r	ℓ
300	10 1	125
300	15 1	56
300	20 1	36
300	20 2	21
300	25 2	12
2000	10 1	477
2000	15 1	278
2000	20 1	197
2000	20 2	93
2000	25 2	56

Fig. 5. Resulting final resolutions under different settings of test numbers and vital lengths under one-round and two-round designs

6   Conclusion

We have presented a novel approach with proven utility which applies DNA synthesis and combinatorial group testing to rapidly identify the location of biological signals. We summarize relevant combinatorial group testing designs, and present designs which implement our prefered approach. Software to construct sequences adhering to our prefered designs is available at: *http://www.algorithm.cs.sunysb.edu/signalSearch.*

Acknowledgements. We would like to thank our experimental collaborators in Wadie Bahou and Eckard Wimmer's groups, and Ian Shields for providing the middle level Hamiltonian cycles which were used to create our group testing designs.

References

1. Balding, D.J., Torney, D.C.: The design of pooling experiments for screening a clone map. Fungal Genetics and Biology 21(3), 302–307 (1997)
2. Bhat, G., Savage, C.: Balanced gray codes. Electronic Journal of Combinatorics 3, R25 (1996)
3. Bruno, W.J., Knill, E., Balding, D.J., Bruce, D.C., Doggett, N.A., Sawhill, W.W., Stallings, R.L., Whittaker, C.C., Torney, D.C.: Efficient pooling designs for library screening. Genomics 26(1), 21–30 (1995)
4. Bugl, H., Danner, J.P., Molinari, R.J., Mulligan, J.T., Park, H.-O., Reichert, B., Roth, D.A., Wagner, R., Budowle, B., Scripp, R.M., Smith, J.A.L., Steele, S.J., Church, G., Endy, D.: Dna synthesis and biological security. Nature Biotechnology 25, 627–629 (2007)
5. Chen, H.B., Hwang, F.K.: Exploring the missing link among d-separable,-separable and d-disjunct matrices. Discrete Applied Mathematics 155(5), 662–664 (2007)
6. Cheng, Y., Du, D.Z.: New constructions of one-and two-stage pooling designs. Journal of Computational Biology 15(2), 195–205 (2008)
7. Colbourn, C.J.: Group testing for consecutive positives. Annals of Combinatorics 3(1), 37–41 (1999)
8. Coleman, J.R., Papamichial, D., Futcher, B., Skiena, S., Mueller, S., Wimmer, E.: Virus attenuation by genome-scale changes in codon-pair bias. Science 320, 1784–1787 (2008)
9. Czar, M.J., Christopher Anderson, J., Bader, J.S., Peccoud, J.: Gene synthesis demystified. Trends in Biotechnology 27(2), 63–72 (2009)
10. Damaschke, P., Sheikh Muhammad, A.: Competitive Group Testing and Learning Hidden Vertex Covers with Minimum Adaptivity. In: Kutyłowski, M., Charatonik, W., Gebala, M. (eds.) FCT 2009. LNCS, vol. 5699, pp. 84–95. Springer, Heidelberg (2009)
11. Damaschke, P., Muhammad, A.S.: Bounds for Nonadaptive Group Tests to Estimate the Amount of Defectives. In: Wu, W., Daescu, O. (eds.) COCOA 2010, Part II. LNCS, vol. 6509, pp. 117–130. Springer, Heidelberg (2010)
12. Du, D., Hwang, F.: Combinatorial group testing and its applications. World Scientific Pub. Co. Inc. (2000)
13. Du, D., Hwang, F.: Pooling Designs and Nonadaptive Group Testing: Important Tools for DNA Sequencing. World Scientific (2006)
14. Eppstein, D., Goodrich, M.T., Hirschberg, D.S.: Improved Combinatorial Group Testing Algorithms for Real-World Problem Sizes. SIAM Journal on Computing 36(5), 1360–1375 (2006)
15. Gray, F.: Pulse code communication. US Patent 2632058, March 17 (1953)
16. Kautz, W., Singleton, R.: Nonrandom binary superimposed codes. IEEE Transactions on Information Theory 10(4), 363–377 (1964)
17. Knuth, D.: The Art of Computer Programming, Volume 4 Fascicle 3: Generating All Combinations and Partitions. Addison Wesley (2005)
18. Lin, Y.-L., Ward, C., Jain, B., Skiena, S.: Constructing Orthogonal de Bruijn Sequences. In: Dehne, F., Iacono, J., Sack, J.-R. (eds.) WADS 2011. LNCS, vol. 6844, pp. 595–606. Springer, Heidelberg (2011)
19. Macula, A.J., Reuter, G.R.: Simplified searching for two defects. Journal of Statistical Planning and Inference 66(1), 77–82 (1998)
20. Mueller, S., Coleman, R., Papamichail, D., Ward, C., Nimnual, A., Futcher, B., Skiena, S., Wimmer, E.: Live attenuated influenza vaccines by computer-aided rational design. Nature Biotechnology 28 (2010)

21. Müller, M., Jimbo, M.: Consecutive positive detectable matrices and group testing for consecutive positives. Discrete Mathematics 279(1-3), 369–381 (2004)
22. Müller, M., Jimbo, M.: Cyclic sequences of k-subsets with distinct consecutive unions. Discrete Mathematics 308(2-3), 457–464 (2008)
23. Pemmaraju, S., Skiena, S.: Computational Discrete Mathematics: Combinatorics and Graph Theory with Mathematica. Cambridge University Press, New York (2003)
24. Savage, C.: A survey of combinatorial gray codes. SIAM Review 39, 605–629 (1997)
25. Schlaghoff, J., Triesch, E.: Improved results for competitive group testing. Combinatorics, Probability and Computing 14(1-2), 191–202 (2005)
26. Shields, I., Shields, B.J., Savage, C.D.: An update on the middle levels problem. Discrete Mathematics 309(17), 5271–5277 (2009)
27. Shimada, M., Amano, K.: A note on the middle levels conjecture. Arxiv preprint arXiv:0912.4564 (2009)
28. Sitaraman, V., Hearing, P., Mueller, S., Ward, C., Skiena, S., Wimmer, E., Bahou, W.: Genetically re-engineered aav rep78 identifies sequence-restricted inhibitory regions affecting adenoviral replication. American Society of Gene and Cell Therapy 13th Annual Meeting (2010)
29. Song, Y., Paul, A., Ward, C., Mueller, S., Futcher, B., Skiena, S., Wimmer, E.: Identification of cis-acting elements in the coding sequence of the poliovirus rna polymerase using computer generated designs and synthetic dna synthesis (manuscript in preparation)
30. Wilf, H.: Combinatorial Algorithms: an update. SIAM, Philadelphia (1989)

The Three-Dimensional Architecture
of a Bacterial Genome and Its Alteration
by Genetic Perturbation

Marc A. Marti-Renom

Structural Genomics Team, Genome Biology Group,
National Center for Genomic Analysis (CNAG), Barcelona

We have determined the three-dimensional (3D) architecture of the *Caulobacter crescentus* genome by combining genome-wide chromatin interaction detection, live-cell imaging, and computational modeling. Using chromosome conformation capture carbon copy (5C), we derive ∼13 kb resolution 3D models of the *Caulobacter* genome. The resulting models illustrate that the genome is ellipsoidal with periodically arranged arms. The *par*S sites, a pair of short contiguous sequence elements known to be involved in chromosome segregation, are positioned at one pole, where they anchor the chromosome to the cell and contribute to the formation of a compact chromatin conformation. Repositioning these elements resulted in rotations of the chromosome that changed the subcellular positions of most genes. Such rotations did not lead to large–scale changes in gene expression, indicating that genome folding does not strongly affect gene regulation. Collectively, our data suggest that genome folding is globally dictated by the *par*S sites and chromosome segregation. Highlight of an article published in Molecular Cell in 2011.

http://www.cell.com/molecular-cell/retrieve/pii/S1097276511007593

B. Chor (Ed.): RECOMB 2012, LNBI 7262, p. 180, 2012.
© Springer-Verlag Berlin Heidelberg 2012

Discovery of Complex Genomic Rearrangements in Cancer Using High-Throughput Sequencing

Andrew McPherson[1], Chunxiao Wu[2], Alexander Wyatt[2], Sohrab Shah[3],
Colin Collins[2], and S. Cenk Sahinalp[1]

[1] School of Computer Science, Simon Fraser University
[2] Vancouver Prostate Centre
[3] Department of Molecular Oncology, BC Cancer Research Centre
andrew.mcpherson@gmail.com, cenk@sfu.ca

Motivation

High-throughput sequencing now allows researchers to identify patterns of rearrangement in tumour genomes, and has led to the discovery of Complex Genomic Rearrangements (CGRs) as a new cytogenetic feature of some cancers. Closed chain breakage and rejoining (CCBR), a generalization of reciprocal translocation, was recently identified in prostate cancer [4]. CCBRs involve the balanced rearrangement of some n loci, and thus have the potential to fuse or interrupt up to n genes. In addition, dispersed throughout some breast cancer genomes are 'genomic shards', small fragments of DNA originating from elsewhere in the genome and inserted at the breakpoints of larger scale rearrangements.

Both CCBRs and genomic shards have implications for the discovery and characterization of gene fusions. A single CCBR can produce multiple gene fusions, and association of these related gene fusions may provide important information about a tumour's biology. Genomic shards have the potential to confound discovery of gene fusions. Discovery of a gene A to gene B rearrangement may falsely nominate an A-B gene fusion, when in fact the complete rearrangement is A-B-C: an A-C gene fusion with a shard of B inserted at the A-C breakpoint. We have developed nFuse (http://n-fuse.sf.net), an algorithmic method for identifying CGRs in WGSS data based on shortest alternating paths in breakpoint graphs. We target our search for CGRs to those that underlie fusion transcripts predicted from matched high-throughput cDNA sequencing (RNA-Seq).

Shortest Alternating Paths in Breakpoint Graphs

Let a breakpoint be defined as a pair of genomic positions, (x^1, x^2), such that x^1 and x^2 are non-adjacent in the reference genome and adjacent in the tumour genome. Let the breakpoint graph for breakpoints B be defined on the vertex set $\{x_i^1, x_i^2 : (x_i^1, x_i^2) \in B\}$ as follows. Breakpoint edges connect x_i^1 and x_i^2 for each breakpoint and are weighted based on the probability the breakpoint exists given the WGSS data (the *breakpoint score*). Adjacency edges connect x_i^u and x_j^v

B. Chor (Ed.): RECOMB 2012, LNBI 7262, pp. 181–182, 2012.
© Springer-Verlag Berlin Heidelberg 2012

if x_i^u and x_j^v are on the same chromosome with opposite orientation. Adjacency edges are weighted based on the chromosomal distance between x_i^u and x_j^v.

A fusion composed of n genomic shards (n-fusion) will involve $n + 2$ separate loci, and will result in $n + 1$ breakpoints in the breakpoint graph forming an alternating path of length $2n - 3$. We identify n-fusions by searching for the shortest alternating path between candidate fused genes. Candidate fused genes are nominated by fusion transcripts predicted from matched RNA-Seq. A CCBR involving n loci will be represented in the breakpoint graph by an alternating $2n$-cycle. Many instances of chromosomal breakage and rejoining involve the loss or duplication of a small amount of chromosomal material. We identify CCBRs by searching for the shortest alternating cycle that includes a candidate breakpoint. Candidate breakpoints are nominated by their association with fusion transcripts predicted from matched RNA-Seq.

Application to Tumour and Cell Line Data

We first applied nFuse to publicly available WGSS (acc. no. ERA010917) and RNA-Seq (acc. no. ERA015355) for HCC1954, a cell line that has been well studied at the molecular level. We compiled a list of 345 validated breakpoints from 3 previous studies of HCC1954 [3,1,2]. Only 3 breakpoints were common to all three studies. The breakpoint detection step of nFuse identifies 296 of the 345 breakpoints for a recall of 0.858. We compared the breakpoint scores for 296 *true positives* to the scores of a random selection of 3000 putative false positive predictions, estimating the area under the curve for the scoring method as 0.96.

In HCC1954, 3 fusion transcripts were associated with n-fusions composed of 2 or more true positive breakpoints, including in-frame, highly expressed PHF20L1-SAMD12. We searched for high confidence n-fusions for which each genomic shard was expressed in at least one fusion transcript, in effect using fusion transcripts as a scaffold for genome reconstruction. The 9 such CGRs we discover are composed of genomic shards that range in size from as 800b to 140kb. Next we applied nFuse to the patient-derived mouse xenograft LTL-313H and primary prostate tumour 963, and PCR validated all breakpoints discussed herein. In LTL-313H, we identified an n-fusion composed of 9 genomic shards scaffolded by 8 fusion transcripts. In 963 we identified 5 CCBRs, 2 involving three loci, 2 involving four loci, and 1 involving 5 loci.

References

1. Bignell, G.R., et al.: Architectures of somatic genomic rearrangement in human cancer amplicons at sequence-level resolution. Genome. Res. 17, 1296–1303 (2007)
2. Stephens, P.J., et al.: Complex landscapes of somatic rearrangement in human breast cancer genomes. Nature 462, 1005–1010 (2009)
3. Galante, P., et al.: Distinct patterns of somatic alterations in a lymphoblastoid and a tumor genome derived from the same indiv. Nucl. Acids Res. 39, 6056–6068 (2011)
4. Berger, M.F., et al.: The genomic complexity of primary human prostate cancer. Nature 470, 214–220 (2011)

Network-Based Prediction and Analysis of HIV Dependency Factors

T.M. Murali[1], Matthew D. Dyer[2], David Badger[1],
Brett M. Tyler[3], and Michael G. Katze[4]

[1] Department of Computer Science, Virginia Tech, Blacksburg, VA
[2] Applied Biosystems, Foster City, CA
[3] Virginia Bioinformatics Institute, Virginia Tech, Blacksburg, VA
[4] Department of Microbiology, University of Washington, Seattle, WA
United States of America

HIV Dependency Factors (HDFs) are a class of human proteins that are essential for HIV replication, but are not lethal to the host cell when silenced. Three previous genome-wide RNAi experiments identified HDF sets with little overlap. We combine data from these three studies with a human protein interaction network to predict new HDFs, using an intuitive algorithm called SinkSource and four other algorithms published in the literature. Our algorithm achieves high precision and recall upon cross validation, as do the other methods. A number of HDFs that we predict are known to interact with HIV proteins. They belong to multiple protein complexes and biological processes that are known to be manipulated by HIV. We also demonstrate that many predicted HDF genes show significantly different programs of expression in early response to SIV infection in two non-human primate species that differ in AIDS progression. Our results suggest that many HDFs are yet to be discovered and that they have potential value as prognostic markers to determine pathological outcome and the likelihood of AIDS development. More generally, if multiple genome-wide gene-level studies have been performed at independent labs to study the same biological system or phenomenon, our methodology is applicable to interpret these studies simultaneously in the context of molecular interaction networks and to ask if they reinforce or contradict each other.

http://www.ploscompbiol.org/article/info:doi/10.1371/
journal.pcbi.1002164

B. Chor (Ed.): RECOMB 2012, LNBI 7262, p. 183, 2012.

Structure-Guided Deimmunization
of Therapeutic Proteins

Andrew S. Parker[1], Karl E. Griswold[2], and Chris Bailey-Kellogg[1]

[1] Department of Computer Science, Dartmouth College,
6211 Sudikoff Laboratory, Hanover, NH 03755, USA
{asp,cbk}@cs.dartmouth.edu
[2] Thayer School of Engineering, Dartmouth College,
8000 Cummings Hall, Hanover, NH 03755, USA
karl.e.griswold@dartmouth.edu

Abstract. Therapeutic proteins continue to yield revolutionary new treatments for a growing spectrum of human disease, but the development of these powerful drugs requires solving a unique set of challenges. For instance, it is increasingly apparent that mitigating potential anti-therapeutic immune responses, driven by molecular recognition of a therapeutic protein's peptide fragments, may be best accomplished early in the drug development process. One may eliminate immunogenic peptide fragments by mutating the cognate amino acid sequences, but deimmunizing mutations are constrained by the need for a folded, stable, and functional protein structure. We develop a novel approach, called EpiSweep, that simultaneously optimizes both concerns. Our algorithm identifies sets of mutations making Pareto optimal trade-offs between structure and immunogenicity, embodied by a molecular mechanics energy function and a T-cell epitope predictor, respectively. EpiSweep integrates structure-based protein design, sequence-based protein deimmunization, and algorithms for finding the Pareto frontier of a design space. While structure-based protein design is NP-hard, we employ integer programming techniques that are efficient in practice. Furthermore, EpiSweep only invokes the optimizer once per identified Pareto optimal design. We show that EpiSweep designs of regions of the therapeutics erythropoietin and staphylokinase are predicted to outperform previous experimental efforts. We also demonstrate EpiSweep's capacity for global protein deimmunization, a case analysis involving 50 predicted epitopes and over 30,000 unique side-chain interactions. Ultimately, EpiSweep is a powerful protein design tool that guides the protein engineer towards the most promising immunotolerant biotherapeutic candidates.

Keywords: structure-based protein design, sequence-based protein design, multi-objective optimization, therapeutic proteints, deimmunization, experiment planning.

1 Introduction

The ever-expanding toolbox of approved therapeutic proteins is one of the crowning achievements of modern biotechnology. Biotherapeutic agents are providing

B. Chor (Ed.): RECOMB 2012, LNBI 7262, pp. 184–198, 2012.

new and efficacious treatment options for common diseases such as cancer, diabetes, rheumatoid arthritis, anemia, heart attacks, strokes, and more. Moreover, they are breaking new ground in helping to treat a broad spectrum of previously intractable illnesses such as multiple sclerosis, cystic fibrosis, congenital lipid and carbohydrate storage disorders, and HIV/AIDS, to name a few [19]. While revolutionizing the treatment of numerous diseases, the rapid growth of the biotherapeutics market has also exposed a range of new challenges in drug design and development. One of the limitations inherent to proteinaceous drugs is the fact that, unlike small molecules, they are subject to surveillance by the human immune system. The recognition of a biotherapeutic as "non-self" can result in the patient's body mounting a concerted anti-biotherapeutic immune response (aBIR). Although in some instances this immune response has no clinical significance, in other cases it manifests a variety of detrimental outcomes ranging from loss of drug efficacy to severe anaphylactic shock [28]. Therapeutic proteins and peptides of non-human origin are especially prone to such an immune response, but the complexity of human immune surveillance can result in immune reactions even against exogenously administered human proteins [16]. Regardless of a protein's origin, the aBIR is fundamentally driven by molecular recognition of antigenic peptide sequences embedded within the full length protein, and removing or manipulating these sequences represents an effective strategy for biotherapeutic deimmunization.

Grafting-based "humanization" strategies simply swap segments of a biotherapeutic candidate for comparable segments of a homologous human protein. While particularly effective for deimmunizing therapeutic antibodies [14,12] these methodologies are predicated on the availability of a homologous human protein as well as detailed knowledge of underlying structure-function relationships. Non-immunoglobulin proteins, which represent a rich but largely untapped reservoir of prospective therapeutic agents, often fail to meet one or both of these criteria. As a result, there exists a growing need for more broadly applicable protein deimmunization methodologies, and the development of such methods will undoubtedly spur further innovations in disease treatment.

As noted above, the aBIR is driven by molecular recognition of immunogenic peptides, hereafter epitopes, which are found within the biotherapeutic's primary sequence. Immune surveillance is initiated when antigen presenting cells internalize the therapeutic protein, which is then hydrolyzed into smaller peptide fragments. Fragments that represent potential antigenic epitopes are loaded into the groove of cognate type II major histocompatibility complex (MHC II) proteins (Fig. 1, left), and the complexes are trafficked to the cell surface for display to the extracellular milieu. There, the peptide-MHC II complexes are free to interact with T cell receptor proteins on the surface of T cell lymphocytes, and true immunogenic epitopes are recognized upon formation of ternary peptide-MHC II-T cell receptor complexes. The subsequent signaling cascade leads to maturation and proliferation of B cell lymphocytes that ultimately secrete immunoglobulin molecules able to bind the original biotherapeutic agent.

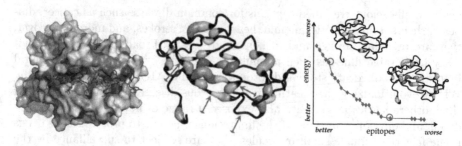

Fig. 1. EpiSweep. **(left)** An immune response to a therapeutic protein is initiated by MHC II (red/white) recognition of an immunogenic peptide epitope (blue) digested from the protein. Our goal is to mutate the protein so that no such recognition will occur. **(center)** MHC/T-cell epitopes are pervasive: exposed on the surface, buried in the core, and covering active sites. Shown is staphylokinase (pdb id 2sak), with black backbone for no epitopes starting at the residue and highlighted sausage for the predicted binding by 8 common MHC II alleles, ranging from thin pale yellow (bind just 1) to thick bright red (bind all 8). Putative active sites are denoted with green arrows. **(right)** EpiSweep explores the Pareto frontier simultaneously optimizing structure (molecular mechanics energy function, y-axis) and immunogenicity (epitope score, x-axis). The left circled plan has fewer predicted epitopes than the right one, but the right one has better predicted energy.

This well-defined immunological pathway suggests that therapeutic proteins might be engineered such that their peptide fragments would evade MHC II/T cell receptor binding and thereby block the undesirable aBIR [3,4]. Indeed, the mutation of key anchor residues within known epitopes of therapeutic proteins has yielded partially deimmunized versions of staphylokinase [34] and erythropoietin [32], among others. It is important to note that unlike antibody epitopes, which localize exclusively to the solvent exposed protein surface, T cell epitopes result from proteolytic processing and may therefore be found anywhere within a protein's primary sequence (Fig. 1, center). As a result, experimental determination of T cell epitopes requires technically challenging immunoassays on large pools of overlapping component peptides that span the entire protein of interest. Upon identifying the immunodominant peptides, the protein engineer then performs scanning alanine or other systematic mutagenesis on each one so as to identify critical binding residues whose substitution ameliorates the undesired immune response. Efficacious mutations must then be engineered back into the full-length protein, where they often affect structural stability or therapeutic function. For example, hydrophobic amino acids typically anchor peptide epitopes within the MHC II binding groove, and these amino acids therefore represent attractive targets for mutagenic epitope deletion. Those same residues, however, can also play a central role in stabilizing the close-packed core of the full-length protein, and substitution at these positions may compromise protein folding. Intuitively, deimmunizing mutations must not undermine the target protein's native structure or function, and consequently protein deimmunization is inherently a multi-objective optimization problem.

To efficiently direct experimental resources towards the most promising sets of mutations, we have developed EpiSweep, a novel approach that integrates validated immunoinformatics and structural modeling methods within a framework for identifying Pareto optimal designs (Fig. 1, right). These designs (sets of mutations) make the best trade-offs between the two objectives of stability and immunogenicity, in that no design is simultaneously better for both objectives. *(Stability)* We compute protein stability using a highly-successful, structure-based, protein design strategy that seeks to optimize side-chain packing [1,20,2]. In this approach, the protein backbone is fixed, and the best side chain conformations (allowing for amino acid subsitutions) are chosen from a discrete set of common, low-energy rotamers. Individual rotamers are selected so as to minimize the total protein energy, calculated with a molecular mechanics energy function. *(Immunogenicity)* To assess immunogenicity, we leverage the well-established development of T-cell epitope predictors that encapsulate the underlying specific recognition of an epitope by an MHC II protein [5]. MHC II proteins from the predominant HLA-DR isotype have a recognition groove whose pockets form energetically favorable interactions with specific side-chains of peptides approximately 9 residues in length (Fig. 1, left). Numerous computational methods are available for identifying peptide epitopes, and studies have shown these methods to be predictive of immunogenicity [33,4]. Here we assess each constituent peptide of our protein and optimize the total.

EpiSweep is the first protein design tool that simultaneously optimizes primary sequence, reducing immunogenicity, and tertiary structure, maintaining stability and function. It significantly extends structure-based protein design by accounting for the complementary goal of immunogenicity. It likewise significantly extends our previous work on Pareto optimization for protein engineering in general [35,11] and for deimmunization in particular, which assessed effects on structure and function only according to a sequence potential [24,25,26]. Inspired by an approach for optimization of stability and specificity of interacting proteins [10], we employ a sweep algorithm that minimizes the energy of the design target at decreasing predicted epitope scores. The sweep reveals an energy-epitope landscape of Pareto-optimal plans (Fig. 1, right) and can also produce near-optimal plans. Although beyond the scope of this paper, EpiSweep promises to inform protein engineering experiments (as our sequence-based algorithms have done [23]) seeking sets of effective deimmunizing mutations for the development of enhanced biotherapeutics.

2 Methods

We seek to make mutations to a target protein so as to reduce its immunogenicity, as evaluated by a sequence-based epitope score, while maintaining its stability and activity, as evaluated by a structure-based effective energy function. We now formalize this as a Pareto optimization problem that extends the standard side-chain packing formulation of structure-based protein design [2] with the complementary / competing epitope score. In general, structure-based protein design problems have been shown to be NP-hard [27].

Problem 1 (Structurally-Guided Deimmunization). We are given a protein sequence A of n amino acids, along with a 3D structure that includes a backbone as well as a rotamer sequence R paralleling A. We are also given a set M of mutable positions (amino acid types allowed to vary) and a set F of flexible positions (side-chains allowed to vary) with $M \subset F$; by default $M = F = \{1, \ldots, n\}$. Finally we are given a mutational load m. Let a *design* be an m-mutation sequence A', with mutations only at positions in M, along with a set R' of selected rotamers, differing from R only at positions in F. Our goal is to determine all Pareto optimal designs, minimizing the two objectives

$$f_\epsilon(A') = \sum_{i=1}^{n-8} \epsilon(A'[i \ldots i+8]) \tag{1}$$

$$f_\phi(R') = \sum_{i \in F} \phi_i(R'[i]) + \sum_{i \in F} \sum_{j \in F, j \neq i} \phi_{i,j}(R'[i], R'[j]) \tag{2}$$

based on the following contributions:

- $\epsilon : \mathcal{A}^9 \to \mathbb{N}$ gives the *epitope score* for a peptide (we assume a 9-mer; see below).
- $\phi_i : \mathcal{R} \to \mathbb{R}$ gives the *singleton energy* capturing the internal energy of a rotamer at position i plus the energy between the rotamer and the backbone structure and side-chains at non-flexible positions
- $\phi_{i,j} : \mathcal{R} \times \mathcal{R} \to \mathbb{R}$ gives the *pairwise energy* between a pair of rotamers at a pair of positions i, j

We use $\mathcal{A}=\{A,C,\ldots,Y\}$ for the set of amino acids and $\mathcal{R}$ for the set of rotamers (from a rotamer library and the input structure). We subscript sequences with brackets, and use the notation $X[i..j]$ to indicate the substring of X from position i to j, inclusive. We will use function $a : \mathcal{R} \to \mathcal{A}$ to obtain the amino acid type of a given rotamer. While our implementation is modular and can support a variety of epitope scores and energy functions, we now summarize those we use for the results.

Epitope score. Our epitope score is the number of the 8 most common HLA-DR alleles (representing a majority of human populations world-wide [30]) predicted by ProPred [29] to recognize a given 9-mer peptide at a 5% threshold. ProPred is an extension of the TEPITOPE "pocket profile" method [31], learning position-specific amino acid scores from experimentally measured binding affinities for individual amino acid types at individual pockets of the MHC II binding groove (Fig. 1, left). Given a 9-mer, the position-specific weights for its amino acids are summed, and the total is compared against a threshold to predict whether or not the given peptide is in a given percentile of the best-recognized peptides. The amino acid in the first pocket is termed the P1 anchor for its significant contribution to binding. ProPred has been successfully employed in a number of experimental studies (e.g., [7,18,22]) and was one of the best predictors in a recent independent assessment [33], achieving an average 0.73 area under the curve in epitope prediction. In our earlier work, we found its predictions to have a strikingly good correspondence with a previously published ELISPOT assay [25], and used it, along with conservation analysis, to optimize a set of mutations for a β lactamase [23].

Molecular mechanics energy function. We assess singleton and pairwise energies according to the AMBER force field as implemented in the Osprey software package for protein redesign [1]. In particular, we use the van der Waals and electrotatics forces as well as an implicit solvation factor and residue reference energies as intended for standard protein redesign (as opposed to active site redesign for change of substrate specificity). We discard rotamers, and thus all conformations containing them, if they have significant overlap of van der Waals radii or otherwise exceptionally high intra- or inter-rotamer energies.

2.1 Sweep Algorithm

We now develop an algorithm to identify *all* Pareto optimal designs. We rely on the fact that epitope scores are discrete at a pre-specified significance level— either the peptide is deemed capable of recognition by an MHC II allele or it isn't. Thus we can "sweep" the epitope score from the wild-type score to successively smaller values, for each epitope score identifying the design with the best energy (Fig. 1, right). However, achieving exactly a specified epitope score might require using a set of substitutions with worse overall energy than that for a set of substitutions for a somewhat smaller epitope score. Thus at each step we find the design with best energy such that the epitope score is at most the current sweep value. A similar sweep approach was employed in developing bZIP partners optimized for stability and specificity, optimizing primarily for stability with a required specificity that is incremented by a fixed amount at each step [10]. Since our epitope scores are discrete, we can safely step by a value of 1 and provide a guarantee to find all and only the Pareto optimal designs.

More formally, we initialize $E = f_\epsilon(A)$, the epitope score of the wild-type protein. Then we repeat the following steps. Let (A', R') be the design minimizing the energy $f_\phi(R')$ using the optimization approach below, with the epitope score $f_\epsilon(A')$ constrained to be $\leq E - 1$. Output (A', R'), update E to $f_\epsilon(A')$, and repeat until there is no solution meeting the smaller epitope constraint. Clearly each identified design is Pareto optimal, and all Pareto optimal designs are identified. In addition, the approach is efficient in an output-sensitive fashion, only requiring D invocations of the optimizer to identify D Pareto optimal designs.

At each step in the sweep, we employ an integer programming (IP) approach to find the global minimum energy conformation with epitope score at most E. Like previous work on side-chain packing, we ensure a consistent rotamer selection that minimizes the singleton and pairwise energy scores [17]. However, we also add a network of constraints for the epitope sweep. Define singleton binary variable $s_{i,r}$ to indicate whether or not the rotamer at position i is rotamer r. Define pairwise binary variable $p_{i,j,r,t}$, derived from $s_{i,r}$ and $s_{j,t}$ to indicate whether or not the rotamers at positions i and j are rotamers r and t, respectively. Finally, define window binary variable $w_{i,X}$, derived from $s_{i,r}$ through $s_{i+8,t}$, to indicate whether or not the amino acids for the rotamers in the window from position i to position $i + 8$ corresponds to peptide $X \subset \mathcal{A}^9$.

We rewrite f_ϕ in terms of these binary variables, and use it as the objective function for the IP:

$$\Phi = \sum_{i,r} s_{i,r} \cdot \phi_i(r) + \sum_{i,j,r,t} p_{i,j,r,t} \cdot \phi_{i,j}(r,t) \tag{3}$$

We then constrain the epitope score according to the current sweep value:

$$\sum_{i,X} w_{i,X} \cdot f_\epsilon(X) \leq E - 1 \tag{4}$$

To guarantee that the variable assignments yield a valid set of rotamers, we impose the following constraints:

$$\forall i : \sum_r s_{i,r} = 1 \tag{5}$$

$$\forall i, r, j > i : \sum_b p_{i,j,r,t} = s_{i,r} \tag{6}$$

$$\forall j, t, i < j : \sum_r p_{i,j,r,t} = s_{j,t} \tag{7}$$

$$\forall i, r \, \forall h \in 1..9 : \sum_{X:X[h]=a(r)} w_{i,X} = s_{i+h-1,r} \tag{8}$$

Eq. 5 ensures that only one rotamer is assigned to a given position. Eq. 6 and Eq. 7 maintain consistency between singleton and pairwise variables, while Eq. 8 maintains consistency between singleton and window variables.

Finally, we enforce the desired mutational load:

$$\sum_{i,r:A[i]\neq a(r)} s_{i,r} = m \tag{9}$$

We implement this IP using the Java API to the IBM ILOG optimization suite. It provides solutions that are guaranteed to be optimal, at the price of having no guarantee on the time required. However, we find that in practice (see results), the globally optimal solutions can in fact be found in at most 8 hours for the size of problems we and others are considering. We can identify increasingly suboptimal solutions by, at each sweep step, repeatedly adding constraints to eliminate identified solutions and solving these extended programs.

Extension: peptide-focused design. Since the "unit" of immunogenicity here (due to MHC binding) is a peptide, initial experimental studies are often peptide-centered. That is, individual immunodominant peptides will be identified by one round of experiments and targeted for epitope deletion in subsequent experiments [32,34,15]. The set M of mutable positions allows us to focus our design effort on such a peptide. However, since only an individual peptide will be assessed for immunogenicity, we modify our epitope scoring and constraints accordingly: we sweep for the epitope score within the peptide, while avoiding the introduction of additional epitopes in the "flanking regions" up- and downstream from that peptide (i.e., 9-mers spanning part of the peptide and part of the protein N- or C-terminal to it).

Let the peptide of interest span from residue p_l to residue $p_h \geq p_l + 9$ (so $M \subset \{p_l, \ldots, p_h\}$). Then there are three sets of starting positions for the 9-mers to be assessed: N-terminal flanking positions $P_l = \{\max\{1, p_l - 8\}, \ldots, \max\{1, p_l - 1\}\}$, core positions $P_c = \{p_l, \ldots, p_h - 8\}$, and C-terminal flanking positions $P_h = \{p_h - 7, \ldots, p_h\}$. We replace Eq. 4 with a score for just the 9-mers in the peptide of interest (i.e., starting at the core positions):

$$\sum_{i \in P_c, X} w_{i,X} \cdot f_\epsilon(X) \leq E \tag{10}$$

and initialize E with a wild-type total that is likewise restricted. We also add flanking region constraints to prevent introducing new epitopes:

$$\forall i \in P_l \cup P_h \; \forall X \; : \; w_{i,X} \cdot f_\epsilon(X) \leq f_\epsilon(A[i..i+8]) \tag{11}$$

2.2 Preprocessing Filters

In practice, it is often helpful to prune the search space in order to focus the combinatorial space of designs, employing additional background knowledge to limit the designs to be considered, as well as eliminating regions of the space guaranteed to be sub-optimal. Our implementation incorporates two filters.

Homology filters. Natural variation provides insights into mutations likely to yield stable, active structures. We can restrict the allowed mutations to be considered at each position, based either on a general substitution matrix such as BLOSUM, or on a position-specific assessment of conservation within a multiple sequence alignment (MSA) of homologs to the target. For BLOSUM, we eliminate an amino acid (and thereby all its rotamers) at a position if its BLOSUM score vs. the wild-type amino acid exceeds a specified threshold. For conservation, we likewise eliminate an amino acid and its rotamers at a position if its frequency in the MSA is below that in a specified background distribution [21].

Dead end elimination. The dead-end elimination (DEE) criterion provides a pruning technique to eliminate rotamers that are provably not part of the global minimum energy conformation (GMEC) [6]. DEE has been generalized and extended in a number of powerful ways (e.g., [20,8,1]); we find the simple Goldstein DEE criterion to be sufficient for our current purposes [9]. However, DEE is correct only for pruning with respect to the overall GMEC; it does not account for our epitope score constraint, which might require making more drastic mutations and thus employ rotamers that would be pruned if relying exclusively on the GMEC criterion. To ensure that we do not eliminate rotamers contributing to Pareto optimal designs, we employ the common technique of performing DEE with respect to a "buffer" energy δ, such that rotamers are pruned only if they provably do not participate in any conformation whose energy is within δ of the GMEC. We do this DEE pruning before initiating the epitope sweep. If during the sweep we find that the solution for a particular epitope value has energy exceeding the GMEC + δ, we must decide either to terminate the sweep (we do not care to produce designs with energy so high) or to re-perform DEE with an increased δ and then re-optimize for that epitope value and continue the sweep.

2.3   Postprocessing Energy Minimization

The molecular mechanics energy function employed during the sweep assesses energies with respect to the fixed backbone and rigid rotamers. Therefore it is standard practice in structure-based protein design to energy-minimize each resulting design, performing a limited relaxation of the conformation in order to improve atomic interactions. Since a suboptimal design may relax to a conformation with a better energy than an optimal one, suboptimal alternatives should be generated and evaluated. In order to ensure that the preprocessing does not *a priori* eliminate rotamers that participate only in sub-optimal rigid designs but could participate in optimal energy-minimized designs, the minDEE variant of DEE should be employed [8].

3   Results

We demonstrate EpiSweep with case study applications to two therapeutic proteins, SakSTAR and Epo, previously targeted for deimmunization by experimental methods. Since the previous experimental work focused on identified immunogenic regions, we first perform analogous peptide-focused design, using the extension discussed in the methods. We then demonstrate that EpiSweep can optimize an entire protein, selecting optimal sets of mutations to hit scattered epitopes. For the results, we focus on the overall trends in the energy-epitope landscape, using EpiSweep to explore the trade-offs between maintaining stability and reducing immunogenicity. Thus we do not list sub-optimal designs and do not perform energy minimization, steps that would be desirable before deciding upon a design for experimentation, but do not provide significant additional insights in these retrospectives.

The initial energy calculations for each peptide case study required 3-6 hours of wall clock time on 60 nodes of a shared cluster, while the full-protein design required several days. Then the actual EpiSweep algorithm took less than an hour on a dedicated 8-core machine for each peptide design problem (target and mutational load), and less than 8 hours for the full-protein design.

3.1   Staphylokinase (SakSTAR) Peptides

Staphylokinase is a fibrin-selective thrombolytic agent with potential therapeutic use in treating heart attacks and strokes. Warmerdam *et al.* [34] sought to deimmunize a variant called SakSTAR, derived from a lysogenic *S. aureus* strain. Based on T-cell profileration assays, they focused on the highly immunogenic C3 region, spanning residues 71–87. They employed alanine scanning mutagensis to identify mutations that reduced immunogenicity. Subsequent proliferation assays then showed that the response was indeed reduced for various designs incorporating 2–4 alanine substitutions. They did not engineer the redesigned peptides back into the whole protein to test stability and activity.

We applied peptide-focused EpiSweep to identify mutations in the C3 region as well as in an additional region, which we name Beta, spanning residues 24–38. We evaluated energies according to the deposited SakSTAR structure (pdb id 2sak) and applied homology filtering according to the family (Pfam PF02821).

The C-terminal end of SakSTAR folds over perpendicular to the C3 peptide, approximately bisecting it and leaving some residues buried, some exposed, and some half-exposed. Warmerdam *et al.* chose substitutions at the positions of C3 underlined in the table in Fig. 2, bottom. They chose not to mutate the large hydrophobics because of potential structural implications, but these same residues tend to anchor MHC II binding. In particular, F76 is a P1 anchor (see Methods, Epitope score) for 4 epitopes. Our energy evaluation predicts that an F76K substitution in fact improves the energy, possibly because it is solvent exposed. Thus our plans (Fig. 2, bottom) uniformly choose it and obtain a better energy than the wild-type. Over the sweep at a fixed mutational load, we see trends from more to less energetically favorable substitutions with decreasing epitope score (left-right in the figure and bottom-up in the table). We are able to delete all the epitopes with either 2 or 3 substitutions.

We chose also to study the Beta peptide because it was predicted to be highly immunogenic in our initial T-cell epitope prediction analysis, and it is also structurally quite different from C3. Beta sits in an anti-parallel beta strand with a pattern of surface-exposed and buried side-chains. The predicted epitope and energy range (Fig. 2, top) is larger than that for C3, and the number of undominated solutions greater except for the 2-mutation curve. In the plans we see that Y24H is commonly taken, as is T30K at the higher mutational loads. We again see trade-offs between stability-preserving and epitope-deleting selections, and note that some of the mutations are predicted to be stability-enhancing (though again, this is before energy minimization).

3.2 Erythropoietin (Epo) Peptides

Erythropoietin (Epo) has therapeutic use in treating anemia but unfortunately induces an immune response in some patients [13]. Based on an intensive analysis of peptides spanning the entire protein, Tangri *et al.* [32] identifed two highly immunogenic regions, spanning 101–115 and 136–150. They engineered four variants targeting the anchor residues of identified T-cell epitopes in these regions: L102P/S146D (named G2), T107D/S146D (G3), L102G/T107D/S146D (G4), and L102S/T107D/S146D (G5). Variants G3 and G4 reduced response in an ELISPOT assay. However, variants G2 and G5 were not bioactive, possibly because of destabilizing substitutions at L102. We target the same two

SakSTAR 24–38 (Beta)

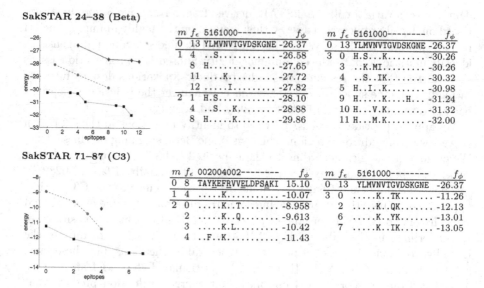

m	f_ϵ	5161000--------	f_ϕ
0	13	YLMVNVTGVDSKGNE	-26.37
1	4	..S............	-26.58
	8	H..............	-27.65
	11	....K.........	-27.72
	12	.....I.........	-27.82
2	1	H.S............	-28.10
	4	..S...K........	-28.88
	8	H.....K........	-29.86

m	f_ϵ	5161000--------	f_ϕ
0	13	YLMVNVTGVDSKGNE	-26.37
3	0	H.S...K........	-30.26
	3	..K.MI.........	-30.26
	4	..S..IK........	-30.32
	5	H..I..K........	-30.98
	9	H.....K...H...	-31.24
	10	H...V.K........	-31.32
	11	H...M.K........	-32.00

SakSTAR 71–87 (C3)

m	f_ϵ	002004002--------	f_ϕ
0	8	TAYKEFRVVELDPSAKI	15.10
1	4	.....K............	-10.07
2	0	.....K..T.........	-8.958
	2	.....K..Q.........	-9.613
	3	.....K.L.........	-10.42
	4	..F..K............	-11.43

m	f_ϵ	5161000--------	f_ϕ
0	13	YLMVNVTGVDSKGNE	-26.37
3	0	.....K..TK........	-11.26
	2	.....K..QK........	-12.13
	6	.....K..YK........	-13.01
	7	.....K..IK........	-13.05

Fig. 2. Pareto-optimal plans in the energy-epitope landscapes for two SakSTAR peptides. In addition to the wild-type (magenta star for Beta, not shown for C3), the plots show three different mutational loads: 1, solid blue and diamond; 2, dash red and circle; 3, dot black and square. The tables detail the mutations from the wild-type. The numbers above the wild-type sequence are the epitope scores for 9-mers starting at those positions.

solvent-exposed peptides, and develop EpiSweep designs based on the structure with pdb id 1buy and Pfam entry PF00758.

In peptide 101–115, EpiSweep avoids the L102 position found to be a problem by Tangri et al. L102 is completely buried and tightly packed in the model structure. Instead, EpiSweep makes an A, G, or L mutation at R103 in every plan, except the energy-optimal plans with 2 of 3 mutations (Fig. 3). While these might not seem conservative substitutions from a purely sequence-based perspective, we find that the energies are favorable. Both EpiSweep and Tangri et al. make substitutions at T107. Small residues are common substitutions here, as they tend not to make good interactions with MHC II pockets, but yet are predicted to be energetically benign. The number of glycines suggests that additional modeling might be appropriate to more fully assess the impact on overall flexibility. We see that the 3-mutation plan has a much nicer deimmunizing trend than the 2-mutation plan, deleting all epitopes without having to take an energetically less favorable choice.

In peptide 136–150, Tangri et al. make a substitution at S146, while we concentrate instead at or around the P1 anchor at F138. As the trends show, mutations are made around, but not at, 138, until the epitope sweep reaches a point where this substitution is required to eliminate as many epitopes as possible. Mutation at 138 is not the most energy favorable, and so we see many plans with nearby R139S and K140E instead. Like the other peptides, as the mutation

Epo 101–115

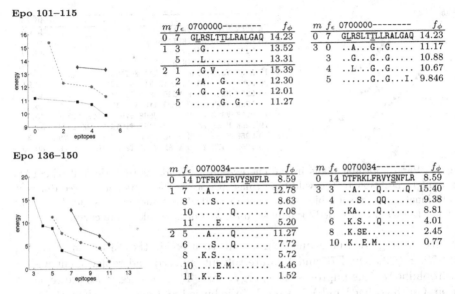

m	f_ϵ	0700000--------	f_ϕ
0	7	GLRSLTTLLRALGAQ	14.23
1	3	..G...........	13.52
	5	..L...........	13.31
2	1	..G.V.........	15.39
	2	..A...G.......	12.30
	4	..G...G.......	12.01
	5	......G..G.....	11.27

m	f_ϵ	0700000--------	f_ϕ
0	7	GLRSLTTLLRALGAQ	14.23
3	0	..A...G..G.....	11.17
	3	..G...G..G.....	10.88
	4	..L...G..G.....	10.67
	5	......G..G...I.	9.846

Epo 136–150

m	f_ϵ	0070034--------	f_ϕ
0	14	DTFRKLFRVYSNFLR	8.59
1	7	..A...........	12.78
	8	...S..........	8.63
	10	.......Q......	7.08
	11	....E.........	5.20
2	5	..A....Q......	11.27
	6	...S...Q......	7.72
	8	.K.S..........	5.72
	10	....E.M.......	4.46
	11	.K..E.........	1.52

m	f_ϵ	0070034--------	f_ϕ
0	14	DTFRKLFRVYSNFLR	8.59
3	3	..A....Q.....Q.	15.40
	4	...S...QQ......	9.38
	5	.KA....Q.......	8.81
	6	.K.S...Q.......	4.01
	8	.K.SE..........	2.45
	10	.K..E.M........	0.77

Fig. 3. Pareto-optimal plans in the energy-epitope landscapes for two Epo peptides. See Fig. 2 for description.

load increases, the sweep has more options to choose substitutions that minimize energy and epitope scores simultaneously.

3.3 SakSTAR Full-Protein Design

Finally, we subjected the entire SakSTAR protein to EpiSweep. Warmerdam *et al.* verified T-cell epitopes exist outside C3 and that the SakSTAR protein requires extensive deimmunization work in other regions. Fig. 4 illustrates some Pareto-optimal designs at mutational loads of 4, 6, and 8. We see the same kinds of trade-offs as in the peptide-focused designs, moving from stability-preserving (or -enhancing) mutations that nominally reduce immunogenicity, to epitope-deleting mutations that might destabilize the protein. However, there is more freedom in the full-protein design so we see more effective epitope-deleting mutations throughout each curve. For example V112K and I128E replace P1 anchors for relatively good energy trade-off.

Previously, we had redesigned SakSTAR based on a 1- and 2-body sequence potential, sampling linear combinations of this sequence potential and the epitope score [25]. It is interesting to compare and contrast the sequence-guided plans with these structure-guided (and Pareto optimal) plans. For example, both tend to choose V112K and I128E, along with some substitution at M26 (M26D for sequence and M26K for structure). These choices are natural, as the wild-type residues are P1 anchors for MHC II binding, and mutation to a charged residue precludes effective binding. However, the sequence-guided approach completely avoided the C3 region; we conjectured that this was to a high degree of covariation among the amino acids in the sequence record. In contrast, many of

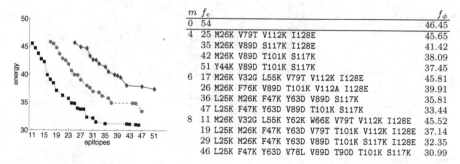

m	f_ϵ							f_ϕ
0	54							46.45
4	25	M26K	V79T	V112K	I128E			45.65
	35	M26K	V89D	S117K	I128E			41.42
	42	M26K	V89D	T101K	S117K			38.09
	51	Y44K	V89D	T101K	S117K			37.45
6	17	M26K	V32G	L55W	V79T	V112K	I128E	45.81
	26	M26K	F76K	V89D	T101K	V112A	I128E	39.91
	36	L25K	M26K	F47K	Y63D	V89D	S117K	35.81
	47	L25K	F47K	Y63D	V89D	T101K	S117K	33.44
8	11	M26K	V32G	L55W	Y62K W66E V79T	V112K	I128E	45.52
	19	L25K	M26K	F47K	Y63D V79T T101K	V112K	I128E	37.14
	29	L25K	M26K	F47K	Y63D V89D T101K	S117K	I128E	32.35
	46	L25K	F47K	Y63D	V78L V89D T90D	T101K	S117K	30.99

Fig. 4. Pareto-optimal plans in the energy-epitope landscape for the full SakSTAR protein. In addition to the wild-type (magenta star), the plots show three different mutational loads: 4, solid blue and diamond; 6, dash red and circle; 8, dot black and square. Details are provided for only sampled sets of plans along the curves.

the structure-guided plans incorporate substitutions in the C3 region at 76, 78, and 79. Specifically, F76 and V79 are solvent exposed and replacement with a hydrophillic residue improves the energy score. These two residues are replaced often. On the other hand V78 side chain is buried and few plans show the more conservative V78L mutation. By directly modeling the energetic interactions, rather than relying just on evolutionary history, we may discover new designs that are favorable for our design goals rather than natural pressures.

4   Conclusion

In order to simultaneously optimize stability/activity and immunogenicity of therapeutic proteins, we have developed a novel Pareto optimization approach that integrates methods from structure-based protein design with immunoinformatics predictors. Our EpiSweep algorithm elucidates the energy-epitope landscape of a protein, identifying all Pareto-optimal plans, along with near-optimal plans as desired. While the underlying design problem is NP-hard, our methods are efficient in practice, requiring only hours to characterize an entire Pareto frontier even for a redesign problem considering an entire protein. We recognize that this speed is due to our reliance on computational models of both stability and immunogenicity that, while extensively validated in numerous retrospective and prospective studies, are imperfect and may yield designs that are unstable or immunogenic in practice. Due to our use of provably correct algorithms, however, this outcome reflects only on the models, and offers an opportunity to improve them. Furthermore, the ability of EpiSweep to characterize the beneficial region of the energy-epitope landscape enables engineers to better identify high-confidence designs worthy of experimental evaluation.

Acknowledgment. This work is supported in part by NSF grant IIS-1017231 to CBK and NIH grant R01-GM-098977 to CBK and KEG.

References

1. Chen, C.Y., Georgiev, I., Anderson, A.C., Donald, B.R.: Computational structure-based redesign of enzyme activity. PNAS 106, 3764–3769 (2009)
2. Dahiyat, B., Mayo, S.: De novo protein design: fully automated sequence selection. Science 278, 82–87 (1997)
3. De Groot, A.S., Knopp, P.M., Martin, W.: De-immunization of therapeutic proteins by T-cell epitope modification. Dev. Biol (Basel) 122, 171–194 (2005)
4. De Groot, A.S., Martin, W.: Reducing risk, improving outcomes: Bioengineering less immunogenic protein therapeutics. Clinical Immunology 131, 189–201 (2009)
5. De Groot, A.S., Moise, L.: Prediction of immunogenicity for therapeutic proteins: State of the art. Curr. Opin. Drug Discov. Devel. 10, 332–340 (2007)
6. Desmet, J., De Maeyer, M., Hazes, B., Lasters, I.: The dead-end elimination theorem and its use in protein side-chain positioning. Nature 356, 539–542 (1992)
7. Dinglasan, R.R., Kalume, D.E., Kanzok, S.M., Ghosh, A.K., Muratova, O., Pandey, A., Jacobs-Lorena, M.: Disruption of Plasmodium falciparum development by antibodies against a conserved mosquito midgut antigen. PNAS 104, 13461–13466 (2007)
8. Georgiev, I., Lilien, R.H., Donald, B.R.: A Novel Minimized Dead-End Elimination Criterion and Its Application to Protein Redesign in a Hybrid Scoring and Search Algorithm for Computing Partition Functions over Molecular Ensembles. In: Apostolico, A., Guerra, C., Istrail, S., Pevzner, P.A., Waterman, M. (eds.) RECOMB 2006. LNCS (LNBI), vol. 3909, pp. 530–545. Springer, Heidelberg (2006)
9. Goldstein, R.F.: Efficient rotamer elimination applied to protein side-chains and related spin glasses. Biophysical J. 66, 1335–1340 (1994)
10. Grigoryan, G., Reinke, A., Keating, A.: Design of protein-interaction specificity gives selective bZIP-binding peptides. Nature 458, 859–864 (2009)
11. He, L., Friedman, A.M., Bailey-Kellogg, C.: A divide and conquer approach to determine the pareto frontier for optimization of protein engineering experiments proteins. Proteins (2011)
12. Hwang, W.Y.K., Foote, J.: Immunogenicity of engineered antibodies. Methods 36, 3–10 (2005)
13. Indiveri, F., Murdaca, G.: Immunogenicity of erythropoietin and other growth factors. Rev. Clin. Exp. Hematol. 1, 7–11 (2002)
14. Jones, P.T., Dear, P.H., Foote, J., Neuberger, M.S., Winter, G.: Replacing the complementarity-determining regions in a human antibody with those from a mouse. Nature 321, 522–525 (1986)
15. Jones, T.D., Phillips, W.J., Smith, B.J., Bamford, C.A., Nayee, P.D., Baglin, T.P., Gaston, J.S.H., Baker, M.P.: Identification and removal of a promiscuous CD4+ T cell epitope from the C1 domain of factor VIII. J. Thromb. Haemost. 3, 991–1000 (2005)
16. Kessler, M., Goldsmith, D., Schellekens, H.: Immunogenicity of biopharmaceuticals. Nephrology, Dialysis, Transplantation 21, v9–v12 (2006)
17. Kingsford, C., Chazelle, B., Singh, M.: Solving and analyzing side-chain positioning problems using linear and integer programming. Bioinf. 21, 1028–1036 (2005)
18. Klyushnenkova, E.N., Kouiavskaia, D.V., Kodak, J.A., Vandenbark, A.A., Alexander, R.B.: Identification of HLA-DRB1*1501-restricted T-cell epitopes from human prostatic acid phosphatase. Prostate 67, 1019–1028 (2007)
19. Leader, B., Baca, Q.J., Golan, D.E.: Protein therapeutics: a summary and pharmacological classification. Nat. Rev. Drug Disc. 7, 21–39 (2008)

20. Lilien, R., Stevens, B., Anderson, A., Donald, B.: A novel ensemble-based scoring and search algorithm for protein redesign, and its application to modify the substrate specificity of the gramicidin synthetase A phenylalanine adenlytaion enzyme. In: Proc. RECOMB, pp. 46–57 (2004)
21. McCaldon, P., Argos, P.: Oligopeptide biases in protein sequences and their use in predicting protein coding regions in nucleotide sequences. Proteins: Structure, Function and Genetics 4, 99–122 (1988)
22. Mustafa, A.S., Shaban, F.A.: Propred analysis and experimental evaluation of promiscuous T-cell epitopes of three major secreted antigens of Mycobacterium tuberculosis. Tuberculosis 86, 115–124 (2006)
23. Osipovitch, D.C., Parker, A.S., Makokha, C.D., Desrosiers, J., DeGroot, A.S., Baiely-Kellogg, C., Griswold, K.E.: Computational design of immune-evading enzymes for ADEPT therapy (in preperation)
24. Parker, A.S., Griswold, K.E., Bailey-Kellogg, C.: Optimization of combinatorial mutagenesis. J. Comput. Biol. 18, 1743–1756 (2011); In: Bafna, V., Sahinalp, S.C. (eds.) RECOMB 2011. LNCS, vol. 6577, pp. 321–335. Springer, Heidelberg (2011)
25. Parker, A.S., Griswold, K.E., Bailey-Kellogg, C.: Optimization of therapeutic proteins to delete t-cell epitopes while maintaining beneficial residue interactions. J. Bioinf. Comput. Biol. 9, 207–229 (2011); conf. ver: Proc CSB, pp.100–113 (2010)
26. Parker, A.S., Zheng, W., Griswold, K.E., Bailey-Kellogg, C.: Optimization algorithms for functional deimmunization of therapeutic proteins. BMC Bioinf. 11, 180 (2010)
27. Pierce, N., Winfree, E.: Protein design is NP-hard. Protein Eng. 15, 779–782 (2002)
28. Schellekens, H.: Bioequivalence and the immunogenicity of biopharmaceuticals. Nature Reviews Drug Discovery 1, 457–462 (2002)
29. Singh, H., Raghava, G.: ProPred: prediction of HLA-DR binding sites. Bioinformatics 17, 1236–1237 (2001)
30. Southwood, S., Sidney, J., Kondo, A., del Guercio, M.F., Appella, E., Hoffman, S., Kubo, R.T., Chesnut, R.W., Grey, H.M., Sette, A.: Several common HLA-DR types share largely overlapping peptide binding repertoires. J. Immunol. 160, 3363–3373 (1998)
31. Sturniolo, T., Bono, E., Ding, J., Raddrizzani, L., Tuereci, O., Sahin, U., Braxenthaler, M., Gallazzi, F., Protti, M.P., Sinigaglia, F., Hammer, J.: Generation of tissue-specific and promiscuous HLA ligand database using DNA microarrays and virtual HLA class II matrices. Nature Biotechnol. 17, 555–561 (1999)
32. Tangri, S., Mothe, B.R., Eisenbraun, J., Sidney, J., Southwood, S., Briggs, K., Zinckgraf, J., Bilsel, P., Newman, M., Chesnut, R., LiCalsi, C., Sette, A.: Rationally engineered therapeutic proteins with reduced immunogenicity. J. Immunol. 174, 3187–3196 (2005)
33. Wang, P., Sidney, J., Dow, C., Mothe, B., Sette, A., Peters, B.: A systematic assessment of MHC class II peptide binding predictions and evaluation of a consensus approach. PLoS Comp. Biol. 4, e1000048 (2008)
34. Warmerdam, P.A.M., Plaisance, S., Vanderlick, K., Vandervoort, P., Brepoels, K., Collen, D., Maeyer, M.D.: Elimination of a human T-cell region in staphylokinase by T-cell screening and computer modeling. J. Thromb. Haemost. 87, 666–673 (2002)
35. Zheng, W., Friedman, A.M., Bailey-Kellogg, C.: Algorithms for joint optimization of stability and diversity in planning combinatorial libraries of chimeric proteins. J. Comput. Biol. 16, 1151–1168 (2009); In: Vingron, M., Wong, L. (eds.) RECOMB 2008. LNCS (LNBI), vol. 4955, pp. 300–314. Springer, Heidelberg (2008)

Evidence for Widespread Association of Mammalian Splicing and Conserved Long-Range RNA Structures

Dmitri Pervouchine[1], Ekaterina Khrameeva[1], Marina Pichugina[2], Olexii Nikolaienko[3], Mikhail Gelfand[4], Petr Rubtsov[5], and Andrei Mironov[1]

[1] Faculty of Bioengineering and Bioinformatics, Moscow State University
[2] Institute of Physics and Technology, Moscow
[3] Institute of Molecular Biology and Genetics NAS of Ukraine, Kyiv
[4] Institute for Information Transmission Problems RAS, Moscow
[5] Engelhardt Institute of Molecular Biology, Russian Academy of Sciences, Moscow

Pre-mRNA structure impacts many cellular processes, including splicing in genes associated with disease. The contemporary paradigm of RNA structure prediction is biased toward secondary structures that occur within short ranges of pre-mRNA, although long-range base-pairings are known to be at least as important. Recently, we developed an efficient method for detecting conserved RNA structures on the genome-wide scale, one that does not require multiple sequence alignments and works equally well for the detection of local and long-range base-pairings. Using an enhanced method that detects base-pairings at all possible combinations of splice sites within each gene, we report a list of RNA structures that could be involved in the regulation of splicing in mammals. We demonstrate statistically that there is a strong association between the occurrence of conserved RNA structures and alternative splicing, where local RNA structures are generally more frequent at alternative donor splice sites, while long-range structures are more associated with weak alternative acceptor splice sites. A fraction of the reported structures is associated with unannotated splicing events that are confirmed by RNA-seq data. As an example, we validated the RNA structure in the human SF1 gene using mini-genes in the HEK293 cell line. Point mutations that disrupted the base-pairing of two complementary boxes between exons 9 and 10 of this gene altered the splicing pattern, while the compensatory mutations that reestablished the base-pairing reverted splicing to that of the wild-type. There is statistical evidence for a Dscam-like class of mammalian genes, in which mutually exclusive RNA structures control mutually exclusive alternative splicing. In sum, we propose that long-range base-pairings carry an important, yet unconsidered part of the splicing code, and that, even by modest estimates, there must be thousands of such potentially regulatory structures conserved throughout the evolutionary history of mammals.

rnajournal.cshlp.org/content/18/1/1.full

B. Chor (Ed.): RECOMB 2012, LNBI 7262, p. 199, 2012.
© Springer-Verlag Berlin Heidelberg 2012

Pathset Graphs: A Novel Approach for Comprehensive Utilization of Paired Reads in Genome Assembly*

Son K. Pham[1,**], Dmitry Antipov[2,**], Alexander Sirotkin[2], Glenn Tesler[3],
Pavel A. Pevzner[1,2], and Max A. Alekseyev[2,4,***]

[1] Dept. of Computer Science and Engineering, University of California, San Diego,
La Jolla, CA, USA
[2] Algorithmic Biology Laboratory, St. Petersburg Academic University,
St. Petersburg, Russia
[3] Dept. of Mathematics, University of California, San Diego, La Jolla, CA, USA
[4] Dept. of Computer Science and Engineering, University of South Carolina,
Columbia, SC, USA

Abstract. One of the key advances in genome assembly that has led
to a significant improvement in contig lengths has been utilization of
paired reads (*mate-pairs*). While in most assemblers, mate-pair information is used in a post-processing step, the recently proposed Paired de
Bruijn Graph (PDBG) approach incorporates the mate-pair information
directly in the assembly graph structure. However, the PDBG approach
faces difficulties when the variation in the insert sizes is high. To address
this problem, we first transform mate-pairs into *edge-pair histograms* that
allow one to better estimate the distance between edges in the assembly
graph that represent regions linked by multiple mate-pairs. Further, we
combine the ideas of mate-pair transformation and PDBGs to construct
new data structures for genome assembly: *pathsets* and *pathset graphs*.

1 Introduction

Current High Throughput Sequencing (HTS) technologies have reduced the time
and costs of genome sequencing and have enabled new experimental opportunities in a variety of applications. However, as sequencing technologies improve,
the challenges in designing software to assemble genomes get harder. Current
genome assemblers face the challenge of assembling billions of short reads. When
the length of a repeat is longer than the read length, correctly matching up its
upstream and downstream flanking regions is difficult. Fortunately, all current
sequencing platforms are able to produce mate-pairs — pairs of reads whose separation in the genome (called the insert size) is approximately known. Because

---

* This work was supported by grants from the National Institutes of Health, USA
(NIH grant 3P41RR024851-02S1) and the Government of the Russian Federation
(grant 11.G34.31.0018).
** These authors contributed equally to this work.
*** Corresponding author.

B. Chor (Ed.): RECOMB 2012, LNBI 7262, pp. 200–212, 2012.
© Springer-Verlag Berlin Heidelberg 2012

insert sizes may be much longer than the read length, mate-pairs may span over long repeats and match up their flanking regions.

Mate-pair information has played an important role in most genome projects and many HTS assemblers incorporate mate-pair information to increase the contigs length in a post-processing step [13,4,17,2,15,3,9]. Most of these assemblers rely on the same basic observation [13]: if there is a unique path of suitable length in the assembly graph that connects the left and right reads of a mate-pair, the gap in the mate-pair can be filled in with the nucleotides spelled by this path (*mate-pair transformation*). However, when multiple paths exist between the left and right reads within a mate-pair, it remains ambiguous which path should be used to fill in the gap. Unfortunately, mate-pairs generated by existing HTS protocols are characterized by rather large variations in insert sizes, leading to multiple paths for a significant fraction of mate-pairs (since the range of suitable path lengths is wide), making it difficult to utilize such mate-pairs.

Recently, [10] introduced the notion of the *paired de Bruijn graph*, which incorporates mate-pair information in the graph structure rather than using it for mate-pair transformations in a post-processing step. Similar methods were also developed independently in [6,7]. Unfortunately, the performance of these methods deteriorates as the variation in the insert size increases.

Below we show that while reducing variation in the insert size remains a difficult experimental problem, it can be addressed computationally by aggregating the mate-pair information for all mate-pairs linking a pair of edges in the condensed de Bruijn graph. This transforms mate-pairs into *edge-pair histograms*, consisting of pairs of edges together with distances aggregated from mate-pair information. Using the collection of edge-pair histograms, we further combine the ideas of mate-pair transformations and paired de Bruijn graphs into new data structures for genome assembly, called *pathsets* and *pathset graphs*. We further compare the performance of our assembler (based on the pathsets approach) to other assemblers on various bacterial datasets.

2 Methods

2.1 De Bruijn Graphs

We represent genomes as circular strings over $\{A, T, G, C\}$, the alphabet of nucleotides. An n-mer is a string of length n. Given a n-mer $s = s_1 \ldots s_n$, we define $\text{PREFIX}(s) = s_1 \ldots s_{n-1}$ and $\text{SUFFIX}(s) = s_2 \ldots s_n$.

Let k be a fixed parameter. Given a set A of k-mers, the standard de Bruijn graph has a directed edge $(\text{PREFIX}(s), \text{SUFFIX}(s))$ for each k-mer $s \in A$. The condensed de Bruijn graph $CG(A, k)$ is obtained from the standard de Bruijn graph by replacing every maximal non-branching path[1] P by a single edge e of length $\ell(e)$ equal to the number of edges in P (see Fig. 1a,b). Below we work with

---

[1] A path is *non-branching* if all its vertices (except possibly the first and last) have indegree and outdegree both equal to 1. The condensed de Bruijn graph may contain loops and multiple edges between the same pair of vertices.

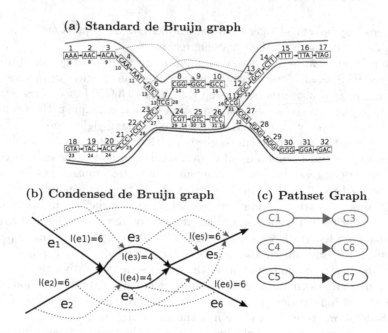

Fig. 1. From de Bruijn graph to pathset graph. (a) A standard de Bruijn graph and the corresponding mapping of mate-pairs. The number on top of each node is the node ID. The smaller blue numbers below/beside each node are the IDs of the corresponding paired right nodes. The bold red, blue and green paths show how the genome traverses the graph. (b) The condensed de Bruijn Graph with edges corresponding to non-branching paths in the standard de Bruijn graph. The dotted red lines indicate edge-pairs. (c) Pathset graph. Initially there are eight pathsets: $C_1 = \{e_1e_3e_5, \mathbf{e_1e_4e_5}\}$, $C_2 = \{e_1e_3\}$, $C_3 = \{e_3e_5\}$, $C_4 = \{\mathbf{e_2e_3e_5}, e_2e_4e_5\}$, $C_5 = \{\mathbf{e_2e_3e_6}, e_2e_4e_6\}$, $C_6 = \{e_4e_5\}$, $C_7 = \{e_4e_6\}$, and $C_8 = \{e_2e_4\}$. Using the edge-pair information, we find phantom paths (indicated in boldface) and remove them. After removal of all prefix pathsets (C_2 and C_8), the pathset graph has six nodes and consists of three edges: $C_1 \rightarrow C_3$ (red path), $C_4 \rightarrow C_6$ (green path), and $C_5 \rightarrow C_7$ (blue path). Each edge in the pathset graph corresponds to a contig; e.g., $C_1 \rightarrow C_3$ spells out the red path (AAACAATCGGCCGCTTTAG).

condensed (rather than standard) de Bruijn graphs and refer to them simply as de Bruijn graphs.

Each k-mer $a \in A$ maps to an edge $e = \text{EDGE}(a)$ in $CG(A, k)$ at position $\text{OFFSET}(a)$ $(1 \leq \text{OFFSET}(a) \leq \ell(e))$. For a path $P = e_1 \ldots e_n$ in a graph $CG(A, k)$, we define $d_P(e_i, e_j) = \Sigma_{t=i}^{j-1} \ell(e_t)$. We further define the length of P as $d_P(e_1, e_n)$, i.e., the total length of its edges, excluding the last edge.

Given a parameter d, a pair of k-mers a, b form a (k, d)-mer $(a|b)$ of string $S = s_1 \ldots s_n$ if there are instances of $a = s_i \ldots s_{i+k-1}$ and $b = s_{i+d} \ldots s_{i+d+k-1}$ in S whose starting positions differ by d nucleotides. Given parameters d and Δ, a pair of k-mers $(a|b)$ is called a (k, d, Δ)-mer of S if it is a (k, d_0)-mer of S for some $d_0 \in [d - \Delta, d + \Delta]$. In particular, error-free pairs of reads of length l with exact insert size s form $(l, s - l)$-mers (Fig. 2).

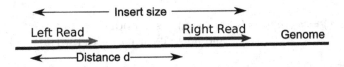

Fig. 2. A mate-pair is a pair of reads with distance d between their starting positions. The insert size is the distance from the start of the left read to the end of the right read. Each platform has reads oriented a particular way, but for presentation purposes, we canonically re-orient them as indicated.

We call a and b the *left* and *right* k-mers of the pair $(a|b)$, respectively, and we refer to the distance between them as the *distance* of $(a|b)$. For a set B of k-mer pairs, we let LEFT(B) (resp., RIGHT(B)) be the set of left (resp., right) k-mers appearing in B.

We transform each pair of reads of length l into a sequence of $l - k + 1$ consecutive k-mer pairs. For the set B of resulting k-mer pairs, we define the *de Bruijn graph of reads* as CG(LEFT$(B) \cup$ RIGHT$(B), k)$.

2.2   From k-mer Pairs to Edge-Pair Histograms

We assume for simplicity that the genome defines a *genomic walk* W that passes through all edges in the de Bruijn graph of reads.[2] Since some edges may appear multiple times in the genomic walk, we distinguish between the edge e and its instance $\tilde{e}$. Every two instances $\tilde{e}_1$ and $\tilde{e}_2$ of edges e_1 and e_2 define a *genomic distance* $d_W(\tilde{e}_1, \tilde{e}_2)$ between edges e_1 and e_2. This in turn yields a triple $(e_1, e_2, d_W(\tilde{e}_1, \tilde{e}_2))$ called a *genomic edge-pair*. Note that a pair of edges may have multiple genomic distances if one of these edges appears multiple times in the genomic walk.

When the genomic walk W is unknown, genomic edge-pairs can be computed from the set of k-mer pairs B when the genomic distance between k-mers in each k-mer pair of B is known exactly. In practice, such distances are approximate rather than exact. Below, we define *edge-pair histograms* and use them for more accurate approximation of genomic edge-pairs.

For a set B of k-mer pairs, a pair of edges (e_1, e_2) in the de Bruijn graph CG(LEFT$(B) \cup$ RIGHT$(B), k)$ is called B-*bounded* if there exists at least one k-mer pair in B whose left (resp., right) k-mer maps to the edge e_1 (resp., e_2).

Below we assume that parameters d_0 (estimated distance between reads within read-pairs) and Δ (maximum error in distance estimates) are fixed. Given a set of read pairs whose distances fall in the range $d_0 \pm \Delta$, one can generate a set B of all (k, d_0, Δ)-mers (extracted from all read-pairs), and then compute the set of B-bounded edge-pairs. For a B-bounded edge-pair (e_1, e_2), we define the *edge-pair histogram* $(e_1, e_2, \mathbf{h})$, where $\mathbf{h}$ is a histogram with $\mathbf{h}(x)$ equal to the number of (k, d_0, Δ)-mers in B that support *genomic distance* x between e_1 and e_2:

---

[2] While this assumption is unrealistic (e.g., gaps in coverage often make the de Bruijn graphs of reads disconnected), the analysis of fragment assembly in this idealized setting can be extended to real sequencing data; see the Results section.

$$\mathbf{h}(x) = \#\{(a|b) \in B \mid \text{EDGE}(a) = e_1, \ \text{EDGE}(b) = e_2,$$
$$\text{and } d_0 + \text{OFFSET}(a) - \text{OFFSET}(b) = x\}.$$

While HTS machines produce large numbers of read pairs (e.g., over 10^7 for the *E. coli* datasets we analyze below), the de Bruijn graph of reads contains a small number of edges (several thousand for our *E. coli* datasets). Thus, edge-pair histograms are typically supported by many k-mer pairs, which allows one to accurately estimate the genomic distance(s) between e_1 and e_2.

Since insert sizes typically follow a Gaussian distribution [5], the histogram $(e_1, e_2, \mathbf{h})$ either represents a sample from a single Gaussian distribution (if e_1 and e_2 appear once in the genomic walk) or a mixture of Gaussian distribution (if e_1 and e_2 appear multiple times in the genomic walk). Inferring each individual Guassian distribution from a sample of mixture distributions represents a challenging problem [11]. Here we sketch a simple approach to address this problem (see [1] for details of a more advanced analysis). We smooth the histogram, choose a threshold, and focus on the regions where the values of the histogram exceed the threshold (above the red line in Fig. 3b). The edge-pair histogram is thus transformed into one or more non-overlapping *edge-pair intervals*, each corresponding to one or more genomic edge-pairs with similar distances.

An edge-pair interval $(e_1, e_2, [a, b])$ is called *correct* if there exists a genomic edge-pair (e_1, e_1, D) such that $a \leq D \leq b$. A properly chosen threshold should maximize the number of correct edge-pair intervals, while still separating the genomic edge-pairs with different distances.

Figure 3a shows a pair of edges (e_1, e_2) that is traversed three times through the paths P_1, P_2, and P_3. The first two traversals, $e_1 P_1 e_2$ and $e_1 P_2 e_2$, have similar lengths while the third, $e_1 P_3 e_2$, has significantly larger length. Setting the threshold to 200 results in two correct edge-pair intervals: $(e_1, e_2, [130, 144])$ and $(e_1, e_2, [156, 164])$. The first interval supports two genomic edge-pairs, $(e_1, e_2, 134)$ and $(e_1, e_2, 140)$, while the second interval supports a single genomic edge-pair, $(e_1, e_2, 160)$.

Edge-pair intervals have two advantages over mate-pairs:

- With proper choice of thresholds, estimates of distances between edges in edge-pair intervals are more accurate than estimates of distances for individual mate-pairs. Better estimates may result in better performance of existing methods for resolving repeats, including mate-pair transformations [14], paired de Bruijn graphs [10], and mate-pair graphs [7].
- Since the edge-pair intervals compactly represent the mate-pair data, many redundant operations can be avoided; for instance, multiple function calls to perform mate-pair transformations for all mate-pairs corresponding to the same edge-pair interval (e.g., in EULER-SR [4]) can be replaced by a single mate-pair transformation of the corresponding edge-pair interval.

Below we use edge-pair intervals to define the notions of *pathset* and *pathset graph*.

(a)

(b)

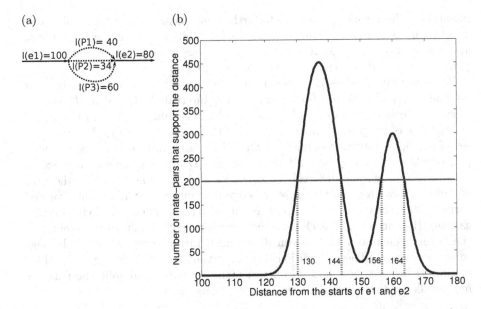

Fig. 3. Estimating genomic distances between edges within edge-pairs using the edge-pair histogram. (a). The genome traverses the graph from e_1 to e_2 through P_1, P_2, and P_3. (b) The smoothed edge-pair histogram (inferred from mate-pairs mapping to e_1 and e_2) supports various genomic distances between e_1 and e_2. Setting the threshold to 200 (red line) splits the histogram into two edge-pair intervals $(e_1, e_2, [130, 144])$ (supporting the traversal via paths P_1 and P_2) and $(e_1, e_2, [156, 164])$ (supporting the traversal through path P_3).

2.3   From Edge-Pair Intervals to Pathsets

We define a *pathset* as any set of paths between a fixed pair of edges.[3] Every edge-pair interval $(e_1, e_2, [a, b])$ corresponds to a pathset $\text{PATHSET}(e_1, e_2, [a, b])$ formed by all paths starting at e_1, ending at e_2, and having lengths in the interval $[a, b]$. For example, in Fig. 1b, $\text{PATHSET}(e_1, e_5, [9, 11]) = \{e_1e_3e_5, e_1e_4e_5\}$.

As a by-product of transforming edge-pair intervals to pathsets, the set of possible genomic distances between edges in each edge-pair interval can be further reduced. Namely, the interval $[a, b]$ in an edge-pair interval $(e_1, e_2, [a, b])$ can be replaced by a list of lengths of paths in the corresponding pathset. This is referred to as an *edge-pair distance set*, or simply an *edge-pair*, and denoted $(e_1, e_2, \mathbf{d})$. If all paths in a pathset have the same length, this length defines the exact distance between e_1 and e_2.

A path in the de Bruijn graph is called a *genomic path* if it corresponds to a substring of the genome (i.e., is a subpath of the genomic walk), and a *phantom path* otherwise. In general, a pathset may contain multiple genomic and phantom paths. Given a collection of pathsets obtained from edge-pairs, our goal is to remove the

---

[3] The term "pathset" is also used in [7] to describe the construction of a different graph that represents mate-pairs.

phantom paths in each pathset and to further split pathsets with t genomic paths into t pathsets consisting from singleton paths. While it is not always possible to accomplish this task (since it is not clear how to separate phantom and genomic paths), below we describe some steps towards this goal.

We note that all edge-pair intervals are *disjoint*, i.e., for every two edge-pair intervals $(e_1, e_2, [a, b])$ and $(e_1, e_2, [a', b'])$, the intervals $[a, b]$ and $[a', b']$ do not overlap. A pair of edges e_1 and e_2 in the genomic walk is called d_0-*bounded* if at least one of its genomic distances falls in the range $[d_0 - \ell(e_2), d_0 + \ell(e_1)]$. A set of edge-pair intervals is *representative* if for each instance of a d_0-bounded pair of edges e_1 and e_2 (separated by distance D in the genome), there exists a *supporting* edge-pair interval $(e_1, e_2, [a, b])$ with $a \leq D \leq b$. For the sake of simplicity, we assume that a set of edge-pair intervals is representative and correct. Then the corresponding pathsets are also (a) disjoint (i.e., do not overlap as sets), (b) correct (i.e., each contains a genomic path), and (c) representative (i.e., every genomic path between d_0-bounded instances of two edges belongs to some pathset). These conditions are important for constructing the pathset graph.[4] Below we show how to remove phantom paths and split the pathsets into smaller pathsets while preserving conditions (a–c).

2.4   Removing Phantom Paths from Pathsets

Since our pathsets are representative (condition (c)), for each instance of a d_0-bounded pair of edges, there exists a supporting edge-pair interval. Therefore, if a path in a pathset violates this condition, it represents a phantom path. Below we describe how to identify some (but not necessarily all) phantom paths.

A path $P = e_1 e_2 \ldots e_n$ is *supported* by a set of edge-pairs EP if for every pair of d_0-bounded edges e and e' in P, there exists an edge-pair $(e, e', \mathbf{d}) \in EP$ such that $d_P(e, e') \in \mathbf{d}$. The path P is *strongly supported* by a set of edge-pairs EP if it can be extended from both ends into a longer path $P' = u \ldots e_1 \ldots e_n \ldots v$ such that (i) P' is supported by EP, (ii) the pair of edges u and e_1 is d_0-bounded, and (iii) the pair of edges e_n and v is d_0-bounded. Paths in a pathset that are not strongly supported are classified as phantom paths and removed. Indeed, if P is a genomic path, then it is a subpath of the genomic walk. In the genomic walk, there exists an edge u preceding P and an edge v succeeding P that satisfy the properties (ii) and (iii). The subpath starting at u and ending at v (denoted P' above) clearly satisfies property (i).

2.5   Splitting Pathsets

A pathset *contains* an edge e if one of the paths in this pathset contains e. For a pathset PS and sets of edges $U = \{u_1, \ldots, u_m\}$ and $V = \{v_1, \ldots, v_n\}$, we define $PS_{u_1, \ldots, u_m, \overline{v_1}, \ldots, \overline{v_n}}$ as the set of all paths in PS that contain all edges from U and no edges from V.

---

[4] In the Results section, we demonstrate that conditions (a–c) are satisfied for the vast majority of pathsets constructed for our bacterial assembly datasets.

Two edges contained in a pathset are called *independent* if no path in this pathset contains them both. An edge e is *essential* for a pathset PS if there exists a genomic path in PS containing e. A set of essential edges in a pathset is called *independent* if every two edges in this pathset are independent. An independent set $A = \{a_1, \ldots, a_t\}$ of essential edges contained in PS defines a *split* of PS into t disjoint pathsets:[5] $PS_{a_1}, \ldots, PS_{a_{t-1}}, PS_{\overline{a_1}, \ldots, \overline{a_{t-1}}}$. We remark that each of these pathsets contains an essential edge from A and thus is correct. It is easy to check that the split operation preserves conditions (a–c).

For example, for the pathset defined by edges e_1 and e_7 in Fig. 4, all edges are essential. Edges e_2 and e_3 (as well as e_5 and e_6) form an independent set.

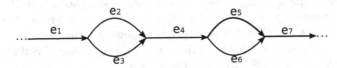

Fig. 4. The pathset $PS = \text{PATHSET}(e_1, e_7, \mathbf{d})$ corresponding to the edge-pair $(e_1, e_7, \mathbf{d})$ contains 4 paths: $PS = \{e_1e_2e_4e_5e_7, e_1e_3e_4e_5e_7, e_1e_2e_4e_6e_7, e_1e_3e_4e_6e_7\}$. Edges e_2 and e_3 (as well as edges e_5 and e_6) form an independent set. Therefore PS can be split into two pathsets: $PS_{e_2} = \{e_1e_2e_4e_5e_7, e_1e_2e_4e_6e_7\}$ and $PS_{\overline{e_2}} = \{e_1e_3e_4e_5e_7, e_1e_3e_4e_6e_7\}$ containing edges e_2 and e_3 respectively.

2.6   Identifying Essential Edges

To split a pathset, one has to identify essential edges. Consider an instance $\tilde{e}$ of an edge e in the genomic walk. A subpath of this walk $e_1 \ldots \tilde{e} \ldots e_2$ is called an $\tilde{e}$-*subpath* if (i) the pair of edges e_1 and $\tilde{e}$ is d_0-bounded, and (ii) the pair of edges $\tilde{e}$ and e_2 is d_0-bounded. A maximal $\tilde{e}$-subpath (i.e., a subpath containing all other $\tilde{e}$-subpaths) is called a *span* of $\tilde{e}$. For an edge e, we define $\text{SPAN}(e)$ as the union of all spans of all instances $\tilde{e}$ of the edge e.

We refer to a pathset as $PS(a, b)$ if all paths in this pathset start at edge a and end at edge b. Given paths $a \ldots e$ and $e \ldots b$ (i.e., starting and ending at the same edge e), we define their *concatenation* as the path $a \ldots e \ldots b$. Given a collection of pathsets and a fixed edge e, consider all pathsets $PS(a, e)$ and $PS(e, b)$ and all concatenations of paths from $PS(a, e)$ with paths from $PS(e, b)$ (for all possible choices of a and b). Define $\text{SPREAD}(e)$ as the set of all supported paths in this set. Obviously, $\text{SPAN}(e) \subseteq \text{SPREAD}(e)$.

An edge e is called (e_1, e_2)-*constrained* if all paths in $\text{SPREAD}(e)$ have the form $\ldots e_1 \ldots e \ldots e_2 \ldots$ and edges e_1 and e_2 in all such path are d_0-bounded. It is easy to see that every (e_1, e_2)-constrained edge e is essential in some pathset $PS(e_1, e_2)$. Indeed, since the genomic walk contains each edge e in the graph, there exists a genomic path P in $\text{SPAN}(e)$. Since $\text{SPAN}(e) \subseteq \text{SPREAD}(e)$ and since e is (e_1, e_2)-constrained, P has the form $\ldots e_1 \ldots e \ldots e_2 \ldots$, where edges e_1 and

---

[5] We remark that the resulting pathsets depend on ordering of elements in A (in particular, on the choice of "last" element a_t); different orderings may result in different splits.

e_2 are d_0-bounded. Therefore, the subpath $e_1 \ldots e \ldots e_2$ of P belongs to some pathset $PS(e_1, e_2)$, implying that e is an essential edge.

2.7 Pathset Graph

Given a collection of pathsets satisfying conditions (a–c), the genome assembly problem becomes similar to traditional genome assembly with each pathset playing a role of a single (long) read. Below, we define the *pathset graph* with nodes corresponding to pathsets and non-branching paths corresponding to assembly contigs.

Path p is a *prefix* (resp., *suffix*) of path q if q can be obtained from p by concatenating some non-empty path to the end (resp., start) of p. Pathset PS is called a *prefix* of pathset PS' if each path of PS is a prefix of a path in PS'. Path q follows path p if they have the following form: $p = e_1 e_2 \ldots e_k$ and $q = e_2 \ldots e_k \ldots e_{k+t}$, where $t \geq 0$. Pathset PS' *follows* pathset PS if there exists paths $q \in PS'$ and $p \in PS$ such that q follows p.

A collection of pathsets is called *prefix-free* if no pathset in this collection is a prefix of another pathset. Given a collection of pathsets, we remove phantom paths, perform splits, and remove prefix pathsets to obtain a prefix-free collection of pathsets. We then construct the *pathset graph* by representing each remaining pathset as a node and forming a directed edge $PS \rightarrow PS'$ if PS' follows PS (similarly to the classical overlap-layout-consensus approach to fragment assembly).

Fig. 1c illustrates the pathset graph for a toy example. After removing phantom paths, we obtain six singleton pathsets. The pathset graph consists of six vertices and three edges (non-branching paths), corresponding to three contigs.

The pathset graph approach can be summarized as follows:

Input: Set of mate-pairs
1: Construct the de Bruijn graph from individual reads in mate-pairs
2: Transform read-pairs into a set of edge-pair histograms
3: Transform edge-pair histograms into edge-pair intervals
4: Transform edge-pairs intervals into pathsets
5: **for each** pathset
6:     Remove phantom paths
7:     Construct an independent edge-set in each pathset
8:     Split the pathset over the independent edge-set
9: Remove prefix pathsets from the resulting collection of pathsets
10: Construct the pathset graph on the resulting prefix-free collection of pathsets
11: Output contigs as non-branching paths in the pathset graph

3 Results

We implemented Pathset as a module in the new assembler SPAdes [1]. The source code is released under the GNU General Public License and is available at http://bioinf.spbau.ru/spades/pathset. To evaluate the performance of Pathset, we compared it with various assemblers on two Illumina *E. coli* datasets.

The first dataset (EMBL-EBI Sequence Read Archive ERA000206, which we refer to as EC215) consists of 28 million paired reads of length 100 bp and mean insert size $\approx$ 215 bp, while the second (EMBL-EBI Sequence Read Archive ERR022075, which we refer to as EC500) consists of 44 million paired reads of length 100 bp and mean insert size $\approx$ 500 bp. The reads in each dataset were error-corrected with Quake [8].

3.1   From Mate-Pairs to Edge-Pair Intervals

For each dataset, we constructed the de Bruijn graph for $k = 55$ and transformed the set of input mate-pairs into a set EP of edge-pair intervals. To evaluate this transformation, we further extracted from the $E.\ coli$ genome a set of k-mer pairs at fixed distance d_0. The mapping of these k-mer pairs to edges of the graph defines a set of genomic edge-pairs, EP_0. Table 1 demonstrates that for insert size $d_0 = 215$, most genomic edge-pairs in EP_0 (97%) are also presented[6] in EP and very few (0.8%) of the edge-pair intervals in EP are incorrect. We also observed that for $d_0 = 500$, the proportion of incorrect edge-pairs in EC500 only slightly increases as compared to EC215.

Table 1. Comparison between the set EP_0 of correct edge-pairs (defined by the known $E.\ coli$ genome) and the set of edge-pairs generated from mate-pairs from the EC215 and EC500 datasets. False positive rate (FPR) is the fraction of edge-pairs intervals in EP that are incorrect. False negative rate (FNR) is the fraction of genomic edge-pairs in EP_0 that are missing in EP.

Insert size	$|EP_0|$	$|EP|$	FPR	FNR
215	6321	6178	0.008	0.030
500	18684	18571	0.017	0.023

3.2   From Edge-Pair Intervals to Pathsets

For each edge-pair interval $(e_1, e_2, [a, b]) \in EP$, we constructed its pathset and further applied phantom path removal and pathset splitting. For EC215 (resp., EC500), we generated 6178 (resp., 18571) pathsets before removing prefix pathsets and 1430 (resp., 2037) pathsets after removing prefix pathsets. Approximately 90% (resp., 74%) of all pathsets that remained represented singletons.

Fig. 5 shows the number of pathsets of each multiplicity before (red) and after (blue) phantom path removal and splitting. Before removing phantom paths and splitting, the largest pathset contained only 4 (resp., 27) paths for the EC215 (resp., EC500) dataset. As Fig. 5 illustrates, most pathsets with multiple paths turn into singletons after phantom path removal and splitting.

---

[6] A genomic edge-pair (e_i, e_j, d_g) is *present (resp., missing)* in EP if there exists (resp., does not exist) an edge-pair interval $(e_i, e_j, [a, b]) \in EP$, such that $a \leq d_g \leq b$.

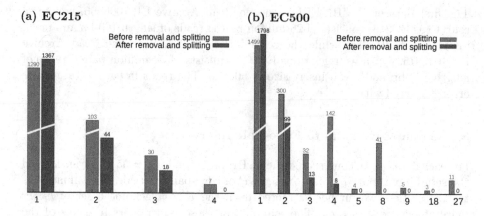

Fig. 5. The number of pathsets of each size in datasets (a) EC215 and (b) EC500. The red columns column counts initially constructed pathsets. The blue columns count pathsets after phantom path removal and splitting.

3.3 Comparing Pathset with Other Genome Assemblers

We compared Pathset to Velvet [17], SOAPdenovo [9], and IDBA [12] assemblers using Plantagora [16], an assembly evaluation tool; see Table 2. For dataset

Table 2. Comparison of different assemblers. *Complete genes* is the number of genes contained completely within assembly contigs (using *E. coli* gene annotations from http://www.ecogene.org). In each column, the best value by that metric is boldfaced.

Assembler	# contigs	N50	N75	Covered (%)	Misassemblies	Substitution Error (per 100 kbp)	Complete Genes
EC215 dataset:							
Velvet	198	82776	42878	**99.93**	4	**1.2**	4223
SOAPdenovo	192	62512	35069	97.72	1	26.1	4141
IDBA	246	48825	25483	99.60	3	1.3	4170
Pathset	360	**91829**	**56830**	99.56	1	2.1	**4249**
EC500 dataset:							
Velvet	169	**105637**	57172	99.28	5	2.5	4095
SOAPdenovo	982	57167	31582	**99.88**	0	**0.2**	4196
IDBA	227	57827	34421	99.26	1	4.2	4158
Pathset	320	97971	**58548**	99.46	2	2.0	**4252**

EC215, Pathset improved on other assemblers in N50, N75, number of misassemblies,[7] and the number of captured complete genes. For dataset EC500, Pathset outperformed the other assemblers in N75 and the number of captured genes. While Velvet had a larger N50 for the EC500 dataset, Velvet's assembly was compromised by the largest number of errors (5). SOAPdenovo produced the most accurate assembly (no assembly errors) but the smallest N50 (57167 as compared to 97971 for Pathset and 105637 for Velvet).

4 Discussion

In this paper, we presented the pathset data structure for assembling genomes using mate-pair data. Instead of using mate-pair transformations on a set of mate-pairs directly, which is computationally expensive and susceptible to failure when the insert size variation is high, we first transform mate-pairs into edge-pairs. We aggregate the distance estimates from all mate-pairs mapping to each pair of edges to make distance estimates more accurate.

As compared with the traditional mate-pair transformation approach, where it is required to have a unique path between paired reads in the de Bruijn graph, our approach stores all suitable paths in a pathset data structure and later uses the paired information to remove phantom paths and further split pathsets. We also introduce the pathset graph, which allows one to construct contigs from the pathsets. Multiple libraries with different insert sizes can be utilized in the pathset data structure. The paired information in different libraries can be used to remove invalid paths in each pathset.

Acknowledgements. As members of the SPAdes assembler developers group [1] at the Algorithmic Biology Laboratory (Russian Academy of Science) and UCSD, the authors would like to thank the rest of the group for productive collaboration throughout the SPAdes and Pathset projects. We are especially indebted to Anton Bankevich, Mikhail Dvorkin, Alexey Gurevich, Alexey Pyshkin, Sergey Nikolenko, Sergey Nurk, Nikolay Vyahhi, and Viraj Deshpande for their support of the Pathset project, helpful comments, and thought-provoking discussions. The authors are grateful to Paul Medvedev for his insightful comments.

References

1. Bankevich, A., Nurk, S., Antipov, D., Gurevich, A., Dvorkin, M., Kulikov, A., Lesin, V., Nikolenko, S., Pham, S., Prjibelski, A., Pyshkin, A., Sirotkin, A., Vyahhi, N., Tesler, G., Alekseyev, M., Pevzner, P.: SPAdes: a New Genome Assembler and its Applications to Single Cell Sequencing (submitted, 2012)
2. Butler, J., MacCallum, I., Kleber, M., Shlyakhter, I., Belmonte, M., Lander, E., Nusbaum, C., Jaffe, D.: ALLPATHS: de novo assembly of whole-genome shotgun microreads. Genome Research 18(5), 810 (2008)

---

[7] A *misassembly* is formed by concatenation of two sequences A and B that both align to the reference genome but with a gap between them larger than 1000 bases.

3. Chaisson, M., Brinza, D., Pevzner, P.: De novo fragment assembly with short mate-paired reads: Does the read length matter? Genome Research 19(2), 336 (2009)
4. Chaisson, M., Pevzner, P.: Short read fragment assembly of bacterial genomes. Genome Research 18(2), 324 (2008)
5. Chen, K., Wallis, J., McLellan, M., Larson, D., Kalicki, J., Pohl, C., McGrath, S., Wendl, M., Zhang, Q., Locke, D., et al.: BreakDancer: an algorithm for high-resolution mapping of genomic structural variation. Nature Methods 6(9), 677–681 (2009)
6. Chikhi, R., Lavenier, D.: Localized Genome Assembly from Reads to Scaffolds: Practical Traversal of the Paired String Graph. In: Przytycka, T.M., Sagot, M.-F. (eds.) WABI 2011. LNCS, vol. 6833, pp. 39–48. Springer, Heidelberg (2011)
7. Donmez, N., Brudno, M.: Hapsembler: An Assembler for Highly Polymorphic Genomes. In: Bafna, V., Sahinalp, S.C. (eds.) RECOMB 2011. LNCS, vol. 6577, pp. 38–52. Springer, Heidelberg (2011)
8. Kelley, D., Schatz, M., Salzberg, S.: Quake: quality-aware detection and correction of sequencing errors. Genome Biology 11(11), R116 (2010)
9. Li, R., Zhu, H., Ruan, J., Qian, W., Fang, X., Shi, Z., Li, Y., Li, S., Shan, G., Kristiansen, K., et al.: De novo assembly of human genomes with massively parallel short read sequencing. Genome Research 20(2), 265 (2010)
10. Medvedev, P., Pham, S., Chaisson, M., Tesler, G., Pevzner, P.: Paired de Bruijn Graphs: A Novel Approach for Incorporating Mate Pair Information into Genome Assemblers. In: Bafna, V., Sahinalp, S.C. (eds.) RECOMB 2011. LNCS, vol. 6577, pp. 238–251. Springer, Heidelberg (2011)
11. Moitra, A., Valiant, G.: Settling the polynomial learnability of mixtures of gaussians. In: 51st Annual IEEE Symposium on Foundations of Computer Science (FOCS), pp. 93–102. IEEE (2010)
12. Peng, Y., Leung, H.C.M., Yiu, S.M., Chin, F.Y.L.: IDBA – A Practical Iterative de Bruijn Graph De Novo Assembler. In: Berger, B. (ed.) RECOMB 2010. LNCS, vol. 6044, pp. 426–440. Springer, Heidelberg (2010)
13. Pevzner, P., Tang, H.: Fragment assembly with double-barreled data. Bioinformatics 17(suppl. 1), S225 (2001)
14. Pevzner, P., Tang, H., Waterman, M.: An Eulerian path approach to DNA fragment assembly. PNAS 98(17), 9748 (2001)
15. Simpson, J., Wong, K., Jackman, S., Schein, J., Jones, S., Birol, İ.: ABySS: a parallel assembler for short read sequence data. Genome Research 19(6), 1117 (2009)
16. Young, S., Barthelson, R., McFarlin, A., Rounsley, S.: PLANTAGORA toolset (2011), http://www.plantagora.org
17. Zerbino, D., Birney, E.: Velvet: algorithms for de novo short read assembly using de Bruijn graphs. Genome Research 18(5), 821 (2008)

Finding Maximum Colorful Subtrees in Practice

Imran Rauf[1,*], Florian Rasche[2],
François Nicolas[2], and Sebastian Böcker[2]

[1] Department of Computer Science, University of Karachi, Karachi, Pakistan
imran.rauf@uok.edu.pk
[2] Lehrstuhl für Bioinformatik, Friedrich-Schiller-Universität Jena, Jena, Germany
{florian.rasche,francois.nicolas,sebastian.boecker}@uni-jena.de

Abstract. In metabolomics and other fields dealing with small compounds, mass spectrometry is applied as sensitive high-throughput technique. Recently, fragmentation trees have been proposed to automatically analyze the fragmentation mass spectra recorded by such instruments. Computationally, this leads to the problem of finding a maximum weight subtree in an edge weighted and vertex colored graph, such that every color appears at most once in the solution.

We introduce new heuristics and an exact algorithm for this MAXIMUM COLORFUL SUBTREE problem, and evaluate them against existing algorithms on real-world datasets. Our tree completion heuristic consistently scores better than other heuristics, while the integer programming-based algorithm produces optimal trees with modest running times. Our fast and accurate heuristic can help to determine molecular formulas based on fragmentation trees. On the other hand, optimal trees from the integer linear program are useful if structure is relevant, e.g., for tree alignments.

1 Introduction

Mass spectrometry is one of the prevalent technologies for the analysis of metabolites and other small compounds with mass below 1000 Da. The study of such compounds is, for example, relevant in drug design and the search for new signaling molecules and biomarkers [11]. Typically, the analyte is fragmented in the mass spectrometer such that fragment masses and their intensities can be recorded in a fragmentation spectrum. Usually, this spectrum is then compared to a database of reference spectra [13]. Unfortunately, such databases are vastly incomplete [6]. Thus, *de novo* approaches, which try to identify the compound solely based on its spectrum, are sought. Whereas *de novo* interpretation of protein spectra makes use of their specific structure, this is not possible for metabolites, as they are not structurally restricted.

In 2008, Böcker and Rasche [2] suggested fragmentation trees for the *de novo* interpretation of metabolite tandem mass spectrometry data. Fragmentation trees can be used to identify the molecular formula of a compound [2], as well as

---

* This work was carried out while the author was a member of the Chair of Bioinformatics at the Friedrich-Schiller-University Jena.

B. Chor (Ed.): RECOMB 2012, LNBI 7262, pp. 213–223, 2012.

to describe the fragmentation process [15]. Fragmentation trees can also be used for the interpretation of multiple mass spectrometry (MS^n) [16]. The automated alignment of such fragmentation trees [14] seems to be a particularly promising approach, allowing a fully automated computational identification of small molecules that cannot be found in any database, and paving the way towards natural product discovery, searching for signaling molecules, biomarkers, novel drugs or ones that are illegal, or other "interesting" organic compounds.

It must be understood that, depending on the application at hand, different aspects of fragmentation tree computation are most relevant: In case we want to identify the molecular formula of an unknown [2] then swift computations are mandatory, as we might have to compute hundreds of trees for a single compound. But if we want to (manually or automatically) interpret the structure of fragmentation trees [14, 15], then it is presumably beneficial to use optimal solutions.

Calculating fragmentation trees leads to the MAXIMUM COLORFUL SUBTREE problem [2]:

Maximum Colorful Subtree Problem. Given a vertex-colored DAG $G = (V, E)$ with colors $\mathcal{C}$ and weights $w : E \to \mathbb{R}$. Find the induced colorful subtree $T = (V_T, E_T)$ of G of maximum weight $w(T) := \sum_{e \in E_T} w(e)$.

This is a special case of the edge-weighted GRAPH MOTIF problem, see [18] for an overview. Scheubert *et al.* [16] present the related COLORFUL SUBTREE CLOSURE problem for analyzing multiple mass spectrometry data. Ljubíc *et al.* [12] presented an Integer Linear Program for the related PRIZE-COLLECTING STEINER TREE problem. The MAXIMUM COLORFUL SUBTREE problem is NP-hard [5] as well as APX-hard [3] even on binary trees. Furthermore, on general trees it has no constant factor approximation [3, 19].

This paper presents an experimental study of various heuristics for the MAXIMUM COLORFUL SUBTREE problem. We implement a new heuristic along with naive greedy approaches, and three exact algorithms for the problem. The various algorithms are then compared with respect to their running times and quality of the generated trees. As a side result, we prove that even after relaxing the color constraints, the resulting MAXIMUM SUBTREE problem remains inapproximable within $O(|V|^{1-\epsilon})$ factor for any $\epsilon > 0$, unless P = NP. Our proof is simpler and improves upon the previous APX-hardness result due to [3] under the same complexity hypothesis. Sikora showed a slightly weaker inapproximability result for unweighted graphs, but maintaining the color constraints [19].

The rest of the paper is organized as follows: The next section defines the MAXIMUM COLORFUL SUBTREE problem and describes how it is applied to analyze tandem mass spectra. In Section 3, we discuss the computational complexity of the closely related MAXIMUM SUBTREE problem, and present exact and heuristic algorithms for MAXIMUM COLORFUL SUBTREE. In Section 4, we evaluate these algorithms on three real-world datasets.

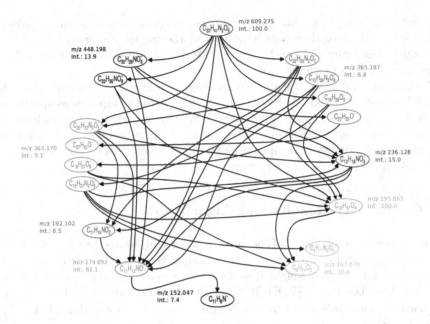

Fig. 1. Input graph calculated from the spectrum of Reserpine (Micromass dataset). Formulas of identically colored nodes have nearly identical masses. Edges are drawn between vertices if the molecular formula of one vertex is a sub-formula of the molecular formula of the other.

2   Computing Fragmentation Trees

Here we outline our method of computing the fragmentation tree of a metabolite compound from its tandem mass spectrum. This leads to a formal definition of the MAXIMUM COLORFUL SUBTREE problem. Let $m_1, \ldots, m_p$ be the peak masses of a spectrum consisting of p peaks. We first compute a list of possible molecular formulas $\{F_i^1, \ldots, F_i^{n_i}\}$ with mass close to m_i, for each $1 \leq i \leq p$. We then construct a directed acyclic graph G, where the set of vertices are the molecular formulas F_i^j of the i-th peak, for $1 \leq j \leq n_i$ and $1 \leq i \leq p$. Next, we label the vertices with colors from $\{1, \ldots, p\}$, where all vertices corresponding to the i-th peak are assigned color i. We connect two vertices u, v with a directed edge uv, if the molecular formula of v is a sub-formula of the formula of u, i. e., v contains at most as many atoms of every chemical element as u. The edge uv thus represents a possible fragmentation step of the molecule associated with u. Note that G is acyclic, since the subset relation used to define edges is a partial order.

Let us assume that the largest peak mass is m_1 which corresponds to the unfragmented compound whose fragmentation tree we want to compute. We further assume without loss of generality that $n_1 = 1$, otherwise we can consider a separate graph for each molecular formula explaining the peak mass m_1. So, we may assume that there is only a single source vertex r associated with the unfragmented compound. We delete all other vertices without incoming edges from the graph, since they cannot be part of the fragmentation process. Consequently, G is a directed acyclic graph (DAG) with unique source r such that every other node is reachable from r.

A subtree of G is *colorful* if any two vertices in the tree have different colors. A colorful subtree of G rooted at r explains each peak in the input mass spectrum at most once. The approach in [15] works by scoring edges by log likelihoods of the fragmentation steps they correspond to. Thus, edges may have negative weights. The maximum colorful subtree rooted at r will then be the most likely explanation of the given mass spectrum.

3   Complexity Results and Algorithms

We now consider a relaxation of the maximum colorful subtree problem obtained by dropping the color constraints. The resulting MAXIMUM SUBTREE problem asks to find a maximum weight tree rooted at a particular vertex in the given edge-weighted DAG $G = (V, E)$. It turns out that this simpler problem is still NP-hard. In fact, this problem is NP-hard to approximate within $O(|V|^{1-\epsilon})$ factor for any $\epsilon > 0$.

Theorem 1. *There is no $O(|V|^{1-\epsilon})$ approximation algorithm for the* MAXIMUM SUBTREE *problem for any $\epsilon > 0$, unless* P = NP.

Note that the inapproximability result from Theorem 1 also holds for the MAXIMUM COLORFUL SUBTREE problem, since it is a generalization of the MAXIMUM SUBTREE problem. We defer the proof of Theorem 1 to the full version of this paper.

Now, we briefly review exact algorithms and heuristics for the MAXIMUM COLORFUL SUBTREE problem. For vertices u and v, let $c(v)$ be the color assigned to v and $w(u, v) \in \mathbb{R}$ the weight of the edge uv. Throughout the rest of the paper we denote the number of colors in $G = (V, E)$ by k. Note that $k \leq p$ can be as large as the number of peaks in the spectrum; but we can also choose a smaller k to decrease running times, limiting our attention to, say, the k most intense peaks.

3.1   Exact Methods

The problem can be solved exactly using dynamic programming over vertices and color subsets [4]. Let $W(v, S)$ be the maximal score of a colorful tree with root v and color set $S \subseteq C$. Now, table W can be computed by the following recurrence [2]:

$$W(v, S) = \max \begin{cases} \max\limits_{u:c(u)\in S\setminus\{c(v)\}, vu\in E} W(u, S \setminus \{c(v)\}) + w(v, u) \\ \max\limits_{(S_1, S_2):S_1\cap S_2=\{c(v)\}, S_1\cup S_2=S} W(v, S_1) + W(v, S_2) \end{cases}$$

where, obviously, we have to exclude the cases $S_1 = \{c(v)\}$ and $S_2 = \{c(v)\}$ from the computation of the second maximum. Using the above recurrence with the initial condition $W(v, \{c(v)\}) = 0$, we can compute a maximum colorful tree in $O(3^k k |E|)$ time and $O(2^k |V|)$ space. The exponential running time and space make the algorithm useful only for small size instances. The running time can be somewhat improved to $O(2^k \cdot poly(|V|, k))$ by using the Möbius transform and the inversion technique of Björklund *et al.* [1]. However, the technique only works for suitably small integer weights.

Guillemot and Sikora [7] suggest a different approach using multilinear detection [10] for input graphs with unit weights. Their algorithm requires $O(2^k \cdot poly(|E|, k))$ time and only polynomial space. The algorithm can be adopted to integer weight graphs in a straight forward manner but the resulting algorithm would be pseudo-polynomial, i.e., its running time would depend polynomially on the integer weights thus making it impractical for our purposes. To the best of our knowledge, neither the above algorithm nor the dynamic programming with Möbius transform of the previous paragraph have been used in implementations.

For small instances a brute-force approach is suggested in [2]. The idea is to find a maximum subtree for each possible combination of vertices forming a colorful set. We then search a maximum subtree in a colorful DAG. Clearly, when all edge weights are positive, then the maximum subtree is a spanning tree. This can be found by a simple greedy algorithm, choosing the maximum weight incoming edge for each vertex but the root. With arbitrary edge weights, the problem becomes NP-hard, see Theorem 1. We solve the problem naively by iterating over all combinations of vertices whose best incoming edge has a negative weight. The brute-force approach is obviously not practical when either the number of combinations is large or when there are many vertices whose maximum incoming edge has negative weight.

Integer Linear Programming. Let us define a binary variable x_{uv} for each edge uv of the input graph. For each color $c \in C$ let $V(c)$ be the set of all vertices in $G = (V, E)$ which are colored with c. Then the following simple Integer Linear Program (ILP) captures the maximum colorful subtree problem:

$$\max \sum_{uv \in E} w(u, v) \cdot x_{uv} \tag{1}$$

$$\text{s.t.} \sum_{u \text{ with } uv \in E} x_{uv} \leq 1 \qquad\qquad \text{for all } v \in V \setminus \{r\}, \tag{2}$$

$$x_{vw} \leq \sum_{u \text{ with } uv \in E} x_{uv} \qquad \text{for all } vw \in E \text{ with } v \neq r, \tag{3}$$

$$\sum_{uv \in E \text{ with } v \in V(c)} x_{uv} \leq 1 \qquad\qquad \text{for all } c \in C, \tag{4}$$

$$x_{uv} \in \{0, 1\} \qquad\qquad \text{for all } uv \in E. \tag{5}$$

Constraint set (2) ensures that the feasible solution is a forest, whereas the constraint set (4) make sure that there is at most one vertex of each color present in the solution. Finally, (3) requires the solution to be connected. Note that in general graphs, we would have to ensure for every cut of the graph to be connected to some parent vertex. That would require an exponential number of constraints [12]. But since our graph is directed and acyclic, a linear number of constraints suffice.

3.2 Heuristics

A simple greedy heuristic has been proposed in [2]. It works by considering the edges according to their weights in descending order. The edge being considered is added to the result, if it does not conflict with the previously picked edges. The algorithm continues until all positive edges are considered and the resulting graph is connected. Note that an edge conflicts with another if they either are incoming edges to the same vertex or are incident edges to different vertices of the same color. Finally, we prune the leaves which are attached by negative weight edges in the resulting spanning tree. We refer to the above heuristic as *greedy* in the rest of the paper.

Another greedy strategy is to consider colors in some ordering and for the current color add an vertex of that color that promises the maximum increase of the score and attaches it to the already calculated tree. The resulting heuristic, called *insertion heuristic* in the rest of the paper, begins with only the root as the current partial solution. The heuristic greedily attaches vertices labeled with unused colors. For every vertex u with unused color, and every vertex v already part of the solution, we calculate how much we gain by attaching u to v. To calculate the gain of attaching u to v, we take into account the score of the edge vu, as well as the possibility of rerouting other outgoing edges of v through u. The vertex with maximum gain is then attached to the solution, and edges are rerouted as required. See Algorithm 1 for details. In our implementation, we choose the colors in Line 2 of the algorithm in the descending order with respect to the intensities of their corresponding peaks in the real-world datasets.

Algorithm 1. Vertex insertion heuristic

Input: An edge weighted and vertex colored DAG G, $r \in V(G)$ and color set C.
Output: A colorful subtree rooted at r.
1: $T := \{r\}$.
2: **for all** $c \in C$ such that c does not appear in T **do**
3:     Let v be a c-colored vertex which maximizes the quantity

$$w(u,v) + \sum_{w(v,x)>w(u,x)} (w(v,x) - w(u,x)),$$

    where u, x are vertices in T.
4:     Add uv to T.
5:     Add vx and remove ux for all $x \in T$ for which $w(v,x) > w(u,x)$ holds.
6: **return** T.

Tree Completion Heuristic. As noted before, the dynamic programming approach of Section 3.1 works only for small inputs. We now present a heuristic that combines DP with the greedy approaches. For a small enough constant b, the heuristic works by first computing the maximum colorful subtree consisting of at most b vertices, which we call *backbone* of a candidate solution. Next, we complete the backbone by using one of the greedy heuristics discussed above.

The tree completion heuristic works with Algorithm 1 by initializing T in Line 1 with the computed backbone. Similarly, the *greedy* heuristic can be used to complete the tree by starting with the backbone and applying the greedy heuristic on the remaining edges. In our experiments, we use the *insertion* heuristic for tree completion since it achieved consistently better scores. This heuristic is referred to as DP_b, where b is the size of the backbone computed exactly.

4   Evaluation Results

In our study, we analyze spectra from three real-world datasets. The analysis of spectra from two artificial datasets (one random, one constructed "computationally hard") is deferred to the full version of this paper. The *Orbitrap* dataset consists of mass spectra of 38 compounds with a mass accuracy of 10 ppm. It includes the 37 compounds used for evaluating fragmentation trees in [15]. The *Micromass* dataset [8] contains spectra of 100 compounds with an accuracy of 50 ppm, while the *QSTAR* dataset [2] consists of 36 mass spectra with 20 ppm accuracy. Note that the later dataset contains only 14.3 peaks per compound on average, whereas the average for Orbitrap and Micromass datasets is 75.4 and 51.6 peaks per compound, respectively. The Orbitrap dataset contains 1.03 vertices per color, the Hill dataset 1.7 and the QSTAR dataset has 1.1 vertices per color. Despite these low ratios, the number of colorful vertex sets grows as large as 10^{55} for certain instances, due to the combinatorial explosion. Experimental details can be found in the corresponding publications.

For each compound in the above three datasets, we assume that we know the correct molecular formula and construct a directed acyclic graph as described in Section 2. We use the scoring from [15] to weight the edges.

We implemented the exact algorithms based on the dynamic program, the integer linear program and the brute-force approach of Section 3.1. We also implemented the *greedy* heuristic and the tree completion heuristic with backbone size 10 and 15 (DP_{10}, DP_{15}) that use *insertion* to complete the backbone. To evaluate an heuristic on an instance of the problem, we consider its *performance ratio*, i. e., the ratio of the weight of generated solutions versus the optimal.

The algorithms are implemented in Java 1.6 by using an adjacency list representation for graphs. In the *DP* algorithm, we use the Java `long` data type to represent sets of colors as bitsets. This limits the maximum possible size of the color set to 64. Memory usage of this algorithm becomes prohibitive long before this number is reached. The experiments were run on a Lenovo T400 laptop powered with dual core Intel P8600 at 2.40 GHz with 2 GB of RAM and running Ubuntu Lucid Lynx as an operating system. Our implementation however

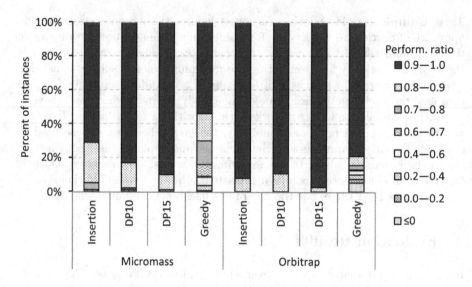

Fig. 2. Performance ratios achieved by different heuristics on *Micromass* and *Orbitrap* datasets. Results on the *QSTAR* dataset were rather uninformative, as nearly all instances reached a performance ratio above 0.9.

is single threaded and does not exploit the availability of multiple cores in the system. The integer linear programming solver, however, uses multiple cores. We run the experiments with default heap size on a Sun Java server virtual machine and use the *Gurobi Optimizer* for solving the Integer Linear Programs.[1]

For our applications, an algorithm is sufficiently fast if it runs in less than ten seconds, since this is usually faster than the data can be acquired. Among the exact algorithms, only the ILP managed to solve all instances of our datasets. Using the ILP the running time stayed under 5.6 minutes per instance while for about 95% of the instances, it terminated in at most 5 seconds.

The brute force algorithm mentioned in Section 3.1 runs fast on most instances of the *Orbitrap* and *QSTAR* dataset. Due to the high mass accuracy and the small compound sizes in these datasets there are only few explanations per peak and thus few vertices with the same color in the input graph. Edges with negative weights are rare. Thus, the algorithm terminates in under a second for all but three instances in *Orbitrap* and *QSTAR* datasets. But for two compounds from the *Orbitrap* dataset and 37 compounds in the *Micromass* dataset, the algorithm does not terminate in 12 hours and a week, respectively. The DP algorithm (Section 3.1) was able to solve the *QSTAR* instances exactly, since the number of colors and vertices is small for this dataset.

In Figure 2, we present performance ratios achieved by several heuristics. The tree completion heuristics (DP_{10} and DP_{15}) work very well with an output tree of weight at least 80 percent of the optimal for the DP_{15} variant. On the other hand, the *greedy* heuristic performs inferior to both DP_{10} and DP_{15}.

---

[1] *Gurobi Optimizer 4.5*. Houston, Texas: Gurobi Optimization Inc., April 2011.

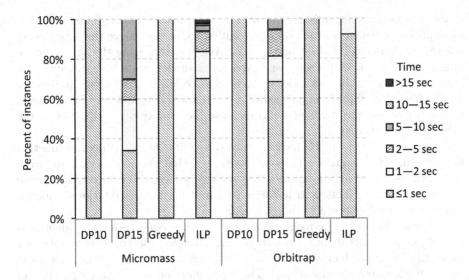

Fig. 3. Running times taken by different heuristics on *Micromass* and *Orbitrap* datasets. Nearly all calculations on the *QSTAR* dataset finished in less than a second.

The *insertion* heuristic performs better than the *greedy* heuristic on real datasets. We also observe improved performance for the tree completion heuristic as we increase the parameter, i. e., the size of the backbone computed exactly. But the performance increase is only marginal for real-world datasets, as can be seen in Figure 2. The tree completion heuristic becomes infeasible when the size of the backbone is ≥ 25. In this case, more than half of the instances from Orbitrap and Micromass datasets fail to terminate in less than a week.

The *insertion, greedy* and DP_{10} heuristics are fast with running times well under a second, whereas the DP_{15} heuristic terminates in less than 8 seconds for all instances. The algorithm based on integer programming also finished in at most 16 seconds while it was actually faster on most of the instances. Figure 3 presents the breakdown of datasets depending on how much time it took to solve them using different algorithms. Running times exclude construction of the graph representations from MS data.

5   Conclusions

We have discussed several exact and heuristic algorithms for the MAXIMUM COLORFUL SUBTREE problem, and also shown its inapproximability even for colorful input graphs. The MAXIMUM COLORFUL SUBTREE problem is relevant for the *de novo* analysis of metabolite mass spectra.

Experiments on five different sets of graphs from actual metabolite spectra reveal that for smaller test instances, the brute force and dynamic programming based algorithms can calculate the optimal solution quickly. When the structure of the fragmentation tree is of interest, for example for tree alignment [14], it is most probably beneficial to find exact solutions. Our tests show that the integer linear program performs best on this task. When determining the sum formula, hundreds of instances have to be solved for a single compound, but only the scores are relevant. In this case, the tree completion heuristic with parameter 10 provides a good performance ratio of 95% on average, and very fast running times. Increasing the parameter did increase the quality of results, but at the price of highly increased running times.

Expert evaluations in [15] indicate that trees with optimal objective function are structurally reasonable. Thus, we conjecture that this is generally the case, but, of course, we cannot guarantee that the tree with optimal objective function is the "true" tree. This is true for most problems in computational biology, however, so we deliberately ignored this fact.

Fast calculation of fragmentation trees is a prerequisite for the identification of unknown metabolites from fragmentation mass spectra. It is required for molecular formula identification [15] and the classification of unknowns using fragmentation tree alignment [14]. The algorithms presented here can play an integral role in this fully-automated pipeline.

In the future, we intend to test whether "advanced" ILP techniques, such as branch-and-cut or column generation will further decrease running times. It also remains open whether a modified version of the ILP can solve the (presumably even harder) COLORFUL SUBTREE CLOSURE instances [17] in reasonable time.

Acknowledgments. IR thanks Tsuyoshi Ito for giving the first idea of the proof of Theorem 1 [9]. Kerstin Scheubert suggested a simplification in the proof. We thank Aleš Svatoš (Max Planck Institute for Chemical Ecology in Jena, Germany), Christoph Böttcher (Institute for Plant Biochemistry in Halle, Germany), and David Grant (University of Connecticut in Storrs, Connecticut) for supplying us with the test data.

References

1. Björklund, A., Husfeldt, T., Kaski, P., Koivisto, M.: Fourier meets Möbius: fast subset convolution. In: Proc. of ACM Symposium on Theory of Computing (STOC 2007), pp. 67–74. ACM Press, New York (2007)
2. Böcker, S., Rasche, F.: Towards de novo identification of metabolites by analyzing tandem mass spectra. Bioinformatics 24, I49–I55 (2008)
3. Dondi, R., Fertin, G., Vialette, S.: Maximum Motif Problem in Vertex-Colored Graphs. In: Kucherov, G., Ukkonen, E. (eds.) CPM 2009. LNCS, vol. 5577, pp. 221–235. Springer, Heidelberg (2009)
4. Dreyfus, S.E., Wagner, R.A.: The Steiner problem in graphs. Networks 1(3), 195–207 (1972)
5. Fellows, M.R., Gramm, J., Niedermeier, R.: On the parameterized intractability of motif search problems. Combinatorica 26(2), 141–167 (2006)
6. Fernie, A.R., Trethewey, R.N., Krotzky, A.J., Willmitzer, L.: Metabolite profiling: from diagnostics to systems biology. Nat. Rev. Mol. Cell Biol. 5(9), 763–769 (2004)

7. Guillemot, S., Sikora, F.: Finding and Counting Vertex-Colored Subtrees. In: Hliněný, P., Kučera, A. (eds.) MFCS 2010. LNCS, vol. 6281, pp. 405–416. Springer, Heidelberg (2010)

8. Hill, D.W., Kertesz, T.M., Fontaine, D., Friedman, R., Grant, D.F.: Mass spectral metabonomics beyond elemental formula: Chemical database querying by matching experimental with computational fragmentation spectra. Anal. Chem. 80(14), 5574–5582 (2008)

9. Ito, T.: Finding maximum weight arborescence in an edge-weighted DAG. Theoretical Computer Science – Stack Exchange, http://cstheory.stackexchange.com/q/4088/189 (retrieved: October 12, 2011)

10. Koutis, I., Williams, R.: Limits and Applications of Group Algebras for Parameterized Problems. In: Albers, S., Marchetti-Spaccamela, A., Matias, Y., Nikoletseas, S., Thomas, W. (eds.) ICALP 2009. LNCS, vol. 5555, pp. 653–664. Springer, Heidelberg (2009)

11. Li, J.W.-H., Vederas, J.C.: Drug discovery and natural products: end of an era or an endless frontier? Science 325(5937), 161–165 (2009)

12. Ljubić, I., Weiskircher, R., Pferschy, U., Klau, G.W., Mutzel, P., Fischetti, M.: Solving the prize-collecting Steiner tree problem to optimality. In: Proc. of Algorithm Engineering and Experiments (ALENEX 2005), pp. 68–76. SIAM (2005)

13. Oberacher, H., Pavlic, M., Libiseller, K., Schubert, B., Sulyok, M., Schuhmacher, R., Csaszar, E., Köfeler, H.C.: On the inter-instrument and inter-laboratory transferability of a tandem mass spectral reference library: 1. results of an Austrian multicenter study. J. Mass Spectrom. 44(4), 485–493 (2009)

14. Rasche, F., Scheubert, K., Hufsky, F., Zichner, T., Kai, M., Svatoš, A., Böcker, S.: Identifying the unknowns by aligning fragmentation trees (October 2011) (manuscript)

15. Rasche, F., Svatoš, A., Maddula, R.K., Böttcher, C., Böcker, S.: Computing fragmentation trees from tandem mass spectrometry data. Anal. Chem. 83, 1243–1251 (2011)

16. Scheubert, K., Hufsky, F., Rasche, F., Böcker, S.: Computing fragmentation trees from metabolite multiple mass spectrometry data. J. Comput. Biol. 18(11), 1383–1397 (2011)

17. Scheubert, K., Hufsky, F., Rasche, F., Böcker, S.: Computing Fragmentation Trees from Metabolite Multiple Mass Spectrometry Data. In: Bafna, V., Sahinalp, S.C. (eds.) RECOMB 2011. LNCS, vol. 6577, pp. 377–391. Springer, Heidelberg (2011)

18. Sikora, F.: An (almost complete) state of the art around the graph motif problem. Technical report, Université Paris-Est, France (2010), http://www-igm.univ-mlv.fr/~fsikora/pub/GraphMotif-Resume.pdf

19. Sikora, F.: Aspects algorithmiques de la comparaison d'éléments biologiques. PhD thesis, Université Paris-Est (2011)

20. Xu, K., Li, W.: Many hard examples in exact phase transitions. Theor. Comput. Sci. 355(3), 291–302 (2006)

Recovering the Tree-Like Trend of Evolution Despite Extensive Lateral Genetic Transfer: A Probabilistic Analysis

Sebastien Roch[1] and Sagi Snir[2]

[1] Department of Mathematics and Bioinformatics Program, UCLA[*]
[2] Institute of Evolution, University of Haifa, Haifa 31905 Israel[**]

Abstract. Lateral gene transfer (LGT) is a common mechanism of non-vertical evolution where genetic material is transferred between two more or less distantly related organisms. It is particularly common in bacteria where it contributes to adaptive evolution with important medical implications. In evolutionary studies, LGT has been shown to create widespread discordance between gene trees as genomes become mosaics of gene histories. In particular, the Tree of Life has been questioned as an appropriate representation of bacterial evolutionary history. Nevertheless a common hypothesis is that prokaryotic evolution is primarily tree-like, but that the underlying trend is obscured by LGT. Extensive empirical work has sought to extract a common tree-like signal from conflicting gene trees. Here we give a probabilistic perspective on the problem of recovering the tree-like trend despite LGT. Under a model of randomly distributed LGT, we show that the species phylogeny can be reconstructed even in the presence of surprisingly many (almost linear number of) LGT events per gene tree. Our results, which are optimal up to logarithmic factors, are based on the analysis of a robust, computationally efficient reconstruction method and provides insight into the design of such methods. Finally we show that our results have implications for the discovery of highways of gene sharing.

Keywords: Phylogenetic Reconstruction, Lateral Gene Transfer, Quartet Reconstruction.

1 Introduction

High-throughput sequencing is transforming the study of evolution by allowing the integration of genome analysis and systematic studies, an area called phylogenomics [EF03, DBP05]. An important step in most phylogenomic analyses is the reconstruction of a tree of ancestor-descendant relationships—a gene tree—for each family of orthologous genes in a dataset. Such analyses have revealed widespread discordance between gene trees [GD08], leading some to question

---

[*] Supported by NSF grant DMS-1007144.
[**] Supported by the USA-Israel Binational Science Foundation and by the Israel Science Foundation. This work was done while SR and SS were visiting the Institute for Pure and Applied Mathematics (IPAM) at UCLA.

B. Chor (Ed.): RECOMB 2012, LNBI 7262, pp. 224–238, 2012.
© Springer-Verlag Berlin Heidelberg 2012

the meaningfulness of the Tree of Life [GDL02, ZLG04, GT05, BSL+05, DB07, Koo07]. In addition to statistical errors in gene tree estimation, various mechanisms commonly lead to incongruences between inferred gene histories, including hybridization events, duplications and losses in gene families, incomplete lineage sorting, and lateral genetic transfers [Mad97].

Here we study specifically lateral gene transfer (LGT), that is, the non-vertical transfer of genes between more or less distantly related organisms (as opposed to the standard vertical transmission between parent and offspring). Estimates of the fraction of genes that have undergone LGT vary widely—with some as high as 99%. See e.g. [DM06, GD08] and references therein. LGT is particularly common in bacterial evolution and it has been recognized to play an important role in microbial adaptation, selection and evolution with implications in the study of infectious diseases [SB05]. As a result, the bacterial phylogeny is usually inferred from genes that are thought to be immune to LGT, typically ribosomal RNA genes. However there is growing evidence that even such genes have in fact experienced LGT [YZW99, vBTP+03, SSJ03, DSS+05]. In any case, LGT appears to be a major source of conflict between gene trees that must be taken into account appropriately in phylogenomic analyses, in particular when building phylogenies. This is the problem we address in this paper.

Despite the confounding effect of LGT, we operate under the prevailing assumption that the evolution of organisms is governed primarily by vertical inheritance. In particular we ask:

1. How much genetic transfer can be coped with before the tree-like signal is completely erased?
2. What phylogenetic reconstruction methods are most effective under this hypothesis?

These questions, and other related issues, have been the subject of some empirical and simulation-based work [BHR05, GWK05, Gal07, PWK09, PWK10, KPW11]. See also [GD08, RB09] for enlightening discussions. In particular there is ample evidence that a strong tree-like signal can be extracted in the presence of extensive LGT (although some debate remains on this question [GDL02]).

In this paper we provide the first (to our knowledge) mathematical analysis of the issues above. We work under a stochastic model of gene tree topologies positing that LGT events occur at more or less random locations on the species phylogeny [Gal07]. In our main result we establish quantitative bounds implying that surprisingly high levels of LGT—almost linear in the number of branches for each gene—can be handled by simple, computationally efficient inference procedures. That amount of genetic transfer appears to be much higher than known empirical estimates of LGT frequency based on genomic datasets in prokaryotes[1]. Hence our results indicate that an accurate, reliable bacterial phylogeny should be reconstructible if the vertical inheritance hypothesis is correct. We prove that our bound on the achievable rate of LGT is tight up to logarithmic

---

[1] Note that such estimates are typically based on small numbers of genomes and, therefore, are probably lower than reality [GD08].

factors. We also show that constraining LGT to closely related species makes the tree reconstruction problem significantly easier.

Our theoretical approach complements simulation-based studies in allowing a broad range of parameters and tree shapes to be considered. Moreover our analysis provides new insights into the design of effective reconstruction methods in the presence of LGT. More precisely we focus on methodologies—both distance-based [KS01] and quartet-based [ZGC+06]—that derive their statistical power from the aggregation of basic topological information across genes.

In addition, we study the effect of so-called highways of gene sharing, roughly, preferred genetic exchanges between *specific* groups of species. Beiko et al. [BHR05] provided empirical evidence for the existence of such highways. To identify highways, they inferred LGT events by reconciling gene trees with a trusted species tree. In subsequent work, Bansal et al. [BBGS11] formalized the problem and designed a fast highway detection algorithm that aggregates conflicting signal across genes rather than solving the difficult LGT inference problem on each gene tree. Similarly to Beiko et al., Bansal et al. rely on a trusted species tree.

Here we show that a species phylogeny can be reliably estimated in the presence of *both* random LGT events and highways of LGT as long as such highways involve a small enough fraction of genes. Under extra assumptions, we also design an algorithm for inferring the location of highways. Because we first recover the species phylogeny, our highway reconstruction algorithm does not require a trusted species tree. In essence, our results on highways indicate that robust phylogeny reconstruction in the presence of random LGT extends to a phylogenetic network setting. For background on phylogenetic networks, see e.g. [HRS10].

We note that there exist related lines of work in phylogenomics addressing the issue of incomplete lineage sorting [DR09] in the presence of gene transfers and hybridization events [TRIN07, JML09, Kub09, MK09, YTDN11, CA11] as well as work on probabilistic models involving gene duplications and losses [ALS09, CM06].

The rest of the paper is organized as follows. In Section 2, we define a stochastic model of LGT and state our main results. A high-level description of our analysis is given in Section 3. Finally in Section 4 we extend our results to highways of gene sharing. For lack of space, proofs are omitted and will appear in the full version of this article.

2   Model and Main Results

Before stating our main results, we present a stochastic model of LGT. Roughly, following Galtier [Gal07], we assume that LGT events occur more or less at random along the species phylogeny. Such a model appears to be consistent with empirical evidence [GD08].

2.1   Stochastic Model of LGT

Gene Trees and Species Phylogeny. A *species phylogeny* (or phylogeny for short) is a graphical representation of the speciation history of a group of organisms. The leaves correspond to extant or extinct species. Each branching indicates a speciation event. Moreover we associate to each edge a positive value corresponding to the time elapsed along that edge. For a tree $T = (V, \mathcal{E})$ with leaf set L and a subset of leaves $X \subseteq L$, we let $T|X$ be the *restriction of T to X*, that is, the subtree of T where we keep only those vertices and edges on paths connecting two leaves in X. We say that T *agrees* (or is *consistent*) with $T|X$.

Definition 1 (Phylogeny). *A (species) phylogeny $T_s = (V_s, E_s, L_s; r, \tau)$ is a rooted tree with vertex set V_s, edge set E_s and n (labelled) leaves $L_s = [n] = \{1, \ldots, n\}$ such that 1) the degree of all internal vertices $V_s - L_s$ is exactly 3 except the root r which has degree 2, and 2) the edges are assigned inter-speciation times $\tau : E_s \to (0, +\infty)$. We assume that T_s includes $n^+ > 0$ extant species L_s^+ and $n^- \geq 0$ extinct species L_s^-, where $n = n^+ + n^-$. We also associate to each edge $e \in E_s$ in T_s a rate of lateral gene transfer $0 < \lambda(e) < +\infty$. We denote by $T_s^+ = (V_s^+, E_s^+, L_s^+; r, \tau^+)$, the subtree of T_s restricted to the extant leaves L_s^+, that is, $T_s^+ = T_s|L_s^+$ rooted at the most recent common ancestor of L_s^+. We further suppress vertices of degree 2 in T_s^+ except the root (in which case we add up the branch lengths to obtain τ^+). We call T_s^+ the extant phylogeny.*

Remark 1 (Ultrametricity). Ultrametricity holds when, from every node, the path length to all its descendant leaves is equal. As the branch lengths in the species phylogeny correspond to inter-speciation times, this property holds for the extant phylogeny T_s^+.

Remark 2 (Extinct species). Although we are ultimately interested in recovering the extant phylogeny, we include extinct species in the model as they can be involved in LGT events that affect the extant restriction of the tree. See e.g. [Mad97].

To infer the species phylogeny, we first reconstruct gene trees, that is, trees of ancestor-descendant relationships for orthologous genes or loci. Phylogenomic studies have revealed extensive discordance between such gene trees (e.g. [BSL+05, DB07]).

Definition 2 (Gene tree). *A gene tree $T_g = (V_g, E_g, L_g; \omega_g)$ for gene g is an unrooted tree with vertex set V_g, edge set E_g and $0 < n_g \leq n$ (labelled) leaves $L_g \subseteq \{1, \ldots, n\}$ with $|L_g| = n_g$ such that 1) the degree of every internal vertex is either 2 or 3, and 2) the edges are assigned branch lengths $\omega_g : E_g \to (0, +\infty)$. We let $T_g = T[T_g]$ be the topology of T_g where each internal vertex of degree 2 is suppressed.*

Remark 3 (Rooted vs. unrooted). Our stochastic model of LGT requires a *rooted* species phylogeny as time plays an important role in constraining valid LGT events. See e.g. [JNST09]. In particular our results rely on the ultrametricity property of the extant phylogeny. In contrast, branch lengths in gene trees correspond to expected numbers of mutations. As mutation rates can change from branch to branch, ultrametricity is not guaranteed and gene trees are typically *unrooted*.

Remark 4 (Taxon sampling). Each leaf in a gene tree corresponds to an extant species in the species phylogeny. However, because of gene loss and taxon sampling, a taxon may not be represented in every gene tree.

Remark 5 (Branch lengths). Each branch e in a gene tree T_g corresponds to a full or partial edge in the species phylogeny T_s. In particular, we allow internal vertices of degree 2 in a gene tree to potentially delineate between two consecutive species edges. We allow the branch lengths $w_g(e)$ to be arbitrary, but one could easily consider cases where the branch lengths are determined by interspeciation times, lineage-specific rates of substitution and gene-specific rates of substitution.

Random LGT. We formalize a stochastic model of LGT similar to Galtier's [Gal07]. See also [KS01, Suc05, JNST06] for related models. The model accounts for LGT events originating at random locations on the species phylogeny with LGT rate $\lambda(e)$ prevailing along edge e. (We come back to highways of LGT in Section 4.)

We will need the following notation. Let $T_s = (V_s, E_s, L_s; r, \tau)$ be a fixed species phylogeny. By a *location* in T_s, we mean any position along T_s seen as a continuous object (also called $\mathbb{R}$-tree), that is, a point x along an edge $e \in E_s$. We write $x \in e$ in that case. We denote the set of locations in T_s by $\mathcal{X}_s$. For any two locations x, y in $\mathcal{X}_s$, we let $\mathrm{MRCA}(x, y)$ be their most recent common ancestor (MRCA) in T_s and we let $\tau(x, y)$ be the length of the path connecting x and y in T_s under the metric naturally defined by the weights $\{\tau(e), e \in E_s\}$, interpolated linearly to locations along an edge. In words $\tau(x, y)$, which we refer to as the τ-distance between x and y, is the sum of times to x and y from $\mathrm{MRCA}(x, y)$. We say that two locations x, y are *contemporaneous* if their respective τ-distance to the root r is identical, that is,

$$\tau(x, r) = \tau(y, r).$$

For $R > 0$, we let

$$\mathcal{C}_x^{(R)} = \{y \in \mathcal{X}_s : \tau(r, x) = \tau(r, y),\ \tau(x, y) \leq 2R\}$$

be the set of locations contemporaneous to x at τ-distance at most $2R$ from x (or in other words with MRCA at τ-distance at most R). In particular, $\mathcal{C}_x^{(\infty)}$ denotes the set of all locations contemporaneous to x. We let $\Lambda(e) = \lambda(e)\tau(e)$, $e \in E_s$. We note that, since $\lambda(e)$ is the LGT rate on e, $\Lambda(e)$ gives the expected number of LGT events along e. Further, we let

$$\Lambda_{\mathrm{tot}} = \sum_{e \in E_s} \Lambda(e),$$

be the *total LGT weight* of the phylogeny and

$$\Lambda = \sum_{e \in \mathcal{E}(T_s | L_s^+)} \Lambda(e),$$

be the total LGT weight of the extant phylogeny, where $\mathcal{E}(T_s | L_s^+)$ is the edge set of $T_s | L_s^+$.

Our model of LGT is the following. Note that an LGT transfer is equivalent to a subtree-prune-and-regraft (SPR) move [SS03]. The recipient location, that is, the location receiving the genetic transfer, is the point of pruning. Similarly, the donor location is the point of regrafting.

Definition 3 (Random LGT). *Let $0 < R \leq +\infty$ possibly depending on n (i.e. not necessarily a constant) and note that we explicitly allow $R = +\infty$. Let $T_s = (V_s, E_s, L_s; r, \tau)$ be a fixed species phylogeny. Let $0 < p \leq 1$ be a sampling effort probability. A gene tree topology T_g is generated according to the following continuous-time stochastic process which gradually modifies the species phylogeny starting at the root. There are two components to the process:*

1. **LGT locations.** *The recipient and donor locations of LGT events are selected as follows:*
    - Recipient locations. *Starting from the root, along each branch e of T_s, locations are selected as recipient of a genetic tranfer according to a continuous-time Poisson process with rate $\lambda(e)$. Equivalently, the total number of LGT events is Poisson with mean Λ_{tot} and each such event is located independently according to the following density. For a location x on branch e, the density at x is $\Lambda(e)/\Lambda_{\text{tot}}$.*
    - Donor locations. *If x is selected as a recipient location, the corresponding donor location y is chosen uniformly at random in $C_x^{(R)}$. The LGT transfer is then obtained by performing an SPR move from x to y, that is, the subtree below x in T_s is moved to y in T_g. Note that we perform genetic transfers chronologically from the root.*
2. **Taxon sampling.** *Each extant leaf is kept independently with probability p. (One could also consider a different probability for each leaf. We use a fixed sampling effort p for simplicity.) The set of leaves selected is denoted by L_g. The final gene tree T_g is then obtained by keeping the subtree restricted to L_g.*

The resulting (random) gene tree topology is denoted by T_g.

Remark 6 (Preferential LGT). When $R < +\infty$ a transfer can only occur between sufficiently closely related species. One could also consider more general donor location distributions. See e.g. [PWK10]. In Section 4, we consider a different form of preferential exchanges, highways of gene sharing.

2.2   Recovering a Tree-like Trend: Main Results

Problem Statement. Let $T_s = (V_s, E_s, L_s; r, \tau)$ be an unknown species phylogeny. Using homologous gene sequences for every gene at hand, we generate N independent gene tree topologies $T_{g_1}, \ldots, T_{g_N}$ as above. Given the gene trees (or their topologies), we seek to reconstruct the topology $T_s^+ = T[T_s^+]$ of the extant phylogeny T_s^+. More precisely we are interested in the amount of LGT that can be sustained without obscuring the phylogenetic signal. To derive asymptotic results about this question, we make some assumptions on the underlying phylogeny. We discuss two cases in detail.

Remark 7 (Gene tree estimation). In practice, one estimates gene trees from sequence data. We come back to gene tree estimation issues below.

Bounded-Rates Model. The following assumption was introduced in [DR10] and is related to a common assumption in the mathematical phylogenetics literature.

Definition 4 (Bounded-rates model). *Let* $0 < \rho_\lambda < 1$ *and* $0 < \rho_\tau < 1$ *be constants. Let further* $0 < \bar{\tau} < +\infty$ *be a constant and* $0 < \bar{\lambda} < +\infty$ *be a value possibly depending on* n^+. *Under the Bounded-rates model, we consider the set of phylogenies* $T_s = (V_s, E_s, L_s; r, \tau)$ *with* $n^+ > 0$ *extant leaves and* $n^- \geq 0$ *extinct leaves and extant phylogeny* $T_s^+ = (V_s^+, E_s^+, L_s^+; r, \tau^+)$ *such that the following conditions are satisfied:*

$$\underline{\lambda} \equiv \rho_\lambda \bar{\lambda} \leq \lambda(e) \leq \bar{\lambda}, \quad \forall e \in E_s,$$

and

$$\underline{\tau} \equiv \rho_\tau \bar{\tau} \leq \tau^+(e^+) \leq \bar{\tau}, \quad \forall e^+ \in E_s^+.$$

Our result in this case is the following. We use $\bar{\lambda}$ to control the amount of LGT in the model.

Theorem 1 (Main result: Bounded-rates model, $R = +\infty$). *Let* $R = +\infty$. *Under the Bounded-rates model, it is possible to reconstruct the topology of the extant phylogeny with high probability (w.h.p.) from* $N = \Omega(\log n^+)$ *gene tree topologies if* $\bar{\lambda}$ *is such that*

$$\Lambda = O\left(\frac{n^+}{\log n^+}\right).$$

In words, we can reconstruct the species phylogeny w.h.p. as long as the expected number of LGT events Λ (as measured on the extant phylogeny) per gene is at most of the order of $\frac{n^+}{\log n^+}$. This result is based on a polynomial-time algorithm we describe in Section 3.

Remark 8 (Number of genes). Note that, in typical phylogenomic studies, the number of genes is much larger than the number of species. Therefore, our assumption that the number of genes should be at least of the order of the logarithm of the number of extant species is mild.

We also show that the bound on Λ in Theorem 1 is close to optimal, up to logarithmic factors.

Theorem 2 (Non-recoverability). *Under the Bounded-rates model as above with* $N = O(\log n^+)$, *the topology of the extant phylogeny cannot, in general, be reconstructed w.h.p. if* $\bar{\lambda}$ *is such that* $\Lambda = \Omega(n^+ \log \log n^+)$.

Theorem 2 is derived by the coupling method [Lin92]. In words, we show that, above $\Lambda = \Omega(n^+ \log \log n^+)$ expected LGT events, there is insufficient signal from the original species phylogeny. More generally, the species phylogeny cannot be reconstructed from N genes if $\Lambda = \Omega(n^+ \log N)$.

Yule Process. Branching processes are commonly used to model species phylogenies [RY96]. In the continuous-time Yule process (or pure-birth process), one starts with two species (representing the two branches emanating from the root). At any given time, each species generates a new offspring at rate $0 < \nu < +\infty$. We stop the process when the number of species is exactly $n + 1$ (and ignore the $n + 1$st species). This process generates a species phylogeny with $n = n^+$ extant species with branch lengths given by the inter-speciation times in the above process. Note that $n^- = 0$ by construction. Let $0 < \rho_\lambda < 1$ be a constant. We also assume that

$$\underline{\lambda} \equiv \rho_\lambda \overline{\lambda} \leq \lambda(e) \leq \overline{\lambda}, \quad \forall e \in E_s,$$

for some $0 < \overline{\lambda} < +\infty$ possibly depending on n. As above, we use $\overline{\lambda}$ to control the amount of LGT in the model.

An advantage of the Yule model is that, unlike the Bounded-rates model, it does not place arbitrary constraints on the inter-speciation times. In particular, the following analog of Theorem 1 shows that our analysis does not hinge on such constraints.

Theorem 3 (Main result: Yule process, $R = +\infty$). *Let $R = +\infty$. Under the Yule model, the following holds with probability arbitrarily close to 1. It is possible to reconstruct the topology of the extant phylogeny w.h.p. from $N = \Omega(\log n)$ gene tree topologies if $\overline{\lambda}$ is such that*

$$\Lambda = O\left(\frac{n}{\log n}\right).$$

Extensions. When $R < +\infty$, that is, when transfers occur only between sufficiently related species, we obtain the following generalization which implies that preferential LGT makes the tree-building problem easier.

Theorem 4 (Preferential LGT). *Let $0 < R < \log n^+$ possibly depending on n^+. Under the Bounded-rates model, it is possible to reconstruct the topology of the extant phylogeny w.h.p. from $N = \Omega(\log n^+)$ gene tree topologies if $\overline{\lambda}$ is such that*

$$\Lambda = O\left(\frac{n^+}{R}\right).$$

A similar result holds under the Yule model.

In Section 4, we also extend our results to cases where highways of LGT are present. The results above rely on a quartet-based method detailed in Section 3.

Sequence Length Issues. We have assumed so far that gene tree topologies are perfectly reconstructed. Of course, this is not the case in practice and the effect of sequence lengths is important to account for. One issue that arises is that LGT events may create very short branches that are difficult to infer. We deal with these issues in the full version of the paper.

3   Probabilistic Analysis

We assume that we are given N independent gene tree topologies $\mathcal{T}_{g_1}, \ldots, \mathcal{T}_{g_N}$ as above. Our goal is to reconstruct the extant phylogeny.

Different algorithms are possible. A simple approach is to take a majority vote over all gene tree topologies. But this approach is problematic under taxon sampling and cannot sustain the high levels of LGT we consider below.

Instead we consider approaches that aggregate partial information over all gene trees. We focus on subtrees over four taxa whose topologies are called quartets [SS03]. We show that computationally efficient quartet-based approaches can sustain high levels of LGT. Although we prove our results for the specific method described below, our analysis is likely to apply to related methods.

3.1   Algorithm

We consider the following approach related to an algorithm of Zhaxybayeva et al. [ZGC+06]. Let X be a four-tuple of extant species The topology $\mathcal{T}|X$ of a tree $\mathcal{T}$ restricted to X can be summarized with a *quartet split* (or *quartet* for short). For $X = \{a, b, c, d\}$ there are three possible (resolved) quartets denoted $q_1 = ab|cd$, $q_2 = ac|bd$, and $q_3 = ad|bc$. We first compute the frequency of each quartet over all gene trees displaying X, that is,

$$f_X(q_1) = \frac{|\{g_i : X \subseteq L_{g_i}, \mathcal{T}_{g_i}|X = q_1\}|}{|\{g_i : X \subseteq L_{g_i}\}|},$$

and similarly for q_2, q_3. (We set the frequency to 0 if the denominator is 0.) For each X, we choose the quartet with highest frequency (breaking ties arbitrarily).

Definition 5. *A set of quartets $Q = \{q_i\}$, with L_{q_i} the leaf set of q_i, is compatible if there is a tree $\mathcal{T}$ with leaf set $L_Q \equiv \cup_{q_i \in Q} L_{q_i}$ such that $\mathcal{T}$ agrees with every q_i.*

Quartet compatibility is, in general, NP-hard [Ste92]. However, when the set Q contains all possible four-tuple of taxa (i.e., exactly $\binom{n}{4}$ quartets), there is a polynomial-time algorithm for compatibility [BD86, Bun71, BG01]. In our procedure, for every four-tuple of taxa, there is a single quartet chosen, so we can check compatibility easily and output the corresponding tree. In practice, if Q is not compatible, one can use instead a heuristic supertree method such as MRP [Rag92, Bau92] or Quartet MaxCut [SR10, SR12].

3.2   A General Formula

Our asymptotic analysis is based on the following claim. Recall that, for a subset of extant species X, we let $\mathcal{T}_s|X$ be the extant phylogeny topology restricted to X with corresponding edge set $\mathcal{E}(\mathcal{T}_s|X)$. Also recall that $\Lambda(e) = \lambda(e)\tau(e)$ is the expected number of LGT events on e which we refer to as the *LGT weight* (or simply *weight*) of e. Let

$$\Lambda_X = \sum_{e \in \mathcal{E}(\mathcal{T}_s|X)} \Lambda(e),$$

be the total weight of the subtree $\mathcal{T}_s|X$ under the weights $\Lambda(e)$, $e \in E_s$. Define the *maximum quartet weight (MQW)* to be

$$\Upsilon^{(4)} = \max\{\Lambda_X : X \subseteq (L_s^+)^4\}.$$

Lemma 1 (Probability of a miss). *Let $\mathcal{T}_g$ be a gene tree topology distributed according to the random LGT model such that $X = \{a, b, c, d\} \subseteq L_g$. Let q_g^X (respectively q_s^X) be the quartet split corresponding to $\mathcal{T}_g|X$ (respectively $\mathcal{T}_s|X$). Then*

$$\mathbb{P}[q_g^X = q_s^X | X \subseteq L_g] \geq \exp\left(-\Upsilon^{(4)}\right).$$

Recall that Λ is the expected number of LGT events (as measured on the extant phylogeny) per gene. As a comparison, note that the probability that a gene tree is LGT-free is $e^{-\Lambda}$, which can be much smaller.

3.3 Bounded-Rates and Yule Models

Next we argue that, under appropriate assumptions on the species phylogeny, the maximum quartet weight is bounded in such a way that the plurality quartet topology for every four-tuple of taxa $X = \{a, b, c, d\}$, which we denote by q_*^X, satisfies $q_*^X = q_s^X$. As a result, our quartet set is compatible and $\mathcal{T}_s^+$ can be reconstructed efficiently.

Bounded-rates Model. We bound the maximum quartet weight $\Upsilon^{(4)}$ in the Bounded-rates model.

Lemma 2 (Bound on quartet weight: Bounded-rates case). *Under the Bounded-rates model it holds that*

$$\Upsilon^{(4)} = O\left(\bar{\lambda} \log n^+\right).$$

Using Lemma 2, we prove Theorem 1.

Yule Process. We now consider the Yule model.

Lemma 3 (Bound on quartet weight: Yule case). *Under the Yule model, it holds that*

$$\Upsilon^{(4)} = O\left(\bar{\lambda} \log n\right),$$

with probability approaching 1 as $n \to +\infty$.

Using Lemma 3, we prove Theorem 3.

4 Highways of LGT

In this section, we add highways of gene sharing to the model. Highways are, in essence, non-random patterns of LGT [BHR05]. These can potentially take different shapes. Following Bansal et al. [BBGS11], we focus on pairs of edges in the phylogeny that undergo an unusually large (that is, more than random) number of LGT events between them. Of course, for highways to be detectable, they must involve a sufficiently large fraction of genes.

We give two results. As long as the frequency of genes affected by highways is low enough, the species phylogeny can be reconstructed using the same approach as in Section 3. Moreover, with extra assumptions on the positions of the highways with respect to each other, the highways themselves can be inferred.

4.1   Model

We generalize our model of LGT as follows.

Definition 6 (Highways of LGT). *Let $T_s = (V_s, E_s, L_s; r, \tau)$ be a species phylogeny with LGT rates $0 < \lambda(e) < +\infty$, $e \in E_s$ and let $0 < p \le 1$ be a taxon sampling probability. For $\beta = 1, \ldots, B$, let $\mathbf{H}_\beta = (e^H_{\beta,0}, e^H_{\beta,1})$ be a pair of edges in T_s which share contemporaneous locations. We call $\mathbf{H}_\beta$ a highway. Let $g_1, \ldots, g_N$ be N genes. Each highway $\mathbf{H}_\beta$ involves a subset $\mathbf{G}^H_\beta$ of the genes. If gene $g_i \in \mathbf{G}^H_\beta$, then it undergoes an LGT event between a pair of contemporaneous locations $x^H_{\beta,i} \in e^H_{\beta,0}$ and $y^H_{\beta,i} \in e^H_{\beta,1}$. We let γ_β be the fraction of genes such that $g_i \in \mathbf{G}^H_\beta$ and we assume that $\gamma_\beta > \underline{\gamma}$ for some $\underline{\gamma}$ (chosen below). In addition, independently from the above, we assume that each gene undergoes LGT events at random locations as described in Definition 3. We denote by $\mathcal{T}_{g_1}, \ldots, \mathcal{T}_{g_N}$ the gene tree topologies so obtained.*

Remark 9 (Deterministic setting). Note that the highways and which genes are involved in them are deterministic in this setting. Only the random LGT events are governed by a stochastic process. Note moreover that we allow highway events to go in either direction, that is, from $e^H_{\beta,0}$ to $e^H_{\beta,1}$ or vice versa.

4.2   Building the Species Tree in the Presence of Highways

We first prove that the species phylogeny can still be reconstructed in the presence of highways as long as the fraction of genes involved in highways is low enough. We only discuss the Bounded-rates model with $R = +\infty$. Similar results hold for the other models we considered in Section 3.

Theorem 5 (Highways of LGT). *Consider the Bounded-rates model with $R = +\infty$ and assume that $B < +\infty$ is constant. Assume further that there is a constant $0 < \overline{\gamma} < 1$ such that*

$$\gamma_\beta < \overline{\gamma}, \quad \beta = 1, \ldots, B.$$

If

$$\overline{\gamma} < \frac{1}{2B},$$

then it is possible to reconstruct the topology of the extant phylogeny w.h.p. from $N = \Omega(\log n^+)$ gene tree topologies if $\overline{\lambda}$ is such that

$$\Lambda = O\left(\frac{n^+}{\log n^+}\right).$$

4.3 Inferring Highways

The problem of inferring the highway locations is essentially a network reconstruction problem. Such problems are often computationally intractable. See
e.g. [HRS10]. Therefore, we require some extra assumptions. Our goal here is
not to provide the most general result but rather to illustrate that our analysis extends naturally to certain network settings. The following assumption is
related to so-called galled trees.

Assumption 1. *We assume that all highway locations lie on the extant phylogeny T_s^+ and that, further, no highway connects two adjacent edges in T_s^+ or
edges adjacent to root edges. (Such cases cannot be reconstructed.) Seen as an
edge superimposed on T_s^+, a highway event $(x_{\beta,i}^H, y_{\beta,i}^H)$ forms a cycle. We assume
that all such cycles are disjoint, that is, they do not share common locations.*

We then prove the following. We use a computationally efficient algorithm described in the proof.

Theorem 6 (Inferring highways). *Consider the Bounded-rates model with
$R = +\infty$ and assume that $B < +\infty$ is constant. Assume further that there are
constants $0 < \underline{\gamma} < \overline{\gamma} < +\infty$ such that*

$$\underline{\gamma} < \gamma_\beta < \overline{\gamma}, \quad \beta = 1, \dots, B.$$

If

$$\overline{\gamma} < \frac{1}{2},$$

*and Assumption 1 holds then it is possible to reconstruct the topology of the
extant phylogeny as well as the highway edges w.h.p. from $N = \Omega(\log n^+)$ gene
tree topologies if $\overline{\lambda}$ is such that*

$$\Lambda = O\left(\frac{n^+}{\log n^+}\right).$$

5 Discussion

We have shown that a species phylogeny or network can be reconstructed despite
high levels of random LGT and we have provided explicit quantitative bounds on
tolerable rates of LGT. Moreover our analysis sheds light on effective approaches
for species tree building in the presence of LGT. Several problems remain open:

- Galtier and Daubin [GD08] speculate that random LGT only becomes a
  significant hurdle when the rate of LGT greatly exceeds the rate of diversi-
  fication. In our setting this would imply that a value of Λ as high as $\Omega(n)$
  may be achievable. Note that branches close to the leaves are particularly
  easy to reconstruct because they lie on small quartet trees that are less likely
  than deep ones to be hit by an LGT event. Is a recursive approach starting
  from the leaves possible here? See [Mos04, DMR11] for recursive approaches
  in a related context.

- In a related problem, we have analyzed a quartet-based method. A better understanding of bipartition-based approaches is needed and may lead to a higher threshold for Λ.
- What can be proved when the number of genes is significantly less than $\log n$?
- In the presence of highways, dealing with more general network settings would be desirable. Also our definition of highways as connecting two edges is somewhat restrictive. In general, one is also interested in preferential genetic transfers between clades.
- On the practical side, the predictions made here should be further tested on real and simulated datasets. We note that there is extisting work in this direction [BHR05, GWK05, Gal07, PWK09, PWK10, KPW11, BBGS11].

References

[ALS09] Arvestad, L., Lagergren, J., Sennblad, B.: The gene evolution model and computing its associated probabilities. J. ACM 56(2) (2009)

[Bau92] Baum, B.R.: Combining trees as a way of combining data sets for phylogenetic inference. Taxon 41, 3–10 (1992)

[BBGS11] Bansal, M.S., Banay, G., Peter Gogarten, J., Shamir, R.: Detecting highways of horizontal gene transfer. Journal of Computational Biology 18(9), 1087–1114 (2011)

[BD86] Bandelt, H.-J., Dress, A.: Reconstructing the shape of a tree from observed dissimilarity data 7, 309–343 (1986)

[BG01] Berry, V., Gascuel, O.: Inferring evolutionary trees with strong combinatorial evidence. Theoretical Computer Science (240), 271–298 (2001)

[BHR05] Beiko, R.G., Harlow, T.J., Ragan, M.A.: Highways of gene sharing in prokaryotes. Proc. Natl. Acad. Sci. USA 102, 14332–14337 (2005)

[BSL+05] Bapteste, E., Susko, E., Leigh, J., MacLeod, D., Charlebois, R.L., Doolittle, W.F.: Do orthologous gene phylogenies really support tree-thinking? BMC Evol. Biol. 5, 33 (2005)

[Bun71] Buneman, P.: The recovery of trees from measures of dissimilarity. In: Hodson, F.R., Kendall, D.G., Tautu, P. (eds.) Anglo-Romanian Conference on Mathematics in the Archaeological and Historical Sciences, pp. 387–395. Edinburgh University Press, Mamaia (1971)

[CA11] Chung, Y., Ane, C.: Comparing two bayesian methods for gene tree/species tree reconstruction: Simulations with incomplete lineage sorting and horizontal gene transfer. Systematic Biology 60(3), 261–275 (2011)

[CM06] Csűrös, M., Miklós, I.: A Probabilistic Model for Gene Content Evolution with Duplication, Loss, and Horizontal Transfer. In: Apostolico, A., Guerra, C., Istrail, S., Pevzner, P.A., Waterman, M. (eds.) RECOMB 2006. LNCS (LNBI), vol. 3909, pp. 206–220. Springer, Heidelberg (2006)

[DB07] Doolittle, W.F., Bapteste, E.: Pattern pluralism and the tree of life hypothesis. Proc. Natl. Acad. Sci. USA 104, 2043–2049 (2007)

[DBP05] Delsuc, F., Brinkmann, H., Philippe, H.: Phylogenomics and the reconstruction of the tree of life. Nat. Rev. Genet. 6(5), 361–375 (2005)

[DM06] Dagan, T., Martin, W.: The tree of one percent. Genome Biology 7(10), 118 (2006)

[DMR11]   Daskalakis, C., Mossel, E., Roch, S.: Evolutionary trees and the ising model on the bethe lattice: a proof of steel's conjecture. Probability Theory and Related Fields 149, 149–189 (2011), doi:10.1007/s00440-009-0246-2

[DR09]   Degnan, J.H., Rosenberg, N.A.: Gene tree discordance, phylogenetic inference and the multispecies coalescent. Trends in Ecology and Evolution 24(6), 332–340 (2009)

[DR10]   Daskalakis, C., Roch, S.: Alignment-Free Phylogenetic Reconstruction. In: Berger, B. (ed.) RECOMB 2010. LNCS, vol. 6044, pp. 123–137. Springer, Heidelberg (2010)

[DSS+05] Dewhirst, F.E., Shen, Z., Scimeca, M.S., Stokes, L.N., Boumenna, T., Chen, T., Paster, B.J., Fox, J.G.: Discordant 16S and 23S rRNA gene phylogenies for the Genus Helicobacter: Implications for phylogenetic inference and systematics. J. Bacteriol. 187(17), 6106–6118 (2005)

[EF03]   Eisen, J.A., Fraser, C.M.: Phylogenomics: Intersection of evolution and genomics. Science 300(5626), 1706–1707 (2003)

[Gal07]   Galtier, N.: A model of horizontal gene transfer and the bacterial phylogeny problem. Systematic Biology 56(4), 633–642 (2007)

[GD08]   Galtier, N., Daubin, V.: Dealing with incongruence in phylogenomic analyses. Philos. Trans. R Soc. Lond. B Biol. Sci. 363, 4023–4029 (2008)

[GDL02]   Peter Gogarten, J., Ford Doolittle, W., Lawrence, J.G.: Prokaryotic evolution in light of gene transfer. Molecular Biology and Evolution 19(12), 2226–2238 (2002)

[GT05]   Peter Gogarten, J., Townsend, J.P.: Horizontal gene transfer, genome innovation and evolution. Nat. Rev. Micro. 3(9), 679–687 (2005)

[GWK05]   Ge, F., Wang, L.S., Kim, J.: The cobweb of life revealed by genome-scale estimates of horizontal gene transfer. PLoS Biol. 3, e316 (2005)

[HRS10]   Huson, D.H., Rupp, R., Scornavacca, C.: Phylogenetic Networks: Concepts, Algorithms and Applications. Cambridge University Press, Cambridge (2010)

[JML09]   Joly, S., McLenachan, P.A., Lockhart, P.J.: A statistical approach for distinguishing hybridization and incomplete lineage sorting. The American Naturalist 174(2), E54–E70 (2009)

[JNST06] Jin, G., Nakhleh, L., Snir, S., Tuller, T.: Maximum likelihood of phylogenetic networks. Bioinformatics 22(21), 2604–2611 (2006)

[JNST09] Jin, G., Nakhleh, L., Snir, S., Tuller, T.: Parsimony score of phylogenetic networks: Hardness results and a linear-time heuristic. IEEE/ACM Trans. Comput. Biology Bioinform. 6(3), 495–505 (2009)

[Koo07]   Koonin, E.V.: The biological big bang model for the major transitions in evolution. Biol. Direct. 2, 21 (2007)

[KPW11]   Koonin, E.V., Puigbo, P., Wolf, Y.I.: Comparison of phylogenetic trees and search for a central trend in the forest of life. Journal of Computational Biology 18(7), 917–924 (2011)

[KS01]   Kim, J., Salisbury, B.A.: A tree obscured by vines: Horizontal gene transfer and the median tree method of estimating species phylogeny. In: Pacific Symposium on Biocomputing, pp. 571–582 (2001)

[Kub09]   Kubatko, L.S.: Identifying hybridization events in the presence of coalescence via model selection. Systematic Biology 58(5), 478–488 (2009)

[Lin92]   Lindvall, T.: Lectures on the Coupling Method. Wiley, New York (1992)

[Mad97]   Maddison, W.P.: Gene trees in species trees. Systematic Biology 46(3), 523–536 (1997)

[MK09]   Meng, C., Kubatko, L.S.: Detecting hybrid speciation in the presence of incomplete lineage sorting using gene tree incongruence: A model. Theoretical Population Biology 75(1), 35–45 (2009)

[Mos04]   Mossel, E.: Phase transitions in phylogeny. Trans. Amer. Math. Soc. 356(6), 2379–2404 (2004)

[PWK09]   Puigbo, P., Wolf, Y., Koonin, E.: Search for a 'tree of life' in the thicket of the phylogenetic forest. Journal of Biology 8(6), 59 (2009)

[PWK10]   Puigbo, P., Wolf, Y.I., Koonin, E.V.: The tree and net components of prokaryote evolution. Genome Biology and Evolution 2, 745–756 (2010)

[Rag92]   Ragan, M.A.: Matrix representation in reconstructing phylogenetic-relationships among the eukaryotes. Biosystems 28, 47–55 (1992)

[RB09]   Ragan, M.A., Beiko, R.G.: Lateral genetic transfer: open issues. Philosophical Transactions of the Royal Society B: Biological Sciences 364(1527), 2241–2251 (2009)

[RY96]   Rannala, B., Yang, Z.: Probability distribution of molecular evolutionary trees: A new method of phylogenetic inference. J. Mol. Evol. 43, 304–311 (1996)

[SB05]   Smets, B.F., Barkay, T.: Horizontal gene transfer: perspectives at a crossroads of scientific disciplines. Nat. Rev. Micro. 3(9), 675–678 (2005)

[SR10]   Snir, S., Rao, S.: Quartets maxcut: A divide and conquer quartets algorithm. IEEE/ACM Trans. Comput. Biology Bioinform. 7(4), 704–718 (2010)

[SR12]   Snir, S., Rao, S.: Quartet maxcut: A fast algorithm for amalgamating quartet trees. Molecular Phylogenetics and Evolution 62(1), 1–8 (2012)

[SS03]   Semple, C., Steel, M.: Phylogenetics. Mathematics and its Applications series, vol. 22. Oxford University Press (2003)

[SSJ03]   Schouls, L.M., Schot, C.S., Jacobs, J.A.: Horizontal transfer of segments of the 16S rRNA genes between species of the Streptococcus anginosus group. J. Bacteriol. 185(24), 7241–7246 (2003)

[Ste92]   Steel, M.: The complexity of reconstructing trees from qualitative characters and subtress. Journal of Classification 9(1), 91–116 (1992)

[Suc05]   Suchard, M.A.: Stochastic models for horizontal gene transfer. Genetics 170(1), 419–431 (2005)

[TRIN07]   Than, C., Ruths, D., Innan, H., Nakhleh, L.: Confounding factors in hgt detection: Statistical error, coalescent effects, and multiple solutions. Journal of Computational Biology 14(4), 517–535 (2007)

[vBTP+03] van Berkum, P., Terefework, Z., Paulin, L., Suomalainen, S., Lindstrom, K., Eardly, B.D.: Discordant phylogenies within the rrn loci of rhizobia. J. Bacteriol. 185(10), 2988–2998 (2003)

[YTDN11]   Yu, Y., Than, C., Degnan, J.H., Nakhleh, L.: Coalescent histories on phylogenetic networks and detection of hybridization despite incomplete lineage sorting. Systematic Biology 60(2), 138–149 (2011)

[YZW99]   Yap, W.H., Zhang, Z., Wang, Y.: Distinct types of rrna operons exist in the genome of the Actinomycete Thermomonospora chromogena and evidence for horizontal transfer of an entire rRNA operon. J. Bacteriol. 181(17), 5201–5209 (1999)

[ZGC+06]   Zhaxybayeva, O., Peter Gogarten, J., Charlebois, R.L., Ford Doolittle, W., Thane Papke, R.: Phylogenetic analyses of cyanobacterial genomes: Quantification of horizontal gene transfer events. Genome Research 16(9), 1099–1108 (2006)

[ZLG04]   Zhaxybayeva, O., Lapierre, P., Gogarten, J.P.: Genome mosaicism and organismal lineages. Trends Genet 20, 254–260 (2004)

RESQUE: Network Reduction Using Semi-Markov Random Walk Scores for Efficient Querying of Biological Networks (Extended Abstract)

Sayed Mohammad Ebrahim Sahraeian and Byung-Jun Yoon

Department of Electrical and Computer Engineering,
Texas A&M University, College Station, TX 77843, USA
msahraeian@tamu.edu, bjyoon@ece.tamu.edu

Abstract. In this work, we present RESQUE, an efficient algorithm for querying large-scale biological networks. The algorithm uses a semi-Markov random walk model to estimate the correspondence scores between nodes across different networks. The target network is iteratively reduced based on the node correspondence scores, which are also iteratively re-estimated for improved accuracy, until the best matching subnetwork emerges. The proposed network querying scheme is computationally efficient, can handle any network query with arbitrary topology, and yields accurate querying results.

1 Introduction

The increasing availability of large-scale biological networks, including protein-protein interaction (PPI) networks, metabolic networks, and co-expression networks, asks for effective tools that can be used to analyze these network data and gain useful biological insights. Network querying is one such example, whose goal is to identify subnetwork regions in a large target network that are similar to a given query network [3]. For instance, a network querying algorithm can be used to search for novel potential pathways in a given biological network that are similar to known pathways, thereby enabling knowledge transfer from well-studied species to less-studied ones. The optimal network querying problem has been shown to be NP-complete, by reduction to the graph isomorphism problem [2], and various approaches have been proposed so far to make it computationally feasible [2,4,1]. Despite recent advances in the field, most of the currently existing algorithms still have a number of practical limitations. For example, most algorithms cannot handle network queries with arbitrary topology, and the structure of the query network is often limited to linear paths or trees [2]. Or alternatively, the query network is simply treated as a collection of nodes by ignoring the network topology [1]. Moreover, many existing algorithms suffer from high computational complexity, which often increases exponentially with the query size, making them unsuitable for performing large queries.

In this work, we propose a novel network querying algorithm called RESQUE (**RE**duction-based scheme using **S**emi-Markov scores for network **QUE**rying)

B. Chor (Ed.): RECOMB 2012, LNBI 7262, pp. 239–240, 2012.

that can effectively address the aforementioned issues. RESQUE takes a probabilistic approach for fast and accurate network querying, which iteratively reduces the target network based on the so-called node correspondence scores computed via a semi-Markov random walk model. These scores provide a probabilistic measure of the similarities between nodes that belong to different networks (i.e., the query and the target networks) and can be efficiently computed by a closed-form formula. At each iteration, the estimated node correspondence scores are used to shrink the search space in the target network by removing the nodes that have minimal correspondence to the query nodes. The node correspondence scores are then re-estimated based on the reduced network, and the network reduction process is repeated until the target network has been sufficiently reduced. The iterative re-estimation of the correspondence scores leads to better querying results with higher expected accuracy. After the iterative reduction process, the final querying result is obtained using two different strategies. The first strategy, called RESQUE-M, uses the maximum weighted matching algorithm to find the best matching subnetwork that maximizes the expected accuracy. The second strategy, called RESQUE-C, finds the largest connected subnetwork in the reduced target network. RESQUE does not restrict the topology of the query network, and it can handle linear paths, trees, and general structures with loops. Furthermore, the query can be connected, partially connected, or simply a collection of nodes with unknown network topology.

To assess the performance of RESQUE, we carried out network queries of 1,184 protein complexes of size $4 \sim 25$ in the PPI networks of fly, yeast, and human, which are the three largest PPI networks that are currently available. We compared our algorithm with Torque [1], a state-of-the-art network querying algorithm, which has been shown to outperform many existing algorithms. Our results show that both RESQUE-M and RESQUE-C yield significantly larger number of hits compared to Torque, with higher functional coherence rates at comparable specificity levels. The computational complexity of RESQUE is polynomial in terms of the query size, and it is able to complete each query within a few seconds for all complexes considered in our experiments, while Torque may need more than an hour for some of them. Further performance assessment based on simulated data shows that RESQUE is also highly robust against distortions in the node similarity score as well as random node deletions and insertions.

References

1. Bruckner, S., Huffner, F., Karp, R.M., Shamir, R., Sharan, R.: Topology-free querying of protein interaction networks. J. Comput. Biol. 17, 237–252 (2010)
2. Dost, B., Shlomi, T., Gupta, N., Ruppin, E., Bafna, V., Sharan, R.: QNet: A tool for querying protein interaction networks. J. Comput. Biol. 15(7), 913–925 (2008)
3. Sharan, R., Ideker, T.: Modeling cellular machinery through biological network comparison. Nat. Biotechnol. 24, 427–433 (2006)
4. Yang, Q., Sze, S.: Path matching and graph matching in biological networks. J. Comput. Biol. 14, 56–67 (2007)

Detecting SNP-Induced Structural Changes in RNA: Application to Disease Studies

Raheleh Salari[1], Chava Kimchi-Sarfaty[2],
Michael M. Gottesman[3], and Teresa M. Przytycka[1,*]

[1] National Center for Biotechnology Information,
National Library of Medicine, NIH,
Bethesda, MD 20894, USA
[2] Laboratory of Hemostasis, Division of Hematology,
Center for Biologics Evaluation and Research,
Food and Drug Administration, Bethesda, MD 20892, USA
[3] Center for Cancer Research, National Cancer Institute,
NIH, Bethesda, MD 20892, USA
przytyck@mail.ncbi.nlm.nih.gov

Genotype-phenotype association studies continue to provide ever larger sets of Single Nucleotide Polymorphisms (SNPs) linked to diseases, or associated with responses to vaccines, medications, and environmental factors. Such associations provide an important step in studies of the genetic underpinnings of human diseases. To gain further insight, a deeper understanding of the molecular mechanisms by which a SNP affects cell function is necessary. When a SNP is localized within a gene or in the close neighborhood of a gene, then it is generally assumed that it affects the phenotype through changes at the expression level, the function, or other properties of this particular gene. However, the molecular mechanisms that lead to the change are usually not obvious. In the case of non-synonymous SNPs, where the underlying mutation occurs in the gene coding region and changes an amino-acid, it is usually expected that this amino-acid change affects protein function, expression, conformation or stability. However, numerous results point to diseases related to occurrences of SNPs in non-translated regions [4, 7] and synonymous SNPs (reviewed in [6]), suggesting that amino-acid change is not the sole explanation. In such cases, pinpointing the mechanism responsible for the functional change is considerably more challenging. Possible causes include, among others, structural changes in the RNA transcript that can in turn influence splicing, expression, stability , or translational regulation. Interestingly, a recent study indicated that a common amino acid deletion that was long thought to be responsible for cystic fibrosis is coincident with a synonymous mutation that results in a change in mRNA structure that may be responsible for the disease [1]. There is also statistical evidence for a conserved, local secondary structure in the coding regions of eukaryotic mRNAs and pre-mRNAs [5] and selection for mRNA stability (see [2] and references within).

---

[*] Corresponding author.

B. Chor (Ed.): RECOMB 2012, LNBI 7262, pp. 241–243, 2012.
© Springer-Verlag Berlin Heidelberg 2012

However, the current understanding of the role of SNP-induced RNA structural alterations on phenotypic changes is limited.

To study the effect of SNP-related changes in RNA structure on cell function, it is necessary to have a rigorous and efficient way of measuring the structural changes caused by SNPs. While there have been several previous attempts to address this question, none of the existing approaches have fully solved the problem. We propose that to measure the impact of a point mutation on RNA structure, one should measure the difference in probability distributions of the wild-type and mutant Boltzmann ensembles. Such differences should be able to capture the emergence of new structural clusters and shifts in energy landscape. In information theory, the well established measure for comparing probability distributions is relative entropy. While theoretically logical, it was not obvious how to efficiently compute relative entropy for ensembles of RNA structures. Note that ensemble size grows exponentially with RNA length and enumerating all the elements of such an ensemble is not usually feasible. Thus, a more sophisticated approach was necessary. In this work, we close this gap and provide *remuRNA*, an efficient method to compute *r*elative *e*ntropy between the wild-type and *mu*tant *RNA* structural ensembles.

We used remuRNA to compare the impact of SNPs naturally occurring in populations to the impact of point mutations randomly inserted into the RNA sequences. If RNA structure is important for gene function, then inserting a point mutation into a position where it naturally does not occur should, on average, generate more significant changes than changes caused by SNPs naturally occurring in the human population. This is indeed what we found. This result indicates there is natural selection against SNPs that significantly change RNA structure. This is also an important factor that needs to be taken into account when such random point mutations are used as a background for estimating the prominence of structural changes caused by disease-associated SNPs.

Subsequently, we applied remuRNA to examine which of the disease-associated SNPs are potentially related to structural changes in RNA. The method applies to any type of RNA sequence. However, the importance of RNA structure, especially in the 5'UTR region is very well established and thus studies of non-coding RNA provide a good test of our approach. While our method pointed to many SNPs previously identified by Halvorsen and colleagues [3], it also identified many additional SNPs in which RNA structure alteration is a potential cause for the disease phenotype.

These results indicate that the proposed approach is not only mathematically well justified and computationally efficient, but also a powerful way to study the impact of SNP-induced RNA structural changes on gene expression and function.

References

[1] Bartoszewski, R.A., Jablonsky, M., Bartoszewska, S., Stevenson, L., Dai, Q., Kappes, J., Collawn, J.F., Bebok, Z.: A synonymous single nucleotide polymorphism in DeltaF508 CFTR alters the secondary structure of the mRNA and the expression of the mutant protein. J. Biol. Chem. 285, 28741–28748 (2010)

[2] Chamary, J.V., Hurst, L.D.: Evidence for selection on synonymous mutations affecting stability of mRNA secondary structure in mammals. Genome Biol. 6, R75 (2005)

[3] Halvorsen, M., Martin, J.S., Broadaway, S., Laederach, A.: Disease-associated mutations that alter the RNA structural ensemble. PLoS Genet. 6, e1001074 (2010)

[4] Kikinis, Z., Eisenstein, R.S., Bettany, A.J., Munro, H.N.: Role of RNA secondary structure of the iron-responsive element in translational regulation of ferritin synthesis. Nucleic Acids Res. 23, 4190–4195 (1995)

[5] Meyer, I.M., Miklos, I.: Statistical evidence for conserved, local secondary structure in the coding regions of eukaryotic mRNAs and pre-mRNAs. Nucleic Acids Res. 33, 6338–6348 (2005)

[6] Sauna, Z.E., Kimchi-Sarfaty, C.: Understanding the contribution of synonymous mutations to human disease. Nat. Rev. Genet. 12, 683–691 (2011)

[7] Villette, S., Kyle, J.A., Brown, K.M., Pickard, K., Milne, J.S., Nicol, F., Arthur, J.R., Hesketh, J.E.: A novel single nucleotide polymorphism in the 3' untranslated region of human glutathione peroxidase 4 influences lipoxygenase metabolism. Blood Cells Mol. Dis. 29, 174–178 (2002)

A Model for Biased Fractionation
after Whole Genome Duplication

David Sankoff, Chunfang Zheng, and Baoyong Wang

Department of Mathematics and Statistics, University of Ottawa

Paralog reduction, the loss of duplicate genes after whole genome duplication (WGD), causes more gene order disruption than chromosomal rearrangements such as inversion or translocation. This is particularly true in flowering plants.

Whether the loss proceeds gene by gene or through deletion of multi–gene DNA segments is controversial, as is the question of fractionation bias, namely whether one of the two homeologous chromosomes is more vulnerable to deletion. The process is subject to the overriding functional constraint that only one gene copy can be deleted.

As a null hypothesis, we assume deletion events, on either homeolog, excise a geometrically distributed number of genes with unknown mean μ. A variable number r of these events overlap to produce deleted runs of length l. There is a fractionation bias $0 \leq \phi \leq 1$ for deletions to occur on one specified homeolog. The random variable r has some distribution $\pi(\cdot)$, which depends on θ, the proportion of remaining genes in duplicate form.

As a preliminary step, we simulate the distribution of run lengths l as a function of μ and ϕ, and show how to estimate these parameters, based on run lengths of single-copy regions. The main part of this work is the derivation of a deterministic recurrence to calculate each $\pi(r)$ as a function of μ, ϕ and θ. The recurrence for π provides a deeper mathematical understanding of fractionation process than simulations.

http://www.biomedcentral.com/1471-2164/13/S1/S8/

B. Chor (Ed.): RECOMB 2012, LNBI 7262, p. 244, 2012.
© Springer-Verlag Berlin Heidelberg 2012

Exact Pattern Matching
for RNA Structure Ensembles

Christina Schmiedl[1],[*], Mathias Möhl[1],[*], Steffen Heyne[1],[*], Mika Amit[2],
Gad M. Landau[2],[3], Sebastian Will[1],[4],[**],
and Rolf Backofen[1],[5],[**]

[1] Bioinformatics, Institute of Computer Science, Albert-Ludwigs-Universität,
Freiburg, Germany
{will,backofen}@informatik.uni-freiburg.de
[2] Department of Computer Science, University of Haifa, Haifa, Israel
[3] Department of Computer Science and Engineering, NYU-Poly, Brooklyn NY, USA
[4] CSAIL and Mathematics Department, MIT, Cambridge MA, USA
[5] Center for Biological Signaling Studies (BIOSS), Albert-Ludwigs-Universität,
Freiburg, Germany

Abstract. ExpaRNA's core algorithm computes, for two fixed RNA structures, a maximal non-overlapping set of maximal exact matchings. We introduce an algorithm ExpaRNA-P that solves the lifted problem of finding such sets of exact matchings in entire Boltzmann-distributed structure ensembles of two RNAs. Due to a novel kind of structural sparsification, the new algorithm maintains the time and space complexity of the algorithm for fixed input structures. Furthermore, we generalized the chaining algorithm of ExpaRNA in order to compute a compatible subset of ExpaRNA-P's exact matchings. We show that ExpaRNA-P outperforms ExpaRNA in BRAliBase 2.1 benchmarks, where we pass the chained exact matchings as anchor constraints to the RNA alignment tool LocARNA. Compared to LocARNA, this novel approach shows similar accuracy but is six times faster.

1   Introduction

Genome-wide transcriptomics [1–3] provides evidence for massive transcription in eukaryotic genomes, going as far as suggesting that most of both genomic strands of human might be transcribed [4]. Most of these transcripts do not code for proteins; furthermore, it has become clear that the majority of them perform primarily regulatory functions [5]. Thus, these RNAs play a crucial role in the living cell. However, their functional annotation is strongly lagging behind; reliable automated annotation pipelines exist only for subclasses of ncRNAs such as tRNAs, microRNAs or snoRNAs [6].

Regulatory RNAs are often structural, and their secondary structures are then usually well-conserved due to their functional importance. This fact is used by a

---

[*] Joint first authors.
[**] Joint corresponding authors.

B. Chor (Ed.): RECOMB 2012, LNBI 7262, pp. 245–260, 2012.

priori RNA-gene finders like QRNA [7], RNAz [8], and Evofold [9], which detect conserved RNA-structures in whole genome alignments. More recently, a strategy towards the automatic annotation of non-coding RNAs has emerged, which identifies RNAs with similar sequence and common secondary structure [10–12] on a genomic scale. This can be used to determine remote members of RNA-*families* as defined in the Rfam-database, or to determine new RNA-*classes* of structural, and hence likely functionally similar, ncRNAs by using clustering approaches. In Rfam, the term RNA-*clan* has been introduced for a collection of RNA-families that share similar structure but little sequence conservation. Prominent examples of RNA-*classes* are microRNAs or snoRNAs.

Albeit this approach is appealing, a wide-spread, or even automated, application of these methods has been hindered by the huge complexity of the underlying sequence/structure alignment approach for detecting similarity in both sequence *and* structure. The first practical approaches for multiple structural alignment, such as RNAforester [13] and MARNA [14], depend on predicted or known secondary structures. In practice, however, these approaches are limited by the low accuracy of non-comparative structure prediction. Sankoff's algorithm [15] provides a general solution to the problem of simultaneously computing an alignment and the common secondary structure of two aligned sequences. In its full form, the problem requires $O(n^6)$ CPU time and $O(n^4)$ memory (for RNA sequences of length n). This complexity is already limiting for most practical problems such as routinely scanning remote members of RNA-*families*. For detecting novel RNA-*classes* in the plethora of newly discovered RNA-transcripts, this complexity becomes plainly prohibitive, since this task requires clustering based on quadratically many all-against-all pairwise RNA comparisons.

For that reason, many variants of the Sankoff algorithm with different optimizations have been introduced. FoldAlign [16] and dynalign [17] implement a full energy model for RNA that is evaluated during the alignment computation. In contrast, PMcomp [18] and LocARNA [10] use a lightweight energy model, which assignes energies to single base pairs. This simplification reduces the computational cost significantly. They achieve their accuracy by precomputing the energy contributions of base pairs from their probabilities in a full-featured energy model [19]. Whereas the approaches [20, 16, 17, 21] have to compensate their computational demands by strong, often sequence-based, heuristics, LocARNA [10] takes advantage of structural sparsity in the RNA structure ensembles to reduce its complexity to $O(n^4)$ time and $O(n^2)$ space. This successful approach is consequently found in other Sankoff-like methods [22–24].

We introduce a strategy that reduces the computational demands further, but differs fundamentally from heuristic improvements, like [24], that restrict the search space based on sequence alignments. It computes the sequence-structure-conserved elements that form highly probable local substructures in the RNA structure ensemble of both input RNAs; subsequently, these elements are used as anchor constraints in a full sequence-structure alignment by LocARNA. In [25], we have proposed a similar strategy, which computes conserved elements in pairs of fixed RNA secondary structures, based on an algorithm with quadratic time and space complexity [26]. Albeit this approach reduces the overall computation time significantly, it faces similar problems as the first generation of RNA aligment methods [13, 14], due to the use of a single predicted input structure for each sequence. Since predicting minimum free energy (MFE) structures from single-sequences is unreliable, this strategy fails frequently and causes severe misalignments.

Overcoming the problems of the previous approach, the novel algorithm for determining exact sequence-structure patterns is based on probabilities in the RNA structure ensembles. We point out that a straight-forward extension of the fixed input structure algorithm to RNA structure ensembles, would result in a complexity of $O(n^4)$ time and $O(n^2)$ space. This complexity is as high as the one of LocARNA, which would nullify the benefits of exact matching.

Thus, our main technical contribution is to solve the ensemble based problem in quadratic time and space; the advancement is comparable to the leap from first generation RNA alignment to efficient Sankoff-style alignment. For this achievement, we introduce a method of sparsification that uses the ensemble properties of the input sequences. Previous sparsification approaches reduced the number of computations required for each entry [27–31] or the number of matrices to be considered [10, 22]. In addition, we identify sparse regions of each matrix *a priori* such that, in total, only quadratically many entries remain; each of these entries is calculated in constant time. The *a priori* identification of sparse regions is based on the joint probability that a sequence position occurs as part of a particular loop. Since the sum of these probabilities is bound by one, we can control the complexity on a global scale by setting a probability threshold. As a further benefit over sparsification methods that filter non-optimal solutions [27–31], our sparsification allows us to enumerate suboptimal solutions.

To evaluate the practical benefits of these algorithmic innovations, we devise a novel pipeline ExpLoc-P for sequence-structure alignment. In its first stage, it enumerates suboptimal exact matchings of local sequence-structure patterns due to the introduced algorithm ExpaRNA-P. In the second stage, the suboptimal matchings are chained to select an optimal subset of compatible matchings that can simultaneously occur in an alignment of RNAs. Finally, these matchings are utilized as anchor constraints in a subsequent Sankoff-style alignment by LocARNA. In benchmarks on BRAliBase 2.1, ExpLoc-P's accuracy is comparable to the unconstrained LocARNA, although it is six times faster.

2   Preliminaries

A RNA sequence A is a string over the alphabet $\{A, C, G, U\}$, the base at the i-th position of A is denoted by A_i, the subsequence from position i to j by $A_{i..j}$ and the length of it by $|A|$. A structure of A is a set P_A of base pairs $p = (i, j)$ such that $1 \le i < j \le |A|$, where A_i and A_j form a complementary Watson-Crick base pair (A-U or C-G) or a non-standard base pair G-U. We denote the left end i of p by p^L and the right end j by p^R. For a single structure, we also assume that each sequence position is involved in at most one base pair (for all $(i, j), (i', j') \in P_A$: ($i = i' \Leftrightarrow j = j'$) and $i \ne j'$) and base pairs do not cross (there are no $(i, j), (i', j') \in P_A$ with $i < i' < j < j'$).

Since non-crossing RNA structures correspond to trees, we define, for any position k of A, the *parent of k* as the $(i, j) \in P_A$ with $i < k < j$ such that there does not exist any $(i', j') \in P_A$ with $i < i' < k < j' < j$. Analogously, the *parent of a base pair (i, j)* is the parent of i (which is also the parent of j). Intuitively, if a base or base pair has a parent (i, j), it is located in the loop closed by (i, j). For external positions k that are not included in any loop, we define the parent to be an additional imaginary base pair $(0, |A| + 1)$ covering the entire sequence.

3 Exact Pattern Matchings in RNA Structure Ensembles

In this section, we formalize the problem that our algorithm ExpaRNA-P solves. We fix sequences A and B.

Definition 1 (EPM). *An Exact Pattern Matching (EPM) is a tuple $(\mathcal{M}, \mathcal{S})$ with $\mathcal{M} \subseteq \{(i \sim k) | i \in \{1, \dots, |A|\}, k \in \{1, \dots, |B|\}\}$ and $\mathcal{S} \subseteq \{(ij \sim kl) | (i, j) \in \{1, \dots, |A|\}^2, i < j, (k, l) \in \{1, \dots, |B|\}^2, k < l\}$ such that*

- *for all $(i \sim k) \in \mathcal{M} : A_i = B_k$*
- *for all $(i \sim k), (j \sim l) \in \mathcal{M} : (i < j \Rightarrow k < l \land i = j \Leftrightarrow k = l)$*
- *$(ij \sim kl) \in \mathcal{S} \Rightarrow \{(i \sim k), (j \sim l)\} \subseteq \mathcal{M}$*
- *the structure $\{(i, j) | (ij \sim kl) \in \mathcal{S}\}$ is non-crossing (with the previous condition this implies $\{(k, l) | (ij \sim kl) \in \mathcal{S}\}$ is non-crossing).*
- *the matching is connected on the sequence or structure level, i.e. the graph $(\mathcal{M}, E)$ with $E = \{(i \sim k, j \sim l) | (i = j + 1 \text{ and } k = l + 1) \text{ or } (ij \sim kl) \in \mathcal{S}\}$ is (weakly) connected.*

EPMs are exact matches that are not necessarily contiguous subsequences, but structure-local in the sense of [32, 33]. We define the set of structure matches as $\mathcal{M}|_{\mathcal{S}} := \{(i \sim k), (j \sim l) | (ij \sim kl) \in \mathcal{S}\}$. Note that the correspondence between nested RNA structures and trees naturally generalizes to EPMs. Therefore, we define the parent of some element of $\mathcal{M} \cup \mathcal{S}$ as

$$\text{parent}_{\mathcal{S}}(i \sim k) = \underset{(i'j' \sim k'l') \in \mathcal{S} \cup \{(0|A|+1 \sim 0|B|+1)\}, i' \leq i \leq j'}{\text{argmin}} |j' - i'| \qquad (1)$$

$$\text{parent}_{\mathcal{S}}(ij \sim kl) = \underset{(i'j' \sim k'l') \in \mathcal{S} \cup \{(0|A|+1 \sim 0|B|+1)\}, i' < i < j < j'}{\text{argmin}} |j' - i'| \qquad (2)$$

Note that every matched element that is not enclosed by matched base pairs has the *pseudo-parent* $(0|A| + 1 \sim 0|B| + 1)$ which is best understood as additional match of pseudo-base pairs outside of the two sequences. Also note that for $\text{parent}_{\mathcal{S}}(ij \sim kl) \in \mathcal{S}$ $\text{parent}_{\mathcal{S}}(ij \sim kl) \neq \text{parent}_{\mathcal{S}}(i \sim k) = \text{parent}_{\mathcal{S}}(j \sim l) = (ij \sim kl)$.

Since we only want to match structures that are probable in the ensemble of the given sequences, we define the notion of significant EPMs. Considering only significant EPMs is crucial for both the quality of the results and the complexity of the algorithm. To define significant EPMs we consider the following probabilities over the Boltzmann ensemble of structures.

- $\Pr\{(i,j)|X\}$ denotes the probability, that a structure in the ensemble of $X \in \{A, B\}$ contains the base pair (i, j),
- $\Pr_{(i,j)}^{\text{loop}}(k|X)$ denotes for $i < k < j$ and $X \in \{A, B\}$ the joint probability that the structure of X contains the base pair (i, j) and the unpaired base k such that (i, j) is the parent of k.
- $\Pr_{(i,j)}^{\text{loop}}((i',j')|X)$ denotes for $i < i' < j' < j$ and $X \in \{A, B\}$ the joint probability that the structure of X contains the base pairs (i, j) and (i', j') and that (i, j) is the parent of (i', j').

In the special case where $(i, j) = (0, |A| + 1)$ we define $\Pr_{(0,|A|+1)}^{\text{loop}}((i',j')|X) :=$ $\Pr\{(i',j')|X\}$ and $\Pr_{(0,|A|+1)}^{\text{loop}}(k|X)$ as the probability that base k of X is unpaired, i.e. $1 - \sum_{j<i} \Pr\{(j,i)|X\} - \sum_{i<j} \Pr\{(i,j)|X\}$.[1] In Sec. 4 we show how to compute the probabilities efficiently.

For significant EPMs we introduce three different thresholds θ_1, θ_2 and θ_3. We require that all matched base pairs have a probability of at least θ_1 and that the probabilities of all matched unpaired bases and matched base pairs to occur as part of the loop of their respective parent is at least θ_2 and θ_3, respectively.

Definition 2 (significant EPM). *Given the thresholds θ_1, θ_2, θ_3, an EPM is significant iff*

- *for all $(ij \sim kl) \in \mathcal{S}$:* $\Pr\{(i,j)|A\} \geq \theta_1$ *and* $\Pr\{(k,l)|B\} \geq \theta_1$
- *for all $(i \sim k) \in \mathcal{M} \setminus \mathcal{M}|_{\mathcal{S}}$ with $(i'j' \sim k'l') = \text{parent}_{\mathcal{S}}(i \sim k)$:* $\Pr_{(i',j')}^{\text{loop}}(i|A) \geq \theta_2$ *and* $\Pr_{(k',l')}^{\text{loop}}(k|B) \geq \theta_2$
- *for all $(ij \sim kl) \in \mathcal{S}$ with $(i'j' \sim k'l') = \text{parent}_{\mathcal{S}}(ij \sim kl)$:* $\Pr_{(i',j')}^{\text{loop}}((i,j)|A) \geq \theta_3$ *and* $\Pr_{(k',l')}^{\text{loop}}((k,l)|B) \geq \theta_3$

The score of an EPM $(\mathcal{M}, \mathcal{S})$ consists of a score $\sigma(i, k)$ for each pair of matched unpaired bases and $\tau(i, j, k, l)$ for each pair of matched base pairs:

$$\text{score}(\mathcal{M}, \mathcal{S}) = \sum_{(i \sim k) \in \mathcal{M} \setminus \mathcal{M}|_{\mathcal{S}}} \sigma(i, k) + \sum_{(ij \sim kl) \in \mathcal{S}} \tau(i, j, k, l) \tag{3}$$

The algorithm described in this paper determines all significant maximally extended EPMs up to a certain score threshold, where maximally extended is defined as follows.

Definition 3 (maximally extended EPMs). *An EPMs $(\mathcal{M}, \mathcal{S})$ is maximally extended, if there does not exist any $(\mathcal{M}', \mathcal{S}')$ with $\mathcal{M} \subset \mathcal{M}'$, $\mathcal{S} \subseteq \mathcal{S}'$ and such that for all $(i \sim k) \in \mathcal{M}$ $\text{parent}_{\mathcal{S}}(i \sim k) = \text{parent}_{\mathcal{S}'}(i \sim k)$.*

---

[1] Note that these probabilities include the cases where (i, j) or k are covered by some base pair. This is reasonable as the EPMs are structurally local; thus, they can be enclosed by other structure or be external.

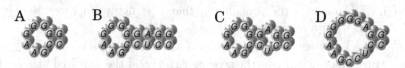

Fig. 1. EPM A is not maximally extended if there exists a larger EPM like B or C. EPMs B, C, and D can all be maximally extended simultaneously since in each case some base matches have different parents.

As shown in Fig. 1, the last condition of this definition is required to ensure that we consider EPMs with different structures as being different. Due to this definition, the set of maximally extended EPMs does not contain proper substructures, such as Fig. 1A depicts a proper substructure of the EPM of Fig. 1B, but contains structural variants of the same set of matched positions. We select a relevant subset of the structural variants in the answer set of our algorithm by considering only significant EPMs.

4   The Algorithm ExpaRNA-P

Precomputing Likely Loops. In a preprocessing step, we compute, separately for each sequence, the probabilities required to determine the significant EPMs. Hence, in clustering scenarios, for example, where all pairs from a set of sequences need to be matched, this preprocessing needs to be done only once for each sequence and not for all quadratically many pairs. To simplify notation, we show how to compute the probabilities for A, the computation for B is identical. While the base pair probabilities $\Pr\{(i,j)|A\}$ are computed by McCaskill's algorithm [19], we extend this algorithm to compute the probabilities $\Pr_{(i,j)}^{\text{loop}}(k|A)$ and $\Pr_{(i,j)}^{\text{loop}}((i',j')|A)$. For this purpose, we utilize the matrices Q_{ij}, Q_{ij}^b, Q_{ij}^m, and Q_{ij}^{m1} of McCaskill's algorithm (details can be found in [19]). For $1 \leq i \leq j \leq |S|$, the entries of these matrices represent the sum over the Boltzmann weights of the following set of structures of $A_{i..j}$

- Q_{ij}: all structures of $A_{i..j}$
- Q_{ij}^b: all structures P of $A_{i..j}$ with $(i,j) \in P$
- Q_{ij}^m: all non-empty structures of $A_{i..j}$ scored as part of a multiloop
- Q_{ij}^{m1}: all structures P of $A_{i..j}$, scored as part of a multiple loop, such that for some k holds $(i,k) \in P$ and for all $(i',j') \in P$ holds $i \leq i' < j' \leq k$.

Intuitively Q_{ij}^{m1} counts the Boltzmann weights of all structures that are part of a multiloop and have exactly one outermost base pair, starting at position i. In addition to the classical McCaskill matrices, we compute a matrix $Q_{ij}^{m2} = \sum_{i<k<j-1} Q_{ik}^m Q_{k+1\,j}^{m1}$, representing parts of a multiloop with at least two outermost base pairs.

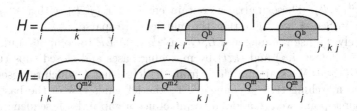

Fig. 2. Recursions of the partition functions for computing the probability that a position k occurs inside a hairpin loop (H), interior loop (I), or multiloop (M) closed by a base pair (i,j)

Given those matrices, we compute $\mathrm{Pr}^{\mathrm{loop}}_{(i,j)}(k|A)$ as

$$\mathrm{Pr}^{\mathrm{loop}}_{(i,j)}(k|A) = \mathrm{Pr}\{(i,j)|A\}\frac{H+I+M}{Q^b_{ij}}, \text{ where} \tag{4}$$

$$H = \exp(-\beta F_1(i,j)) \tag{5}$$

$$I = \sum_{\substack{i',j' \\ i<k<i'<j'<j}} \exp(-\beta F_2(i,j,i',j'))Q^b_{i'j'} + \sum_{\substack{i',j' \\ i<i'<j'<k<j}} \exp(-\beta F_2(i,j,i',j'))Q^b_{i'j'} \tag{6}$$

$$M = Q^{m2}_{k+1j-1}\exp(-\beta(a+(k-i)c)) + Q^{m2}_{i+1k-1}\exp(-\beta(a+(j-k)c)) \tag{7}$$
$$+ Q^m_{i+1k-1}Q^m_{k+1j}\exp(-\beta(a+c))$$

The formulas are visualized in Fig. 2. H, I, and M represent the cases where k is contained in a hairpin, interior loop, and multiloop, respectively. The constants a, c, k, and T and the energy functions F_1 and F_2 are defined as in McCaskill [19], where $\beta := (kT)^{-1}$. The three sums in the computation of M cover the cases where k is in the leftmost, the rightmost, and any other unpaired region of the loop, respectively. Note that Q^{m2} is required to ensure, without increasing complexity, that M considers only multiloops (with at least two inner base pairs). In a similar way we compute $\mathrm{Pr}^{\mathrm{loop}}_{(i,j)}((i',j')|A)$. All these joint probabilities are computed within the same asymptotic complexity as the McCaskill algorithm.

Computing the Significant EPMs. The algorithm computes table entries $D((ij),(kl))$, which store the best EPM enclosed by each base pair match ($ij \sim kl$). More precisely, $D((ij),(kl))$ has entries for each $(i,j) \in P_A$ and $(k,l) \in P_B$ with $\mathrm{Pr}\{(i,j)|A\} \geq \theta_1$ and $\mathrm{Pr}\{(k,l)|B\} \geq \theta_1$ and $D((ij),(kl))$ denotes the maximum score of a significant EPM $(\mathcal{M},\mathcal{S})$ of $A_{i..j}$ and $B_{k..l}$ with $(ij \sim kl) \in \mathcal{S}$. The entries of D are computed in increasing order with respect to their size such that during the computation of some $D((ij),(kl))$ any $D((i'j'),(k'l'))$ with $i < i' < j' < j$ and $k < k' < l' < l$ is already computed. For the computation of each $D((ij),(kl))$ we compute matrices $L^{ijkl}(j',l')$, $G^{ijkl}_A(j',l')$, $G^{ijkl}_{AB}(j',l')$, and $LR^{ijkl}(j', l')$. These matrices contain entries for (j', l') with $i < j' < j$, $k < l' < l$. In Sec. 4, we argue that the matrices are sparse.

Intuitively, we use the matrices L, G_A, G_{AB}, and LR to compute a matching of the loops below (i, j) and (k, l) by matching bases and closed substructures from left to right. We start matching from the left using L which represents the part of the matching that is connected to the left ends i, k of the base pairs. Then at some point we are allowed to introduce a gap in both sequences using matrices G_A and G_{AB} and then start matching the part that is connected to the right ends j, l of the base pairs using matrix LR.

The matrices are computed according to the recursions visualized in Fig. 3. The base cases are $L^{ijkl}(i, k) = 0$, $L^{ijkl}(j', k) = -\infty$ for all $j' > i$, $L^{ijkl}(i, l') = -\infty$ for all $l' > k$, $LR^{ijkl}(i, l') = G_A^{ijkl}(i, l') = G_{AB}^{ijkl}(i, l') = 0$ for all $l' > k$ and $LR^{ijkl}(j', k) = G_A^{ijkl}(j', k) = G_{AB}^{ijkl}(j', k) = 0$ for all $j' > i$. Intuitively, the recursion for L always matches the last positions j' and l' or assigns $-\infty$ if they don't match. Left of this match of the last positions can either be a matched unpaired position of the loops (second case) or a match of two base pairs (third case). The recursion for LR is analogous to L except that it considers the additional case that the gap has just been ended at positions $j' - 1$ and $l' - 1$. The gap itself, computed in G_A and G_{AB}, simply allows to skip once an arbitrary number of positions in both sequences when going from the left matched part to the right matched part. To avoid ambiguity, the recursion enforces to first skip the positions in A (using G_A) and after that the positions of B (using G_{AB}). This is necessary for the suboptimal traceback which would otherwise enumerate the same solutions more than once. Also note that in the computation of $D((ij), (kl))$ not only LR^{ijkl} but also L^{ijkl} is considered. Here, LR^{ijkl} represents the situations where the best EPM contains a gap, and L^{ijkl} the situation where the best matching has no gap, i.e. the parts matched at the left and right ends are connected.

After the matrix D has been computed, a final matrix F is computed where for $0 \leq j' \leq |A|$ and $0 \leq l' \leq |B|$ each $F(j', l')$ denotes the maximum score of a significant EPM of $A_{1..j'}$ and $B_{1..l'}$ which ends at (j', l') (i.e. with $(j \sim l) \in \mathcal{M}$). The base cases are $F(j', 0) = F(0, l') = 0$ for all j', l'. The recursion for F (Fig. 3) is almost identical to the recursion for L, except for the first case, which is 0 instead of $-\infty$, since the EPMs in F are (similar to local sequence alignments) allowed to start at any point. Also, since the base pairs of F are external (i.e. not enclosed by some other base pair of the EPM), the check for the second and third condition of significant EPMs (Def. 2) are discarded.

The suboptimal maximally extended EPMs are obtained by doing standard suboptimal tracebacks enumerating all EPMs up to a given score threshold. Since the recursions are all unambiguous (i.e. the cases do not overlap) no EPM is enumerated more than once. To enumerate only maximally extended EPMs, we start tracebacks only from entries $F(j', l')$ for which $A_{j'+1} \neq B_{l'+1}$.

Lemma 1. *A maximally extended EPM $(\mathcal{M}, \mathcal{S})$ of $A_{1..j'}$ and $B_{1..l'}$ with $(j' \sim l') \in \mathcal{M}$ is also a maximally extended EPM of A and B, iff $A_{j'+1} \neq B_{l'+1}$.*

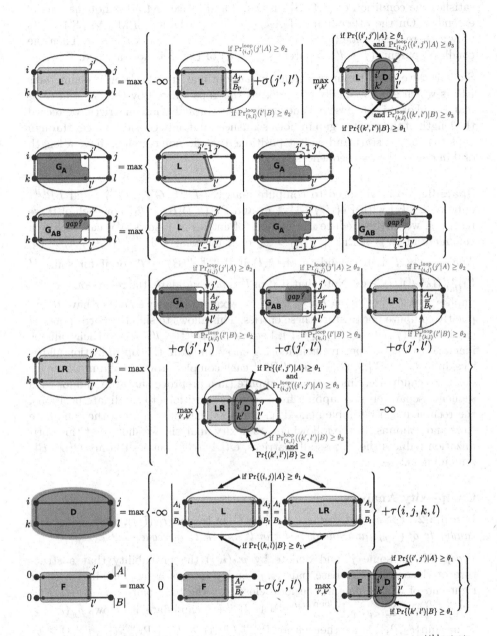

Fig. 3. Visualization of the recursions to compute the matrix entries $L^{ijkl}(j',l')$, $G_A^{ijkl}(j',l')$, $G_{AB}^{ijkl}(j',l')$, $LR^{ijkl}(j',l')$, $D((ij),(kl))$, and $F(j',l')$

Proof. Obviously, if $A_{j'+1} = B_{l'+1}$ the larger EPM $(\mathcal{M} \cup \{(j'+1 \sim l'+1)\}, \mathcal{S})$ satisfies the condition of $(\mathcal{M}', \mathcal{S}')$ in Def. 3 and hence $(\mathcal{M}, \mathcal{S})$ is not maximally extended. On the other hand, if $A_{j'+1} \neq B_{l'+1}$ a larger EPM $(\mathcal{M}', \mathcal{S}')$ exists only if there exist some $(ij \sim kl) \in \mathcal{S}'$ with $i \leq j' < j$ and $k \leq l' < l$. Then the condition $\text{parent}_\mathcal{S}(j' \sim l') = \text{parent}_{\mathcal{S}'}(j' \sim l')$ of Def. 3 is not satisfied.

Similarly, we need to ensure that the suboptimal traceback only enumerates EPMs which are maximally extended at the gaps created by the G_A and G_{AB} matrices. For this purpose, whenever we trace through these matrices, we record the length of the created gap in both sequences and only consider traces through gaps that either start and end at positions that do not match or have a length of 0 in one of the two sequences.

Sparsification. We need to compute matrices L^{ijkl}, G_A^{ijkl}, G_{AB}^{ijkl}, and LR^{ijkl} only for i, j, k, l with $\Pr\{(i, j)|A\} \geq \theta_1$ and $\Pr\{(k, l)|B\} \geq \theta_1$. For each of these matrices, we further reduce the number of entries as follows. We call each j' a *candidate* of (i, j) if $\Pr_{(i,j)}^{\text{loop}}(j'|A) \geq \theta_2$ or if for some i' $\Pr_{(i,j)}^{\text{loop}}((i', j')|A) \geq \theta_3$. Analogously, l' is a candidate of (k, l) if $\Pr_{(k,l)}^{\text{loop}}(l'|B) \geq \theta_2$ or if for some k' $\Pr_{(k,l)}^{\text{loop}}((k', l')|B) \geq \theta_3$. Note that if j' or l' is no candidate, the recursion directly implies that $L^{ijkl}(j', l') = LR^{ijkl}(j', l') = -\infty$ and hence we neither have to explicitly compute nor to store these entries. This allows to skip the corresponding entries $G_A^{ijkl}(j', l')$ and $G_{AB}^{ijkl}(j', l')$, because for $L^{ijkl}(j', l') = -\infty$ their value is identical to their respective neighboring entry. In total, this optimization allows to skip in L^{ijkl}, G_A^{ijkl}, G_{AB}^{ijkl}, and LR^{ijkl} each complete row or column whose index is no candidate. Since we can compute (in a preprocessing step and for each sequence separately) a mapping from sequence positions to candidate positions, the recursion can be implemented on matrices that only contain the candidate rows and columns. In the following complexity analysis, we show that this optimization reduces the, across all matrices, $O(|A|^3|B|^3)$ entries to only $O(|A||B|)$ remaining entries.

Complexity Analysis

Lemma 2. *For a fixed j', there are only $O(1)$ base pairs (i, j), such that j' is a candidate of (i, j) (and analogously for l' and (k, l) in sequence B).*

Proof. We fix some j' and denote by $p_{j'}(i, j)$ the probability that a structure of A contains the base pair (i, j) and j' occurs as an unpaired base or right end of a base pair in the loop closed by the base pair (i, j): $p_{j'}(i, j) := \Pr_{(i,j)}^{\text{loop}}(j'|A) + \sum_{i < i' < j'} \Pr_{(i,j)}^{\text{loop}}((i', j')|A)$. If j' is a candidate, it follows $p_{j'}(i, j) \geq \theta^* := \min\{\theta_2, \theta_3\}$, since then either $\Pr_{(i,j)}^{\text{loop}}(j'|A) \geq \theta_2$ or $\Pr_{(i,j)}^{\text{loop}}((i', j')|A) \geq \theta_3$ for some i'. Note that for different (i, j) the events of probabilities $p_{j'}(i, j)$ are disjoint, since in any structure j' can occur in just one loop. Therefore $\sum_{i,j} p_{j'}(i, j) \leq 1$. Hence there are at most $\frac{1}{\theta^*} \in O(1)$ base pairs (i, j) for which $p_{j'}(i, j) \geq \theta^*$ and only for those j' can be a candidate.

Note that for this lemma it is crucial to consider the probabilities within the single loops and not only general base pair and unpaired probabilities. Considering these probabilities is the key insight of this new way of sparsification.

Theorem 1. *There are only $O(n^2)$ entries $L^{ijkl}(j', l')$, $G_A^{ijkl}(j', l')$, $G_{AB}^{ijkl}(j', l')$, and $LR^{ijkl}(j', l')$ such that j' is a candidate of (i, j) and l' is a candidate of (k, l).*

Proof. Due to Lem. 2 there are $O(n)$ many combinations i, j, j'. Analogously there are $O(n)$ combinations k, l, l' and therefore $O(n^2)$ combinations i, j, k, l, j', l' satisfying the conditions.

Corollary 1. *The time and space complexity of computing all entries $L^{ijkl}(j', l')$, $G_A^{ijkl}(j', l')$, $G_{AB}^{ijkl}(j', l')$, and $LR^{ijkl}(j', l')$, D and F is $O(n^2)$.*

Consequently, the preprocessing step, namely computing the base pair probabilities using McCaskill's algorithm is the dominating factor in the complexity.

Chaining. Furthermore, we implemented a chaining algorithm that selects from the computed suboptimal EPMs a non-crossing and non-overlapping subset that can be extended to an alignment. It generalizes the chaining of ExpaRNA [25] to cope with more than one EPM ending at the same position. The algorithm recursively fills the gaps of all EPMs with other EPMs. For each of the gaps a matrix of size $O(|A||B|)$ is computed (for details see [25]). At each of its entries all EPMs are considered that end at this position. Since each EPM ends at exactly one position, the complexity is $O(H \cdot (|A||B| + E))$, where E is the number of input EPMs and H the total number of their gaps.

If we guarantee that E is $O(|A||B|)$, i.e. there is only a constant number of EPMs ending at each position, the complexity of the chaining is $O(H|A||B|)$ (as in ExpaRNA). Whereas the suboptimal traceback does not *guarantee* $E \in O(|A||B|)$, we also evaluated a heuristic strategy that satisfies the assumption by considering only the best EPM ending at each position.

5   Evaluation

We implemented ExpaRNA-P together with the chaining algorithm in C++. Furthermore we implemented two versions of the traceback: the suboptimal traceback and a heuristic version that, for each match $i \sim j$, considers only the optimal EPM ending at that match. We instantiated the scoring of ExpaRNA-P (see Eq. 3) by $\sigma(i, k) = 1$ and $\tau(i, j, k, l) = 5(\Pr\{(i, j)|A\} + \Pr\{(k, l)|B\}) + 2$. In addition to the presented scoring, we add a reward of $5\Pr\{(i, j)|(i + 1, j - 1)|X\}$ ($X \in \{A, B\}$) for each stacking in the EPM. In the suboptimal traceback, we enumerate EPMs that have a score of at least 90 and a score difference of less than 20 to the optimal EPM. Furthermore, we set $\theta_1 = \theta_2 = 0.01$ and $\theta_3 = 0$.

In order to assess the performance of ExpaRNA-P in comparison to other alignment tools, we designed the following pipeline: In a first step we compute the significant EPMs with ExpaRNA-P and use the chaining algorithm to extract from these EPMs an optimal non-overlapping and non-crossing subset.

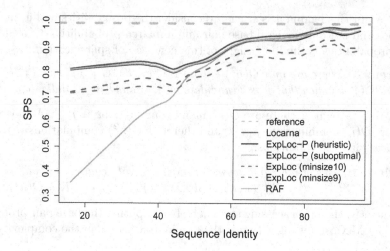

Fig. 4. Alignment quality vs. sequence identity on the k2 dataset of BRAliBase 2.1

Then we compute a sequence structure alignment that includes all matches of the chained EPMs. For this purpose, we apply LocARNA using the EPMs as anchor constraints [25]. This is faster than computing an unconstrained alignment since each anchor reduces the alignment space. We refer to this pipeline, i.e. the combination of ExpaRNA-P and LocARNA, as ExpLoc-P.

We did a benchmark test on the k2 dataset of BRAliBase 2.1 which contains only pairwise alignments [34, 35]. To measure the quality of the calculated alignment in comparison to the reference alignment, we utilized the compalign score which refers to a sum-of-pairs score (SPS) introduced in this specific form with BRAliBase 2.1 [34]. Besides the quality of the results, we also compared the runtime of the different methods. We compared our new approach ExpLoc-P with three other approaches: LocARNA without any anchor constraints, ExpLoc [25], and RAF [24]. ExpLoc is similar to our new ExpLoc-P except that it identifies EPMs in the MFE structure (using ExpaRNA). RAF is the currently fastest Sankoff-style sequence structure alignment approach due to its heuristic filtering based on sequence alignments. Fig. 4 shows the compalign score with respect to the sequence identity on the k2 dataset of BRAliBase 2.1. LocARNA achieves the best results at the expense of the highest computation time. Tab. 1 lists the speedups of the other approaches compared to LocARNA. Our novel combined approach ExpLoc-P achieves with both the heuristic and the suboptimal traceback almost the same quality as LocARNA but is 6 and 4.9 times faster, respectively. The best alignment quality that could be obtained with ExpLoc in [25] was achieved with parameter minsize = 10. Even for this optimal setting the quality of the result is significantly lower than the one for LocARNA alone and ExpLoc-P. Additionally, the speedup for this setting is only 4.4 which is also less than both speedups for ExpLoc-P. With minsize = 9, the speedup of ExpLoc is comparable to ExpLoc-P but the quality declines much more. RAF achieves the best speedup of 15.6 but the drawback of the sequence alignment

Table 1. Runtime comparison of the different approaches. The speedup factor is measured relative to the speed of LocARNA. The runtime is the total runtime for computing the entire benchmark dataset on a single Opteron 2356 processor (2.3 GHz). For ExpLoc-P the first value in brackets is the time for computing and chaining the EPMs and the second one the runtime for the subsequent LocARNA alignments.

	LocARNA	ExpLoc-P (heuristic)	ExpLoc-P (suboptimal)	ExpLoc (minsize 10)	ExpLoc (minsize 8)	RAF
speedup	1	6.0	4.9	4.4	5.4	15.6
total time	14.3h	2.4h (0.4h+2h)	2.9h (0.4h+2.5h)	3.2h	2.6h	0.9h

based heuristic filtering which causes this speedup is clearly visible: For sequence similarities below 50% the quality drops tremendously. This indicates that RAF is only successful on instances where sequence information alone is sufficient to get already reasonable alignments. In summary this means that our novel tool ExpLoc-P finds the best tradeoff between alignment quality and speedup and is robust regarding the alignment quality for the whole range of sequence identities.

To analyze the quality of ExpLoc-P further, we investigated whether the compalign scores of ExpLoc-P and unconstraint LocARNA do correlate well. We found a high correlation of 0.85. This indicates that the six-times faster ExpLoc-P pipeline can replace LocARNA in clustering approaches such as [10–12].

Notably, ExpLoc-P significantly outperforms LocARNA for a prominent cluster of the RNA family IRES_HCV. In contrast to most other RNA families, this cluster shows only local conservation. This suggests that ExpLoc-P can improve cluster-based approaches for genome-wide prediction of structural RNA-families, where typically the boundaries of ncRNAs are loosely defined.

6   Conclusion

We introduced the algorithm (ExpaRNA-P) to identify exact pattern matches (EPMs) in RNA structure ensembles. Using a novel sparsification technique, the complexity of the algorithm can be reduced to quadratic complexity, which is the same as the complexity of the algorithm for the corresponding, simpler case of fixed RNA structures (ExpaRNA). Our evaluation demonstrates that EPMs from structure ensembles outperform EPMs from fixed structures when utilized as anchor constraints for structure alignments in our new pipeline ExpLoc-P.

As future work, we will investigate relaxations of the notion of exact patterns to further improve the results. In particular, the same recursions can be used to detect patterns that allow mismatches of base pairs or unpaired bases. Furthermore, EPM based anchor constraints could be used to improve other alignment tools, like RAF. While for LocARNA the constraints yield a considerable speed-up, in RAF they could improve the quality which is poor for low sequence similarity. The score of the chained EPMs could also be used as a distance measure for clustering approaches. This would speed up the clustering process since the expensive computation of full structure alignments can be avoided.

Acknowledgement. This work was partially supported by the German Research Foundation (BA 2168/3-1, MO 2402/1-1, and WI 3628/1-1) and by the German Federal Ministry of Education and Research (S.H., BMBF grant 0313921 FRISYS to R.B.).

References

1. The FANTOM Consortium: The transcriptional landscape of the mammalian genome. Science 309(5740), 1559–1563 (2005)
2. Cheng, J., Kapranov, P., Drenkow, J., Dike, S., Brubaker, S., Patel, S., Long, J., Stern, D., Tammana, H., Helt, G., Sementchenko, V., Piccolboni, A., Bekiranov, S., Bailey, D.K., Ganesh, M., Ghosh, S., Bell, I., Gerhard, D.S., Gingeras, T.R.: Transcriptional maps of 10 human chromosomes at 5-nucleotide resolution. Science 308, 1149–1154 (2005)
3. Bertone, P., Stoc, V., Royce, T.E., Rozowsky, J.S., Urban, A.E., Zhu, X., Rinn, J.L., Tongprasit, W., Samanta, M., Weissman, S., Gerstein, M., Snyder, M.: Global identification of human transcribed sequences with genome tiling arrays. Science 306, 2242–2246 (2004)
4. Kapranov, P., Willingham, A.T., Gingeras, T.R.: Genome-wide transcription and the implications for genomic organization. Nat. Rev. Genet. 8(6), 413–423 (2007)
5. Mattick, J.S., Taft, R.J., Faulkner, G.J.: A global view of genomic information - moving beyond the gene and the master regulator. Trends in Genetics (2009)
6. Consortium, A.F.B., Backofen, R., Bernhart, S.H., Flamm, C., Fried, C., Fritzsch, G., Hackermuller, J., Hertel, J., Hofacker, I.L., Missal, K., Mosig, A., Prohaska, S.J., Rose, D., Stadler, P.F., Tanzer, A., Washietl, S., Will, S.: RNAs everywhere: genome-wide annotation of structured RNAs. J. Exp. Zoolog. B. Mol. Dev. Evol. 308(1), 1–25 (2007)
7. Rivas, E., Eddy, S.R.: Noncoding RNA gene detection using comparative sequence analysis. BMC Bioinformatics 2(1), 8 (2001)
8. Washietl, S., Hofacker, I.L.: Identifying structural noncoding RNAs using RNAz. In: Curr. Protoc. Bioinformatics, ch.12, Unit 12.7 (2007)
9. Pedersen, J.S., Bejerano, G., Siepel, A., Rosenbloom, K., Lindblad-Toh, K., Lander, E.S., Kent, J., Miller, W., Haussler, D.: Identification and Classification of Conserved RNA Secondary Structures in the Human Genome. PLoS Comput. Biol. 2(4), e33 (2006)
10. Will, S., Reiche, K., Hofacker, I.L., Stadler, P.F., Backofen, R.: Inferring non-coding RNA families and classes by means of genome-scale structure-based clustering. PLOS Computational Biology 3(4), e65 (2007)
11. Kaczkowski, B., Torarinsson, E., Reiche, K., Havgaard, J.H., Stadler, P.F., Gorodkin, J.: Structural profiles of human miRNA families from pairwise clustering. Bioinformatics 25(3), 291–294 (2009)
12. Parker, B.J., Moltke, I., Roth, A., Washietl, S., Wen, J., Kellis, M., Breaker, R., Pedersen, J.S.: New families of human regulatory RNA structures identified by comparative analysis of vertebrate genomes. Genome Res. (2011)

13. Höchsmann, M., Töller, T., Giegerich, R., Kurtz, S.: Local similarity in RNA secondary structures. In: Proceedings of Computational Systems Bioinformatics (CSB 2003), vol. 2, pp. 159–168. IEEE Computer Society (2003)
14. Siebert, S., Backofen, R.: MARNA: multiple alignment and consensus structure prediction of RNAs based on sequence structure comparisons. Bioinformatics 21(16), 3352–3359 (2005)
15. Sankoff, D.: Simultaneous solution of the RNA folding, alignment and protosequence problems. SIAM J. Appl. Math. 45(5), 810–825 (1985)
16. Havgaard, J.H., Lyngso, R.B., Stormo, G.D., Gorodkin, J.: Pairwise local structural alignment of RNA sequences with sequence similarity less than 40%. Bioinformatics 21(9), 1815–1824 (2005)
17. Mathews, D.H., Turner, D.H.: Dynalign: an algorithm for finding the secondary structure common to two RNA sequences. Journal of Molecular Biology 317(2), 191–203 (2002)
18. Hofacker, I.L., Bernhart, S.H., Stadler, P.F.: Alignment of RNA base pairing probability matrices. Bioinformatics 20(14), 2222–2227 (2004)
19. McCaskill, J.S.: The equilibrium partition function and base pair binding probabilities for RNA secondary structure. Biopolymers 29(6-7), 1105–1119 (1990)
20. Gorodkin, J., Heyer, L., Stormo, G.: Finding the most significant common sequence and structure motifs in a set of RNA sequences. Nucleic Acids Res. 25(18), 3724–3732 (1997)
21. Bradley, R.K., Pachter, L., Holmes, I.: Specific alignment of structured RNA: stochastic grammars and sequence annealing. Bioinformatics 24(23), 2677–2683 (2008)
22. Torarinsson, E., Havgaard, J.H., Gorodkin, J.: Multiple structural alignment and clustering of RNA sequences. Bioinformatics 23(8), 926–932 (2007)
23. Bauer, M., Klau, G.W., Reinert, K.: Accurate multiple sequence-structure alignment of RNA sequences using combinatorial optimization. BMC Bioinformatics 8, 271 (2007)
24. Do, C.B., Foo, C.S., Batzoglou, S.: A max-margin model for efficient simultaneous alignment and folding of RNA sequences. Bioinformatics 24(13), i68–i76 (2008)
25. Heyne, S., Will, S., Beckstette, M., Backofen, R.: Lightweight comparison of RNAs based on exact sequence-structure matches. Bioinformatics 25(16), 2095–2102 (2009)
26. Backofen, R., Siebert, S.: Fast detection of common sequence structure patterns in RNAs. Journal of Discrete Algorithms 5(2), 212–228 (2007)
27. Wexler, Y., Zilberstein, C., Ziv-Ukelson, M.: A study of accessible motifs and RNA folding complexity. Journal of Computational Biology 14(6), 856–872 (2007)
28. Havgaard, J.H., Torarinsson, E., Gorodkin, J.: Fast pairwise structural RNA alignments by pruning of the dynamical programming matrix. PLoS Comput. Biol. 3(10), 1896–1908 (2007)
29. Ziv-Ukelson, M., Gat-Viks, I., Wexler, Y., Shamir, R.: A Faster Algorithm for RNA Co-folding. In: Crandall, K.A., Lagergren, J. (eds.) WABI 2008. LNCS (LNBI), vol. 5251, pp. 174–185. Springer, Heidelberg (2008)
30. Backofen, R., Tsur, D., Zakov, S., Ziv-Ukelson, M.: Sparse RNA Folding: Time and Space Efficient Algorithms. In: Kucherov, G., Ukkonen, E. (eds.) CPM 2009. LNCS, vol. 5577, pp. 249–262. Springer, Heidelberg (2009)
31. Salari, R., Möhl, M., Will, S., Sahinalp, S.C., Backofen, R.: Time and Space Efficient RNA-RNA Interaction Prediction via Sparse Folding. In: Berger, B. (ed.) RECOMB 2010. LNCS, vol. 6044, pp. 473–490. Springer, Heidelberg (2010)

32. Backofen, R., Will, S.: Local sequence-structure motifs in RNA. Journal of Bioinformatics and Computational Biology (JBCB) 2(4), 681–698 (2004)
33. Otto, W., Will, S., Backofen, R.: Structure local multiple alignment of RNA. In: Proceedings of German Conference on Bioinformatics (GCB 2008). LNI, Gesellschaft für Informatik (GI), vol. P-136, pp. 178–188 (2008)
34. Wilm, A., Mainz, I., Steger, G.: An enhanced RNA alignment benchmark for sequence alignment programs. Algorithms Mol. Biol. 1, 19 (2006)
35. Gardner, P.P., Wilm, A., Washietl, S.: A benchmark of multiple sequence alignment programs upon structural RNAs. Nucleic Acids Research 33(8), 2433–2439 (2005)

Reconstructing Boolean Models of Signaling

Roded Sharan[1] and Richard M. Karp[2]

[1] Blavatnik School of Computer Science, Tel Aviv University, Tel Aviv 69978, Israel
roded@post.tau.ac.il
[2] International Computer Science Institute, 1947 Center st., Berkeley CA, 94704
karp@icsi.berkeley.edu

Abstract. Since the first emergence of protein-protein interaction networks, more than a decade ago, they have been viewed as static scaffolds of the signaling-regulatory events taking place in the cell and their analysis has been mainly confined to topological aspects. Recently, functional models of these networks have been suggested, ranging from Boolean to constraint-based ones. However, learning such models from large-scale data remains a formidable task and most modeling approaches rely on extensive human curation. Here we provide a generic approach to learning Boolean models automatically from data. We apply our approach to growth and inflammatory signaling systems in human and show how the learning phase can improve the fit of the model to experimental data, remove spurious interactions and lead to better understanding of the system at hand.

Keywords: Signaling network, Boolean modeling, Integer linear programming.

1 Introduction

A fundamental question in biology is how a certain network of interacting genes and proteins gives rise to a specific cellular function. Most studies to date, particularly in the protein-protein interaction domain, aim to answer this question by analyzing a static topological description of the network. The most notable exception is the large-scale analysis of metabolic networks which relies on constraint-based models that quantitatively describe the network's fluxes under a steady-state assumption. The advantage of the latter models is that they allow simulating the process of interest under different genetic and environmental perturbations. Recently, it was suggested that similar models could be applied to signaling networks. While metabolic-like models of signaling are only beginning to emerge [4,10], a large body of work exists on Boolean network modeling dating back to the sixties and seventies [3].

Our focus here is on learning Boolean models from experimental data. In contrast to the rich literature on Boolean modeling frameworks, the learning and application of these frameworks to protein networks is very recent. Saez-Rodriguez et al. developed the CNO algorithm to optimize a Boolean model against experimental measurements on the involved proteins [7]. Their algorithm is based

B. Chor (Ed.): RECOMB 2012, LNBI 7262, pp. 261–271, 2012.
© Springer-Verlag Berlin Heidelberg 2012

on starting with an initial model and learning using heuristic genetic algorithms a compact representation of the model that fits the data well. In a follow-up work, Mitsos et al. presented an integer linear programming (ILP) formulation of the problem that allows learning a subset of the initial model interactions that will yield optimal fit to the observed data [5]. Both modeling frameworks were applied to growth and inflammatory signaling systems in human and were shown to agree well with available experimental data [7,5,6,8].

Despite their relative success, these previous learning algorithms had several shortcomings: (i) they relied on the availability of an annotation of activation/repression signs to the edges of the signaling network, thus limiting the search to logical functions that are monotone in an appropriate set of inputs; (ii) they relied on having an initial model in which each Boolean function is associated with a superset of the actual terms in its DNF representation – moreover, in some cases very simple Boolean functions were assumed (e.g., in [5] it is assumed that a function evaluates to TRUE iff all its member activators (at most two) are active and all its member repressors (at most one) are inactive); and (iii) in most cases (except in [5]) heuristic rather than exact search algorithms were used.

Here we suggest a novel algorithm for learning a Boolean model with two key advantages over previous work: (1) the algorithm allows learning general Boolean functions, with a particularly efficient learning scheme for symmetric threshold functions. In particular, it can deal with functions whose DNF terms are unknown and does not necessarily need an annotation of edge signs. (2) The algorithm is guaranteed to find an optimum solution. We apply our algorithm to learn a Boolean model for two well-studied growth and inflammatory systems: epidermal growth factor receptor (EGFR) signaling and interleukin 1 (IL-1) signaling. We compare our learned models to state-of-the-art manual models for these systems. We show that our algorithm produces accurate reconstructions and can successfully pinpoint possible modifications to a model that will improve its fit to the experimental data.

2   Preliminaries

We assume we are given a directed acyclic signaling network $G = (V, E)$ for the process in question. Such a network can either be gathered from the literature as in the case of the EGFR and IL-1 systems described below or learned from data using network reconstruction algorithms [12,11]. The network may be signed with activation/repression effects on its edges or not. We treat each vertex (molecular species) in G as being in one of two states: active (1) or inactive (0). We further assume that the state of a vertex v is a Boolean function of the states of its direct predecessors in the network $P(v)$. We denote this function by $f(v)$. Adopting the terminology used in [9], we will refer to each term in the DNF representation of f as a *reaction*.

If the given network is signed, we will assume that each of the "hidden" Boolean functions is monotone non-decreasing in an appropriately modified set of input variables. Specifically, given a vertex v and one of its predecessors $u \in P(v)$, the function $f(v)$ that determines the value of v is monotone non-decreasing in u if the sign of (u, v) is $+1$ and, otherwise, $f(v)$ is monotone non-decreasing in $\bar{u}$ – the negation of u.

The goal is to learn the Boolean logic of the network, i.e., the truth table of every Boolean function associated with a vertex of G. In order to learn the logic of the network we are given a set of experiments, in each of which some subset of vertices is perturbed and the states of another subset of vertices are observed. Any suggested model gives rise to a state assignment to the nodes of G (under a steady state assumption). We will aim to derive a model that fits best (under a least squares criterion) to the observed data.

3   Logic Learning via ILP

In this section we present our algorithm for learning a Boolean model with no assumptions on the functions involved. Denote the number of nodes in the network by $N = |V|$. Let δ be the maximum in-degree of a node in G.

A Boolean model is specified by giving, for each non-input node v, a Boolean function $f(v)$ on $n(v) = |P(v)|$ inputs $u(v)_1, u(v)_2, \cdots, u(v)_{n(v)}$ specifying how the value of the node depends on its inputs. The function $f(v)$ is specified by its truth table, a collection of $2^{n(v)}$ binary variables $x(v)_1, x(v)_2, \cdots, x(v)_{2^n(v)}$, where $x(v)_i$ denotes the value of v under the i-th input combination in lexicographic order.

Our goal is to find a Boolean model that minimizes the sum, over all experiments, of the number of experimental observations that differ from the predictions of the model. We formulate this as an integer programming problem. To do so, we derive integer linear constraints that determine the state $a(v)$ of every node v, under some experimental condition (where the index of the condition is omitted for clarity), from the states of its input nodes under that condition, i.e., from $a(u(v)_1), \cdots, a(u(v)_{n(v)})$. When v is one of the perturbed nodes, $a(v)$ is fixed to its perturbed value. Otherwise, we derive its state using the following auxiliary constants and variables:

1. $b(v, i, j)$ – a constant which is equal to the value of the j-th input variable in the input combination corresponding to the i-th row of the truth table of $f(v)$. In other words, $b(v, i, j)$ is the j-th bit of the number $i - 1$. The purpose of the constants $b(v, i, j)$ is to select the correct row of the truth table.
2. $y(v)_i$ – a variable which is 1 iff i is the row of the truth table that is selected by the inputs to node v and $x(v)_i = 1$.

The variables $y(v)_i$ are determined by the following inequalities:

$$y(v)_i \leq x(v)_i \tag{1}$$

$$y(v)_i \leq [1 - b(v, i, j)] + a(u(v)_j)[2b(v, i, j) - 1] \quad \forall j = 1, \ldots, n(v) \tag{2}$$

$$y(v)_i \geq x(v)_i + \sum_j [2b(v, i, j)a(u(v)_j) - a(u(v)_j) - b(v, i, j)] \tag{3}$$

The first constraint ensures that $y(v)_i$ will evaluate to TRUE only if the truth value of the i-th row is TRUE. The second constraint ensures that $y(v)_i$ will evaluate to TRUE only if the activity value of each input matches its designated value. Finally, the third constraint ensures that $y(v)_i$ will evaluate to FALSE only if one of the previous constraints was not satisfied.

Finally, $a(v)$ is 1 iff any of the $y(v)_i$ is 1, as expressed by the following constraints:

$$y(v)_i \leq a(v) \leq 1 \ \forall i \tag{4}$$

$$a(v) \leq \sum_i y(v)_i \tag{5}$$

In case the input network is signed and the monotonicity assumption holds, the ILP formulation can be made more efficient by noting that: (i) for any given row no constraints on the variables that should attain a value of 0 are needed; and (ii) the monotonicity requirements can be forced by appropriate inequality constraints on truth table variables: $\forall i \in C(j) : x(v)_i \geq x(v)_j$, where $C(j)$ is the set of at most $n(v)$ indices whose binary representation has exactly one more '1' than that of j.

The above constraints can be used to derive the state of all nodes given the input perturbed states. It remains to specify the objective of the ILP which measures the agreement between the model-derived states and the experimental data at hand. We use a least squares criterion which becomes linear on binary variables – see, e.g., [5]. Formally, let O be the set of nodes whose output is experimentally observed. For a given condition, let $e(v)$ denote the experimentally measured state of node $v \in O$. Then the non-constant contribution of this condition to the objective function is $\sum_{v \in O} a(v) - 2 \cdot a(v) \cdot e(v)$.

The overall number of variables in the above ILP is $O(N2^\delta)$ and the same bound applies to the number of constraints. The overall size of the ILP is $O(N2^{2\delta})$.

4   Symmetric Threshold Functions

In practice, for known signed models, most or all of the pertaining Boolean functions are of simple structure: single AND or OR gates or, more generally, symmetric threshold functions, where the function evaluates to TRUE iff sufficiently many of its activators (resp. repressors) are present (resp. absent). For example, in the EGFR system that we study below 98 of its 112 species (87.5%) are associated with a single AND or OR gate. Similarly, in the IL-1 system all species whose gates are known (110 of 121) are associated with a single AND or OR gate.

Here we provide a more efficient formulation for symmetric threshold functions. The main improvement is obtained by representing each function using a *single* integer variable that ranges from 0 to the fan-in of the function plus one. Given a node v with incoming nodes $u(v)_1, \ldots, u(v)_{n(v)}$, we let $x(v)$ be the threshold variable representing its underlying Boolean function $f(v)$. The activity status of v can be derived from the following set of constraints:

$$a(v) \geq \frac{\sum_j a(u(v)_j) - x(v) + 1}{n(v) + 1} \tag{6}$$

$$a(v) \leq \frac{\sum_j a(u(v)_j) - x(v)}{n(v) + 1} + 1 \tag{7}$$

Notably, a symmetric threshold function contains AND and OR as special cases, with $x(v)$ equal to the fan-in in the case of an AND gate and $x(v) = 1$ in the case of an OR gate. Furthermore, a value of $n(v) + 1$ allows us to remove this function from consideration, as it will always evaluate to FALSE. This is advantageous when some of the given reactions may be redundant (see Section 6). In addition, this setting allows the discovery of redundant components of an OR gate. Last, a value of 0 allows setting the output of the function to 1 regardless of the inputs. This can be used to search for redundant components of an AND gate (see Section 6).

The overall number of variables in this special ILP is at most twice the number of species, and this is also the bound on the number of constraints. Thus, the overall size of the ILP is quadratic in the number of species and linear when the maximum fan-in is bounded.

5   Experimental Design

5.1   Learning Read-Once Functions

There are interesting connections to be made between the reconstruction of Boolean models of signaling circuits and the branch of computational learning theory involving the identification of Boolean functions using queries [1]. In this theoretical framework one is given black-box access to a Boolean function of n variables known to be drawn from a specified class of functions. The input to a query is one of the 2^n input combinations to the unknown function and the output is the corresponding function value. The theory studies the worst-case number of queries required to identify a function in the given class.

The setting of the current paper differs in several ways from the standard model of learning via queries. Most importantly, instead of black-box access to the unknown Boolean function we assume knowledge of the wiring diagram of the network being analyzed and of the possible Boolean functions that can be associated with the gates within it. Also, our networks may have multiple outputs rather than a single output, and there may be technological limitations on the input combinations that can be applied (i.e, on the feasible combinations of perturbations of the state of the network). Finally, it may be out of reach to determine the network exactly; instead, we seek a network model that has high agreement with the observed experimental outcomes.

Motivated by these differences, we concentrate on algorithms for learning an n-variable Boolean function realized by a network with a known wiring diagram but unknown gates. We allow arbitrary queries. We show that in the case of monotone read-once functions knowing the wiring diagram gives a great advantage over black-box learning. A Boolean function is called *monotone read-once* if it is realized by a Boolean formula in which each connective is AND or OR and each input occurs exactly once. Such a formula can be represented as a tree of AND and OR gates, with the edges directed toward the root, such that each input variable occurs at exactly one leaf. We assume that the structure of the tree is given but the identities of the gates are unknown. We show how to identify the Boolean function with at most n queries (one query per gate), whereas $\Omega(n^2)$ queries seem to be required in the black-box model [2].

Our result follows from three observations:

1. We can set the Boolean value on any edge e of the tree to a Boolean value a by setting to a each input variable from which edge e is reachable.
2. Given any set of edges, no two of which are reachable from the same input variable, we can simultaneously set the values on those edges to any desired combination of values. This is done by applying the above construction simultaneously to each edge in the set.
3. Let g be a gate such that the types (AND or OR) of all gates on the path from g to the output are known. To find the type of g we can set to 0 one of the inputs to g, set to 1 all the other inputs to g, and cause the output of g to "propagate" to the final output by: (i) placing 1 on every edge that is not on the path from g to the output but is directed into an AND gate on the path; and (ii) placing 0 on every edge that is not on the path from g to the output but is directed into an OR gate on the path. Then g is an AND gate if the final output is 0, and an OR gate if the final output is 1.

Using these observations we can work backward from the root to the leaves, determining the type of each gate with one query.

By an extension of this construction we can show the following. Let K be a class of non-constant monotone non-decreasing Boolean functions learnable in the black-box query model with $f(n)$ queries, where n is the number of inputs. Consider the class of Boolean functions representable by read-once tree networks with gates drawn from the class K. Given the wiring diagram of such a network one can identify the gates, and hence learn the Boolean function, with $\sum_i f(d_i)$ queries, where i ranges over the gates in the network and d_i is the fan-in to the i-th gate. For example, given the wiring diagram of a read-once network of bounded fan-in symmetric threshold functions, the Boolean function represented by the network can be learned with $O(n)$ queries.

5.2   An Information-Theoretic Experimental Design Algorithm

Thus far we have assumed that the set of experiments to be applied to the network is externally specified. In this subsection we present an algorithm for adaptively choosing experiments in order to efficiently acquire information about the Boolean circuit in question. Our algorithm draws inspiration from the field of genetic algorithms, and is information-theoretic in nature. At a general step we execute a feasible experiment whose outcome is least predictable according to an entropy measure, thus maximizing the information gain at each step. In selecting the next experiment we assume as given a specification of the inputs and outputs of each experiment performed thus far, a population P of p high-scoring models relative to these experiments (where p is a parameter of the experimental design algorithm) and a set F of feasible experiments from which the next experiment will be drawn. We define a *mutation* of a model as any single change in the truth table of a gate that would alter the output of that gate under some past experiment.

The initial collection of N networks can be learned using the integer programming formulation of Section 3, while instructing the ILP solver to produce multiple optimal (or near-optimal) solutions. A step of the algorithm consists of the following substeps:

1. Enlarge the population P by applying all possible mutations to its members.
2. Restrict the enlarged population P to the p models within it having the highest scores relative to the set of experiments performed so far; this entails simulating each circuit in the enlarged set under every past input.
3. Simulate each experiment in F on each of the p circuits to compute the output it would produce, and compute the entropy of the distribution of these outputs, assuming the uniform distribution over the p models in P.
4. Perform the candidate experiment of maximum entropy (i.e., the most informative candidate experiment).

6   Results

To test the ability of our ILP-based approach to provide an accurate logical model of a signaling system, we applied it to the well-studied EGFR and IL-1 signaling systems. These systems served as ideal test cases as their topologies and underlying logics have been extensively researched and large scale models have been manually constructed for them [9,6]. In the following we describe our implementation, evaluation criteria and the results we attained with respect to each of these systems.

6.1   Implementation and Evaluation

We implemented the algorithm for learning symmetric threshold functions described in Section 4. To deal with the problem of multiple equally-good solutions, we also implemented a variant of the algorithm in which a secondary objective is used to choose among the best performing solutions. In detail, the application of this variant is done as follows: (i) first, we run our algorithm to attain some optimal solution and record its value; (ii) then, we add a constraint to the ILP which restricts its solutions to those attaining the optimal value identified in the previous step; (iii) last, we optimize the ILP w.r.t. the second objective. In the experiments reported below we used as a secondary objective the similarity (number of identical gates) of the output model to the original (manually curated) model. In case a curated model is not available, we may use as a secondary objective some measure of the complexity of the resulting circuit, like the number of non-constant gates.

The implementation was written in C using the CPLEX callable library version 12.1. All runs reported below were conducted on a single core of a Xeon 3.06 GHz server with 8GB memory and were completed in a few seconds.

To evaluate our modeling framework we applied it for the modeling of two signaling systems in human: (i) EGFR, which regulates cellular growth, proliferation, differentiation and motility; and (ii) IL-1, which is involved in coordinating the immune response upon bacterial infection and tissue injury. Both systems are well studied and detailed manual models exist for them. In particular, Samaga et al. have constructed a comprehensive Boolean model of the EGFR system which contains 112 molecular species and their associated Boolean functions [9]; Ryll et al. have created a Boolean model of the IL-1 system with 121 molecular species [6]. We retrieved both models from the CellNetAnalyzer repository (http://www.mpi-magdeburg.mpg.de/projects/cna/repository.html).

In order to learn logical models for these systems we used data published by the above authors on the activity (phosphorylation) levels of certain proteins under different cellular conditions. Specifically, Samaga et al. measured within the EGFR system the activity levels of 11 proteins under 34 distinct conditions in Hep2G cells (Figure 8 in [9]). Similarly, Ryll et al. measured within the IL-1 system the activity levels of 9 proteins under 14 distinct conditions in primary hepatocytes (Figure 6 in [6]). In both cases the cells were stimulated with different ligands and treated with different inhibitors, thus simulating different conditions. Following [9,6], we focused our analysis on the measurements at the 30min time point, representing the early response of the system. Reconstructed models were scored by their fit, according to a least squares criterion, to the observed data.

6.2   The EGFR System

Our first and main test case is the well-studied EGFR system, whose model contains 112 species, 157 reactions (excluding input/output ones) and the maximum fan-in is 5. Experimental data for this system includes the activity levels of 11 proteins under 34 different perturbations. When testing the fit of the curated model of [9] to the observed experimental data, 278 out of 366 predictions (76%) matched the observed activities. In an effort to understand the erroneous cases, the authors have identified 12 gates for which the underlying logic was not clear (along with one more reaction that does not have an effect on the measured proteins and another reaction that happens at a later time; hence, both could be removed). In addition, they have suggested 4 modifications to existing gates that improve the fit of the model (along with three more modifications that involve the introduction of new reactions into the model and, hence, could not be captured by our ILP framework). Overall, there were 2^{33} possible models to choose from when constraining the 16 gates in question to symmetric threshold functions. Full enumeration of these models was prohibitively expensive as each model had to be simulated in order to compare its fit to the experimental data.

We applied our algorithm to reconstruct these unknown logical functions, while aiming to maximize the fit to the experimental data. The program finished in less than a second producing a solution that matched 330 of the data points (90%). We then searched for a model with the same score (fit to data) that is closest to the original model. The reconstructed functions and their curated counterparts appear in Table 1. The proposed changes to the model and the respective reconstructions are summarized in Table 2. An in-depth analysis of the reconstructed functions revealed that: (i) 11 of the 12 reconstructed functions matched the curated description, where an additional one was reconstructed as a majority function on three variables while its curated description was not a symmetric threshold function but was deemed "closest" to a majority one (line 9 in Table 1); and (ii) in 3 of the 4 proposed changes, the algorithm correctly predicted the suggested modification; the fourth modification was rejected by the algorithm.

Quite strikingly, when testing the performance of the curated model with the suggested modifications, the same fit (330 of 366) was observed.

Table 1. Performance evaluation of the reconstruction algorithm. Reconstructions that match the curated functions appear in bold.

Curated function	Reconstructed function
sos1_eps8_e3b1 OR vav2 → rac_cdc42	**OR**
mekk1 OR raf1 → mek12	**OR**
mkk3 OR mkk6 OR mkk4 → p38	**OR**
mekk1 OR mekk4 OR mlk3 → mkk4	**OR**
pak1 OR csrc → pak1crscd	**OR**
p90rsk OR mk2 → creb	**OR**
!akt AND !pak1 → bad	**AND**
!akt AND !p90rsk → gsk3	**AND**
jnk OR (erk12 AND p90rsk) → jnkerkp90rskd	MAJ
erk12 OR jnk → p70s6_1	**OR**
erbb11 AND eps8r → rntre	**AND**
dag AND ca → dagcad	**AND**

Table 2. Performance evaluation of the reconstruction algorithm with respect to proposed modifications. Reconstructions that match the suggested modification appear in bold.

Original function	Proposed modification	Reconstructed function
erb11 AND (pip3 OR pi34p2) → vav2	erb11 → vav2	**erb11 → vav2**
sos1eps8e3b1 → raccdc42	REMOVE	sos1eps8e3b1 → raccdc42
erb11 AND csrc → stat3	REMOVE	**REMOVE**
mk2 → hsp27	REMOVE	**REMOVE**

6.3   IL-1 Signaling

As a second test case we applied our method to automatically learn the logic of the IL-1 circuit [6]. This circuit contains 121 species and 112 reactions (excluding input/output ones) with a maximum fan-in of 6. The logic of 11 of the reactions is not known, but this has no effect on the measured proteins under the available conditions (and assuming monotonicity). Overall, the model successfully explains 104 of the 118 experimental points (88%). In an attempt to improve this fit, Ryll et al. manually inspected the model and data and suggested the addition of seven reactions and the removal of one, albeit achieving only a slight increase in performance (89%).

Ryll et al. have classified the reactions in their model according to the literature support for them. The least reliable categories included those reactions that had evidence under different stimulations than IL-1/6. Hence, we thought to improve the fit of the model by learning the logic of these reactions (focusing on a subset of those that have at least two inputs) in addition to the 11 unknown ones. Overall, we applied our framework to learn the logic of 26 reactions, amounting to a search space of 2^{60} models. Among the solutions obtained, we chose the one that is closest (in terms of the number of identical gates) to the original model. By modifying two reactions (removing one and changing another from AND to OR), the algorithm managed to find a solution with a slightly better fit to the data: 106 of the points agreed with the experimental measurements (90%).

7   Conclusions

Transforming topological networks into working functional models of signaling is a fundamental problem with vast applications. Here we make a step toward achieving this goal by providing an algorithm to learn a Boolean model of a given signaling system automatically from data. We provide a general variant that is applicable to all Boolean functions and does not require knowledge on the activation/repression properties of the network's edges. In addition, we provide a specialized variant for learning symmetric threshold functions that are very common in known models. Our algorithms are based on reducing the learning problem to an integer linear program which is solved to optimality in seconds on current systems. We demonstrate the power of our approach by applying it to two well-annotated signaling systems involved in growth and inflammatory response. The produced models allow completing information gaps, improving the fit to the experimental data and pinpointing redundant reactions and components of reactions.

While our approach is generic, we focused our evaluation and experimentation on learning constrained models in which the initial network is signed and the pertaining Boolean functions are assumed to be symmetric threshold functions. Further experimentation is needed to test the effectiveness of the full model and the accuracy of its predictions when relaxing these assumptions. In addition, it would be interesting to test the utility of our experimental design scheme in prioritizing current experiments and suggesting new ones.

Acknowledgments. Part of this work was done while RS was on sabbatical in ICSI, Berkeley. RS was supported by a research grant from the Israel Science Foundation (grant no. 241/11).

References

1. Angluin, D.: Queries and concept learning. Machine Learning 2, 319–342 (1988)
2. Angluin, D., Hellerstein, L., Karpinski, M.: Learning read-once formulas with queries. J. ACM 40, 185–210 (1993)
3. Kauffman, S.: Metabolic stability and epigenesis in randomly constructed genetic nets. J. Theoret. Biol. 22, 437–467 (1969)
4. Klamt, S., Saez-Rodriguez, J., Lindquist, J., Simeoni, L., Gilles, E.: A methodology for the structural and functional analysis of signaling and regulatory networks. BMC Bioinformatics 7, 56 (2006)
5. Mitsos, A., et al.: Identifying drug effects via pathway alterations using an integer linear programming optimization formulation on phosphoproteomic data. PLoS Computational Biology 5, e1000591 (2009)
6. Ryll, A., Samaga, R., Schaper, F., Alexopoulos, L., Klamt, S.: Large-scale network models of il-1 and il-6 signalling and their hepatocellular specification. Molecular Biosystems (2011)
7. Saez-Rodriguez, J., Alexopoulos, L., Epperlein, J., Samaga, R., Lauffenburger, D., Klamt, S., Sorger, P.: Discrete logic modelling as a means to link protein signalling networks with functional analysis of mammalian signal transduction. Mol. Sys. Biol. 5, 331 (2009)
8. Saez-Rodriguez, J., Alexopoulos, L., Zhang, M., Morris, M., Lauffenburger, D., Sorger, P.: Comparing signaling networks between normal and transformed hepatocytes using discrete logical models. Cancer Research 71, 5400–5411 (2011)
9. Samaga, R., Saez-Rodriguez, J., Alexopoulos, L., Sorger, P., Klamt, S.: The logic of EGFR/ErbB signaling: theoretical properties and analysis of high-throughput data. PLoS Computational Biology 5, e1000438 (2009)
10. Vardi, L., Ruppin, E., Sharan, R.: A linearized constraint based approach for modeling signaling networks. J. Computational Biology 19, 232–240 (2012)
11. Yeger-Lotem, E., Riva, L., Su, L.J., Gitler, A.D., Cashikar, A.G., King, O.D., Auluck, P.K., Geddie, M.L., Valastyan, J.S., Karger, D.R., Lindquist, S., Fraenkel, E.: Bridging high-throughput genetic and transcriptional data reveals cellular responses to alpha-synuclein toxicity. Nat. Genet. 41, 316–323 (2009)
12. Yosef, N., Ungar, L., Zalckvar, E., Kimchi, A., Kupiec, M., Ruppin, E., Sharan, R.: Toward accurate reconstruction of functional protein networks. Mol. Syst. Biol. 5, 248 (2009)

Alignment-Free Sequence Comparison Based on Next Generation Sequencing Reads: Extended Abstract

Kai Song[1], Jie Ren[1], Zhiyuan Zhai[2], Xuemei Liu[3], Minghua Deng[1,*], and Fengzhu Sun[4,5,*]

[1] School of Mathematics, Peking University, Beijing, P.R. China
[2] School of Mathematics, Shandong University, P.R. China
[3] School of Physics, South China University of Technology, Guangzhou, P.R. China
[4] TNLIST/Department of Automation, Tsinghua University, Beijing, P.R. China
[5] Molecular and Computational Biology Program, University of Southern California, Los Angeles, California, USA
dengmh@pku.edu.cn, fsun@usc.edu

Abstract. Next generation sequencing (NGS) technologies have generated enormous amount of shotgun read data and assembly of the reads can be challenging, especially for organisms without template sequences. We study the power of genome comparison based on shotgun read data without assembly using three alignment-free sequence comparison statistics, D_2, D_2^*, and D_2^S, both theoretically and by simulations. Theoretical formulas for the power of detecting the relationship between two sequences related through a common motif model are derived. It is shown that both D_2^* and D_2^S outperform D_2 for detecting the relationship between two sequences based on NGS data. We then study the effects of length of the tuple, read length, coverage, and sequencing error on the power of D_2^* and D_2^S. Finally, variations of these statistics, d_2, d_2^* and d_2^S, respectively, are used to first cluster 5 mammalian species with known phylogenetic relationships and then cluster 13 tree species whose complete genome sequences are not available using NGS shotgun reads. The clustering results using d_2^S are consistent with biological knowledge for the 5 mammalian and 13 tree species, respectively. Thus, the statistic d_2^S provides a powerful alignment-free comparison tool to study the relationships among different organisms based on NGS read data without assembly.

Keywords: NGS, HMM, statistical power, normal approximation, word count statistics.

1 Introduction

Next generation sequencing (NGS) technologies are producing unprecedented volumes of sequence data and are being applied to study many biological and

---

* Corresponding authors.

B. Chor (Ed.): RECOMB 2012, LNBI 7262, pp. 272–285, 2012.
© Springer-Verlag Berlin Heidelberg 2012

biomedical problems such as *de novo* sequencing, RNA expression and alternative splicing, transcription factor binding site (TFBS) identification, etc. The initial step of most currently available methods for the analysis of NGS data is to map the reads onto the known genomes or RNA sequences. However, for genomes without template sequences, it is generally challenging to assemble the shotgun reads because the reads are usually short and there may be a large number of repeats within the genomes. Thus, genome comparison based on NGS shotgun reads can be difficult and no methods have been developed to compare genomes based on shotgun read data directly without assembly.

Alignment-free methods for the comparison of two long sequences have recently received increasing attention because they are computationally efficient and can potentially offer better performance than alignment-based methods for gene regulatory sequence comparison [1,2,3,4,5,6,7,8,9,10,11]. For the comparison of long sequences, one widely used alignment-free statistic is D_2 [1], an uncentered correlation between the number of occurrences of k-words for two sequences of interest. However, it was shown that D_2 was dominated by the noise caused by the randomness of the sequences and has low statistical power to detect potential relationship between two sequences [6,8,11]. Two new variants, D_2^* and D_2^S, were developed by standardizing the k-tuple counts with their means and standard deviations [8,11]. These two statistics are more powerful than the D_2 statistic for the detection of relationship between sequences related through a common motif model that the two sequences share instances of one or multiple motifs [8,11]. The calculations of D_2^* and D_2^S depend only on the numbers of occurrences of k-tuples in the two sequences of interest and the exact long molecular sequences are not needed. Thus, we expect that they can equally be adapted for genome comparison based on NGS shotgun read data.

However, no such studies are yet available and new statistics based on NGS shotgun read data need to be developed. In this study, we address the following questions. 1). How do we modify the D_2, D_2^* and D_2^S statistics so that they can be applicable for genome comparison based on NGS shotgun read data? 2). What are their approximate distributions under the null model that the two sequences are independent and both are generated by independent identically distributed (iid) models? 3). What is the power of these statistics for detecting the relationships between sequences when they are related? In particular, we will study their power using both simulation and theoretical studies when they are related through a common motif model as in [8,11]. 4). What are the effects of the length of the tuple, read lengths, coverage, sequencing errors, and the distribution of reads along the genome on the power of these statistics? 5) How do these statistics perform on whole genome shotgun read data from multiple genomes?

The current study differs from our previous studies [8,11,7] in the following aspects. First, two random processes need to be considered to study the distribution of the number of occurrences of word patterns from shotgun read data. One is that the long genome sequences are random and they can be modeled by a hidden Markov model as in [11,12]. The other randomness comes from the stochastic sampling of the reads from the long genome sequences. A mathematical model similar to that in [13] for the random sampling of the reads is developed. Second, NGS shotgun reads can come from either the forward or the reverse strand of the genomes and it is not known which strand the reads come from. Thus, the reads together with their complements need to be considered simultaneously when counting the numbers of occurrences of word patterns for NGS reads. The inclusion of both strands further complicates our mathematical analysis for the distribution of these statistics. Third, we study the distributions of the statistics D_2, D_2^* and D_2^S under the null and the alternative models based on the stochastic models for the long sequences and the sampling of the reads. The key challenges include the calculation of covariance for the numbers of occurrences of different word patterns from the shotgun reads within one long sequences and between the genome sequences. The major difficulty comes from the random sampling of the reads from the genomes and the consideration of double strands of the genome.

2   Materials and Methods

2.1   Extending the D_2, D_2^* and D_2^S Statistics to the NGS Read Data

The D_2, D_2^* and D_2^S statistics were originally developed for the comparison of two long sequences [8]. Here we extend them so that they can be applicable to NGS shotgun read data. Consider two genome sequences taking L letters $(0, 1, \cdots, L-1)$ at each position. Suppose that M reads of length β are sampled from a genome of length n. Since the reads can come from either the forward strand or the reverse strand of the genome in NGS, we supplement the observed reads by their complements and refer to the joint set of the reads and the complements as the read set. Let $X_{\mathbf{w}}$ and $Y_{\mathbf{w}}$ be the numbers of occurrences of word pattern $\mathbf{w}$ in the M pairs of reads from the first genome and the second genome, respectively. For the null model, we assume that the two genomes are independent and both are generated by iid models with p_l being the probability of taking state l, $l = 0, 1, \cdots, L - 1$. It can be easily shown that

$$EX_{\mathbf{w}} = EY_{\mathbf{w}} = M(\beta - k + 1)(p_{\mathbf{w}} + p_{\bar{\mathbf{w}}}),$$

where $\mathbf{w} = \mathbf{w}_1 \mathbf{w}_2 \cdots \mathbf{w}_k$, $p_{\mathbf{w}} = p_{\mathbf{w}_1} p_{\mathbf{w}_2} \cdots p_{\mathbf{w}_k}$, and $\bar{\mathbf{w}}$ is the complement of word $\mathbf{w}$.

Similar to the definitions of D_2, D_2^* and D_2^S for the comparison of long sequences in [8,11], we define them for the shotgun read data as follows,

$$D_2 = \sum_{\mathbf{w} \in \mathcal{A}^k} X_{\mathbf{w}} Y_{\mathbf{w}}, \quad D_2^* = \sum_{\mathbf{w} \in \mathcal{A}^k} \frac{\tilde{X}_{\mathbf{w}} \tilde{Y}_{\mathbf{w}}}{M(\beta - k + 1)(p_{\mathbf{w}} + p_{\bar{\mathbf{w}}})}, \quad D_2^S = \sum_{\mathbf{w} \in \mathcal{A}^k} \frac{\tilde{X}_{\mathbf{w}} \tilde{Y}_{\mathbf{w}}}{\sqrt{\tilde{X}_{\mathbf{w}}^2 + \tilde{Y}_{\mathbf{w}}^2}},$$

where $\tilde{X}_{\mathbf{w}} = X_{\mathbf{w}} - M(\beta - k + 1)(p_{\mathbf{w}} + p_{\bar{\mathbf{w}}})$ and $\tilde{Y}_{\mathbf{w}}$ is defined analogously. We test the alternative hypothesis, H_1, that the two genome sequences are related against the null hypothesis, H_0, that they are independent. The more specific hypotheses are given in Subsection 2.2. For a type I error α, we find thresholds z_α, z_α^*, and z_α^S such that

$$P(D_2 \geq z_\alpha) = P(D_2^* \geq z_\alpha^*) = P(D_2^S \geq z_\alpha^S) = \alpha,$$

where P indicates the probability distribution under the null model that the two sequences are independent. The null hypothesis is rejected if the statistics are larger than the corresponding thresholds.

2.2 Modeling the Long Underlying Sequences and the Sampling of Reads Using NGS

We model the long genome sequences related through a common motif model as in [8,11]. Each long genome sequence is modeled by three components: 1) the background model for describing the generation of the long sequence, 2) the foreground model for the motif using a position weight matrix (PWM), and 3) the distribution of motif instances along the sequence of interest. First, the background sequence is modeled by iid random variables taking L different states $(0, 1, \cdots, L-1)$ with p_l being the probability of taking state l, for example, $L = 4$ for nucleotide sequences with states (A, G, C, T), and $L = 20$ for amino acid sequences. Second, for a motif of length K, let $p_l^{(k)}, k = 1, 2, \cdots, K$ be the probability that the k-th position of the motif takes value l. We also assume that the motif positions are independent. Third, for a position along the background, the next K positions are replaced with a motif instance with probability $1 - \rho$ and we refer to $1 - \rho$ as motif intensity throughout the paper. Under this model, the null hypothesis corresponds to $H_0 : \rho = 1$, and the alternative hypothesis corresponds to $H_1 : \rho < 1$.

Next we model the sampling of reads from NGS. Recent studies have shown that the distribution of reads from NGS along the genomic region of interest is not homogeneous. Instead, the read distribution is biased by the base composition of the sequences for most of the current NGS technologies [13,14,15]. To model the read distribution heterogeneity along the genome, we assume that a read generated by NGS starts from position i with probability λ_i, where $\sum_{i=1}^{n-\beta+1} \lambda_i = 1$ and n is the length of the sequence. If a read is generated from the sequence, we also consider its complement. Assume that a total of M pairs of reads of length β are generated from NGS.

2.3 The Mean and Covariance of the Numbers of Occurrences of Word Patterns in a Read Set

For a fixed word pattern $\mathbf{w}$ of length k (note that k does not have to equal K, the length of the motif), let $X_{\mathbf{w}}$ be the number of occurrences of $\mathbf{w}$ within the M pairs of reads as described in Subsection 2.2. It can be seen that the expectation of $X_{\mathbf{w}}$ is given by,

$$\mathbf{E}X_{\mathbf{w}} = M(\beta - k + 1)(P_\rho(\mathbf{w}) + P_\rho(\bar{\mathbf{w}})),$$

where P_ρ indicates the probability distribution for the forward strand and can be calculated based on the hidden Markov model in [11,12].

To calculate $\mathbf{E}X_\mathbf{u}X_\mathbf{v}$ where $\mathbf{u}$ and $\mathbf{v}$ are two words of length k, we note that $X_\mathbf{u} = \sum_{i=1}^{M} C_\mathbf{u}(i)$, where $C_\mathbf{u}(i)$ is the number of occurrences of word $\mathbf{u}$ in the i-th read and it's complement. Thus,

$$\mathbf{E}X_\mathbf{u}X_\mathbf{v} = \mathbf{E}\left(\sum_{i=1}^{M} C_\mathbf{u}(i) \sum_{i=1}^{M} C_\mathbf{v}(i)\right) = M\mathbf{E}C_\mathbf{u}(1)C_\mathbf{v}(1) + M(M-1)\mathbf{E}C_\mathbf{u}(1)C_\mathbf{v}(2).$$

For the first term, we have $\mathbf{E}C_\mathbf{u}(1)C_\mathbf{v}(1) = \mathbf{E}\big(X_\mathbf{u}[1,\beta]X_\mathbf{v}[1,\beta]\big) = \mathbf{E}_{\beta,0}(\mathbf{u},\mathbf{v})$ assuming that both sequences start from the stationary distribution, where $X_\mathbf{u}[\gamma,\gamma']$ is the number of occurrences of word $\mathbf{u}$ in the sequence from γ to γ' at the forward strand and its complement, and $\mathbf{E}_{\beta,\eta}(\mathbf{u},\mathbf{v}) = \mathbf{E}(X_\mathbf{u}[1,\beta]X_\mathbf{v}[1 + \eta, \beta + \eta])$

For the second term, we have

$$\mathbf{E}C_\mathbf{u}(1)C_\mathbf{v}(2)$$
$$= \mathbf{E}\left(\sum_{i=1}^{n-\beta+1} \lambda_i X_\mathbf{u}[i, i + \beta - 1] \sum_{j=1}^{n-\beta+1} \lambda_j X_\mathbf{v}[j, j + \beta - 1]\right)$$
$$= \mathbf{E}_{\beta,0}(\mathbf{u},\mathbf{v}) \sum_{i=1}^{n-\beta+1} \lambda_i^2 + \sum_{i=1}^{n-\beta+1} \lambda_i \sum_{\eta=1}^{n-i-\beta+1} \lambda_{i+\eta}\left(\mathbf{E}_{\beta,\eta}(\mathbf{u},\mathbf{v}) + \mathbf{E}_{\beta,\eta}(\mathbf{v},\mathbf{u})\right).$$

Therefore

$$\mathbf{E}X_\mathbf{u}X_\mathbf{v} = \left(M + M(M-1) \sum_{i=1}^{n-\beta+1} \lambda_i^2\right)\mathbf{E}_{\beta,0}(\mathbf{u},\mathbf{v})$$
$$+ M(M-1) \sum_{i=1}^{n-\beta+1} \lambda_i \sum_{\eta=1}^{n-i-\beta+1} \lambda_{i+\eta}\left(\mathbf{E}_{\beta,\eta}(\mathbf{u},\mathbf{v}) + \mathbf{E}_{\beta,\eta}(\mathbf{v},\mathbf{u})\right).$$

The method for calculating $\mathbf{E}_{\beta,\eta}(\mathbf{u},\mathbf{v})$ will be given in the full paper. The following proposition gives the approximate covariance between $X_\mathbf{u}$ and $X_\mathbf{v}$.

Proposition 1. *Consider the models for the long genome sequences and the sampling of reads described in Subsection 2.2. Assume* $\lim_{n\to\infty}(n - \beta - \eta + 1)$ $\sum_{i=1}^{n-\beta-\eta+1} \lambda_i\lambda_{i+\eta} = r_\eta$ *and M depends on n such that* $\lim_{n\to\infty} M/n = \theta$ *where θ is a constant. Then*

$$\sigma_\rho(\mathbf{u},\mathbf{v}) = \lim_{n\to\infty} \frac{\mathrm{Cov}(X_\mathbf{u}, X_\mathbf{v})}{M}$$
$$= (1 + \theta r_0)\left(\mathbf{E}_{\beta,0}(\mathbf{u},\mathbf{v}) - (\beta - k + 1)^2(P_\rho(\mathbf{u}) + P_\rho(\bar{\mathbf{u}}))(P_\rho(\mathbf{v}) + P_\rho(\bar{\mathbf{v}}))\right)$$
$$+ \theta \sum_{\eta=1}^{\infty} r_\eta\left(\mathbf{E}_{\beta,\eta}(\mathbf{u},\mathbf{v}) + \mathbf{E}_{\beta,\eta}(\mathbf{v},\mathbf{u}) - 2(\beta - k + 1)^2(P_\rho(\mathbf{u}) + P_\rho(\bar{\mathbf{u}}))(P_\rho(\mathbf{v}) + P_\rho(\bar{\mathbf{v}}))\right).$$

For simplicity of notation, we also denote $\sigma_\rho^2(\mathbf{u}) = \sigma_\rho(\mathbf{u}, \mathbf{u})$. The following proposition gives the normal approximation for $X_\mathbf{u}, \mathbf{u} \in \mathcal{S}$, where $\mathcal{S}$ is a subset of words of length k.

Proposition 2. *Let $\mathcal{S}$ be a subset of words of length k such that $\Sigma_\rho = (\sigma_\rho(\mathbf{u}, \mathbf{v}))_{\mathbf{u}, \mathbf{v} \in \mathcal{S}}$ is non-degenerate. Then*

$$\lim_{M \to \infty} \sqrt{M} \left(\frac{X_\mathbf{u}}{M} - (\beta - w + 1)(P_\rho(\mathbf{u}) + P_\rho(\bar{\mathbf{u}})) \right)_{\mathbf{u} \in \mathcal{A}^k} = \mathcal{MN}(0, \Sigma_\rho),$$

where $\mathcal{MN}(0, \Sigma_\rho)$ is a multinormal distribution.

2.4 The Approximate Distributions of D_2, D_2^* and D_2^S under the Null and the Alternative Models

The following theorems give the approximate distributions of D_2, D_2^* and D_2^S under the null and the alternative models, respectively. These theorems are then used to give the thresholds for a type I error α under the null model $\rho = 1$, and to derive the approximate power formulas for detecting the relationships between two sequences under the alternative model $\rho < 1$ in Theorem 5. These theorems are extensions of the corresponding results for the limiting distributions of D_2, D_2^* and D_2^S for long sequences without NGS in [11]. For the theorems in this subsection, we assume that both the sequence length n and the number of reads M tend to infinity such that $\lim_{n \to \infty} \frac{M}{n} = \theta$, where θ is a constant. We also assume that the alphabet size, motif length, and word length are kept fixed. The conditions in Propositions 1 and 2 should also be satisfied. All the limits in Theorems 2-4 are in distribution.

Theorem 1. *Under the models for the long sequences and the NGS sampling of sequence reads as in Subsection 2.2, the means of D_2 and D_2^* and the approximate mean of D_2^S are given by*

$$\mathbf{E}D_2 = M^2(\beta - k + 1)^2 \sum (P_\rho(\mathbf{w}) + P_\rho(\bar{\mathbf{w}}))^2,$$

$$\mathbf{E}D_2^* = M(\beta - k + 1) \sum \frac{(P_\rho(\mathbf{w}) + P_\rho(\bar{\mathbf{w}}) - (p_\mathbf{w} + p_{\bar{\mathbf{w}}}))^2}{p_\mathbf{w} + p_{\bar{\mathbf{w}}}},$$

$$\lim_{M \to \infty} \frac{\mathbf{E}D_2^S}{M} = \frac{\beta - k + 1}{\sqrt{2}} \sum |(P_\rho(\mathbf{w}) + P_\rho(\bar{\mathbf{w}})) - (p_\mathbf{w} + p_{\bar{\mathbf{w}}})|.$$

Theorem 2. *Assume that in the background model for the long sequences not all letters are equally likely.*
a). Suppose $\rho = 1$ (the null model that the sequences are iid). Then

$$\lim_{M \to \infty} \sqrt{M} \left(\frac{D_2}{M^2} - (\beta - k + 1)^2 \sum_{\mathbf{w} \in \mathbf{A}^k} (p_\mathbf{w} + p_{\bar{\mathbf{w}}})^2 \right) = Z_1,$$

where Z_1 has normal distribution $\mathcal{N}(0, 2(\Sigma_1)^2)$.

b). Suppose $0 < \rho < 1$. *Then*

$$\lim_{M \to \infty} \sqrt{M}\left(\frac{D_2}{M^2} - (\beta - k + 1)^2 \sum_{\mathbf{w} \in \mathcal{A}^k} (P_\rho(\mathbf{w}) + P_\rho(\bar{\mathbf{w}}))^2\right) = Z_\rho,$$

where Z_ρ *has normal distribution* $\mathcal{N}(0, 2(\Sigma_\rho)^2)$, *and* $(\Sigma_\rho)^2$ *is given by*

$$(\beta - k + 1)^2 \Bigg\{ \sum_{\mathbf{w} \in \mathcal{A}^k} (P_\rho(\mathbf{w}) + P_\rho(\bar{\mathbf{w}}))^2 \sigma_\rho^2(\mathbf{w})$$

$$+ \sum_{\mathbf{w} \neq \mathbf{w}'} (P_\rho(\mathbf{w}) + P_\rho(\bar{\mathbf{w}}))(P_\rho(\mathbf{w}') + P_\rho(\bar{\mathbf{w}}'))\sigma_\rho(\mathbf{w}, \mathbf{w}') \Bigg\}.$$

Theorem 3. a). *Suppose* $\rho = 1$. *Then the limit distribution of* D_2^* *is given by*

$$\lim_{M \to \infty} D_2^* = Z_1^* = \sum_{\mathbf{w} \in \mathcal{A}^k} \frac{Z_\mathbf{w}^{(1)} Z_\mathbf{w}^{(2)}}{(\beta - k + 1)(p_\mathbf{w} + p_{\bar{\mathbf{w}}})},$$

where $\{Z_\mathbf{w}^{(1)}, \mathbf{w} \in \mathcal{A}_k\}$ *and* $\{Z_\mathbf{w}^{(2)}, \mathbf{w} \in \mathcal{A}_k\}$ *are independent and have mean 0 normal distributions.*

b). *Suppose* $0 < \rho < 1$ *and that* $\frac{(P_\rho(\mathbf{w}) + P_\rho(\bar{\mathbf{w}}) - (p_\mathbf{w} + p_{\bar{\mathbf{w}}}))}{p_\mathbf{w} + p_{\bar{\mathbf{w}}}}$ *is not constant in* $\mathbf{w}$. *Then*

$$\lim_{M \to \infty} \sqrt{M}\left(\frac{D_2^*}{M} - (\beta - k + 1) \sum_{\mathbf{w} \in \mathcal{A}^k} \frac{(P_\rho(\mathbf{w}) + P_\rho(\bar{\mathbf{w}}) - (p_\mathbf{w} + p_{\bar{\mathbf{w}}}))^2}{p_\mathbf{w} + p_{\bar{\mathbf{w}}}}\right) = Z_\rho^*,$$

where Z_ρ^* *has normal distribution* $\mathcal{N}(0, 2(\Sigma_\rho^*)^2)$, *and* $(\Sigma_\rho^*)^2$ *is given by*

$$\sum_{\mathbf{w} \in \mathcal{A}^k} \frac{(P_\rho(\mathbf{w}) + P_\rho(\bar{\mathbf{w}}) - (p_\mathbf{w} + p_{\bar{\mathbf{w}}}))^2}{(p_\mathbf{w} + p_{\bar{\mathbf{w}}})^2} \sigma_\rho^2(\mathbf{w})$$

$$+ \sum_{\mathbf{w} \neq \mathbf{w}'} \frac{(P_\rho(\mathbf{w}) + P_\rho(\bar{\mathbf{w}}) - (p_\mathbf{w} + p_{\bar{\mathbf{w}}}))(P_\rho(\mathbf{w}') + P_\rho(\bar{\mathbf{w}}') - (p_{\mathbf{w}'} + p_{\bar{\mathbf{w}}'}))}{(p_\mathbf{w} + p_{\bar{\mathbf{w}}})(p_{\mathbf{w}'} + p_{\bar{\mathbf{w}}'})} \sigma_\rho(\mathbf{w}, \mathbf{w}').$$

Theorem 4. a) *Suppose* $\rho = 1$. *Then*

$$\lim_{M \to \infty} \frac{D_2^S}{\sqrt{M}} = Z_1^S = \sum_{\mathbf{w} \in \mathcal{A}^k} \frac{Z_\mathbf{w}^{(1)} Z_\mathbf{w}^{(2)}}{\sqrt{(Z_\mathbf{w}^{(1)})^2 + (Z_\mathbf{w}^{(2)})^2}},$$

where $\{Z_\mathbf{w}^{(1)}, \mathbf{w} \in \mathcal{A}_k\}$ *and* $\{Z_\mathbf{w}^{(2)}, \mathbf{w} \in \mathcal{A}_k\}$ *are independent and have mean 0 normal distributions.*

b). *Suppose* $0 < \rho < 1$, *and* $(P_\rho(\mathbf{w}) + P_\rho(\bar{\mathbf{w}}) - (p_\mathbf{w} + p_{\bar{\mathbf{w}}}))$ *have different signs in* $\mathbf{w}$. *Then*

$$\lim_{M \to \infty} \sqrt{M}\left(\frac{D_2^S}{M} - (\beta - k + 1) \sum_{\mathbf{w} \in \mathcal{A}^k} \frac{|P_\rho(\mathbf{w}) + P_\rho(\bar{\mathbf{w}}) - (p_\mathbf{w} + p_{\bar{\mathbf{w}}})|}{\sqrt{2}}\right) = Z_\rho^S,$$

where Z_ρ^S has normal distribution $\mathcal{N}(0, 2(\Sigma_\rho^S)^2)$, and $(\Sigma_\rho^S)^2$ is given by

$$\frac{1}{8}\Bigg\{ \sum_{\mathbf{w}\in\mathcal{A}^k} \sigma_\rho^2(\mathbf{w}) +$$

$$\sum_{\mathbf{w}\neq\mathbf{w}'} \operatorname{sign}(P_\rho(\mathbf{w})+P_\rho(\bar{\mathbf{w}})-(p_\mathbf{w}+p_{\bar{\mathbf{w}}}))\operatorname{sign}(P_\rho(\mathbf{w}')+P_\rho(\bar{\mathbf{w}}')-(p_{\mathbf{w}'}+p_{\bar{\mathbf{w}}'}))\sigma_\rho(\mathbf{w},\mathbf{w}') \Bigg\}.$$

The following theorem gives the theoretical formulas for the power of D_2, D_2^* and D_2^S to detect the relationship between two sequences.

Theorem 5. *Assume that* $\frac{(P_\rho(\mathbf{w})+P_\rho(\bar{\mathbf{w}})-(p_\mathbf{w}+p_{\bar{\mathbf{w}}}))}{p_\mathbf{w}+p_{\bar{\mathbf{w}}}}$ *and* $(P_\rho(\mathbf{w})+P_\rho(\bar{\mathbf{w}})-(p_\mathbf{w}+$ $p_{\bar{\mathbf{w}}}))$ *are not constant in* $\mathbf{w}$*. Then, for any given type I error* α*, the power of detecting the relationship between two sequences under the common motif model in Subsection 2.2 against the null model that* $\rho = 1$ *using* D_2, D_2^* *and* D_2^S *can be approximated by* $1 - \Phi(C(\rho)), 1 - \Phi(C^*(\rho))$ *and* $1 - \Phi(C^S(\rho))$*, respectively, where*

$$C(\rho) = -(\beta - k + 1)\sqrt{M}B(\rho) + z_\alpha/\sqrt{2}\Sigma_\rho,$$
$$C^*(\rho) = -(\beta - k + 1)\sqrt{M}B^*(\rho) + z_\alpha^*/\sqrt{2M}\Sigma_\rho^*,$$
$$C^S(\rho) = -(\beta - k + 1)\sqrt{M}B^S(\rho) + z_\alpha^S/\sqrt{2}\Sigma_\rho^S,$$

and

$$B(\rho) = \frac{\beta - k + 1}{\sqrt{2}\Sigma_\rho} \sum_{\mathbf{w}\in\mathcal{A}^k} \left((P_\rho(\mathbf{w}) + P_\rho(\mathbf{w}))^2 - (p_\mathbf{w} + p_{\bar{\mathbf{w}}})^2\right),$$

$$B^*(\rho) = \frac{1}{\sqrt{2}\Sigma_\rho^*} \sum_{\mathbf{w}\in\mathcal{A}^k} \frac{(P_\rho(\mathbf{w}) + P_\rho(\bar{\mathbf{w}}) - (p_\mathbf{w} + p_{\bar{\mathbf{w}}}))^2}{p_\mathbf{w} + p_{\bar{\mathbf{w}}}},$$

$$B^S(\rho) = \frac{1}{2\Sigma_\rho^S} \sum_{\mathbf{w}\in\mathcal{A}^k} |P_\rho(\mathbf{w}) + P_\rho(\bar{\mathbf{w}}) - (p_\mathbf{w} + p_{\bar{\mathbf{w}}})|.$$

where, z_α, z_α^**, and* z_α^S *are upper* α *quantiles of* Z_1, Z_1^**, and* Z_1^S *from Theorems 2, 3, and 4, respectively.*

2.5 New Distance Measures for Clustering Genome Sequences Based on k-tuple Distributions

The statistics D_2, D_2^*, and D_2^S cannot be used directly to cluster genome sequences as the ranges of the statistics depend on several factors such as the nucleotide frequencies, the length of the reads, and the number of reads. To avoid these problems, we define the following distance measures d_2, d_2^* and d_2^S such that they range from 0 to 1, an interval not depending on these factors.

$$d_2 = \frac{1}{2}\left(1 - \frac{D_2}{\sqrt{\sum_{\mathbf{w}\in\mathcal{A}^k} X_{\mathbf{w}}^2}\sqrt{\sum_{\mathbf{w}\in\mathcal{A}^k} Y_{\mathbf{w}}^2}}\right),$$

$$d_2^* = \frac{1}{2}\left(1 - \frac{M(\beta - w + 1)D_2^*}{\sqrt{\sum_{\mathbf{w}\in\mathcal{A}^k} \tilde{X}_{\mathbf{w}}^2/p_{\mathbf{w}}}\sqrt{\sum_{\mathbf{w}\in\mathcal{A}^k} \tilde{Y}_{\mathbf{w}}^2/p_{\mathbf{w}}}}\right),$$

$$d_2^S = \frac{1}{2}\left(1 - \frac{D_2^S}{\sqrt{\sum_{\mathbf{w}\in\mathcal{A}^k} \tilde{X}_{\mathbf{w}}^2/\sqrt{\tilde{X}_{\mathbf{w}}^2 + \tilde{Y}_{\mathbf{w}}^2}}\sqrt{\sum_{\mathbf{w}\in\mathcal{A}^k} \tilde{Y}_{\mathbf{w}}^2/\sqrt{\tilde{X}_{\mathbf{w}}^2 + \tilde{Y}_{\mathbf{w}}^2}}}\right).$$

When the genome sequences of interest are closely related, the values of d_2, d_2^S and d_2^* are close to 0. Therefore, we can use them to measure the distances among different genome sequences.

To evaluate the validity of these distance measures in clustering genome sequences, we first use them to classify human, rabbit, mouse, opossum and chicken based on pseudo-simulated shotgun reads using MetaSim [16]. The phylogenetic relationship among the five species are clearly known. We then use d_2, d_2^* and d_2^S to cluster whole genome NGS data in [17] including 8 tree species of Fagaceae (primarily of the stone oaks, Lithocarpus) and 5 tree species of Moraceae (Ficus), mostly tropical Asian trees.

3   Results

We use simulations and theoretical analysis to study factors affecting the power of different statistics D_2, D_2^* and D_2^S. The details of the simulation results are omitted here and we summarize them as follows. 1). The power of D_2^* and D_2^S is generally higher than that of D_2; 2). Under the common motif model, the power of D_2^* is the highest when the length of the tuples k is the same as the length of the motif inserted in the background, which is consistent with our intuition; 3). For both D_2^* and D_2^S, the power first decreases and then increases with the read length β; 4). The power of D_2^* and D_2^S increases with the coverage of the reads over the genome and is generally lower than the corresponding power when the complete genome sequences are known. The randomness during sampling of the reads in NGS introduces noise into these statistics making them less powerful; 5). The power of D_2^* and D_2^S under the homogeneous sampling of the reads is higher than the corresponding power under the heterogeneous sampling of the reads; 6). When the lengths of the background sequences are very long, the power of D_2^S can be higher than that of D_2^*; 7). The above conclusions are insensitive to sequencing errors when the error rate is less than 0.005.

3.1 Clustering of Five Mammalian Species Using d_2, d_2^* and d_2^S Based on Pseudo-NGS Reads

In order to see the validity of clustering different species using NGS short reads based on d_2, d_2^* and d_2^S, we simulate NGS short reads using MetaSim [16] from five mammalian species: human, rabbit, mouse, opossum, and chicken, whose phylogenetic relationships are well established [18].

We first download their complete genome sequences from UCSC Genome Browser and Ensembl.org. Next, we use MetaSim to simulate NGS reads from each of the five species under the "empirical error model", which is derived from empirical studies of the Illumina Sequencing Technology. The read length is set at 62bp and the coverage is set to 1. Finally, we calculate the distances between any pair of the species using d_2, d_2^* and d_2^S for $k = 7, 9, 11$ based on the simulated reads, and use UPGMA in PHYLIP (http://evolution.genetics.washington.edu/phylip.html) to cluster them. Unfortunately, none of the resulting clustering is consistent with the known phylogenetic relationships of the five species (data not shown).

We reason that the large fraction of repeat regions along the genomes may make the k-tuple frequencies along the complete genomes significantly different from the k-tuple frequencies along the non-repeat regions. Thus, we take the following approach to eliminate or mitigate the effects of repeat regions. The basic idea is that if the number of occurrences of a k-tuple in the reads is much higher than expected, we eliminate the k-tuple from consideration when we calculate d_2, d_2^*, and d_2^S. For every $\mathbf{w}$, we calculate $T_{\mathbf{w}} = X_{\mathbf{w}}/EX_{\mathbf{w}}$. When $T_{\mathbf{w}}$ is larger than a threshold T_0, we set $X_{\mathbf{w}}$ to 0 in the calculation of d_2, and set $X_{\mathbf{w}}$ to $EX_{\mathbf{w}}$ in the calculation of d_2^*, and d_2^S.

When $k = 7$ and $T_0 = 2$, we observe that the clustering using d_2^S is consistent with the true underlying evolutionary tree while the clusterings using d_2 and d_2^* are not. This indicates that d_2^S identifies the relationships between species more efficiently than d_2 and d_2^*. We also note that the clusterings using any of the three distance measures for $k = 9$ and $k = 11$ are consistent with the true phylogenetic tree of the five species.

3.2 Applications to the Detection of the Relationship among Different Tree Species Using NGS Data

We then use the distance measures d_2, d_2^* and d_2^S defined in Subsection 2.5 to cluster the 13 tree species based on the NGS shotgun read data sets in [17]. Note that the number of tree species we study here is more than the 9 tree species in the original paper [17] because more data are now available. The 13 tree species can be generally classified into two groups: 5 tree species from Moraceae and 8 tree species from Fagaceae. Using the data set, we answer the following questions:

- Can the three distance measures d_2, d_2^* and d_2^S clearly separate the two groups of tree species based on the shotgun read data?

– How does the tuple size k affect the clustering of the tree species?
– How does the sequence depth γ affect the clustering of the tree species?

To answer these questions, we first use the complete shotgun read data to calculate the distances, d_2, d_2^* and d_2^S, between any pair of tree species from the 13 species for different values of tuple size $k = 7, 9, 11$. Taking the distance matrix as input, we apply the UPGMA program to cluster the tree species. Figure 1 shows the resulting clusterings using d_2, d_2^* and d_2^S, respectively, with $k = 9$. The clusterings of the tree species using $k = 7, 11$ are given in the full version of the paper.

Second, in order to see whether the clustering of the tree species can be correctly inferred using only a portion of the shotgun read data, we use $\gamma = 5\%$ of the total read data for each tree species to cluster them. Since the γ percent of the reads can be sampled randomly from the original read data, the resulting clustering of the tree species can be different. To study the variation of the clusters due to random sampling of the reads, we repeat the sampling process of the reads 100 times and calculate the frequencies that each internal branch of the clustering using all the reads occurs among the 100 clusterings. The frequencies are given in Figure 1 for $k = 9$ and $\gamma = 5\%$.

It can be seen from the figure that the two groups of tree species can be completely separated using the distance measures d_2^* and d_2^S for any $k = 7, 9, 11$. However, the tree species cannot be distinguished using the distance measure d_2. These two tree groups are quite far apart, as they are in different orders and probably separated by at least 50 million years, if not considerably longer (Cannon, personal communication). A good clustering should be able to separate the two groups of trees. This result indicates that d_2^* and d_2^S are more sensitive to distinguish the tree species than d_2. With d_2^S, most of the resulting clusters can be completely recovered with only 5% of the reads. However, the clustering with d_2^* is less stable when a small fraction of the data are available.

In addition, Ficus altissima and Ficus microcarpa cluster together using all three distance measures, which is consistent with the fact that both are large trees and are closely related while the other three Moraceae species are small dioecious shrubs. Similarly, the two Castanopsis species within the Fagaceae group also cluster together separate from the others using d_2^S. Thus, the clustering based on d_2^S is the most reasonable among the three distance measures we study. Finally, we note that the clustering by d_2^S is not perfect. F. Trigonobalanus is an ancestral genus that is very divergent from the rest of the family and has undergone considerable sequence evolution. It should not group within Lithocarpus (Cannon, personal communication).

4   Discussion

We modified the original D_2, D_2^* and D_2^S statistics for alignment-free sequence comparison of two long sequences to the scenario of genome sequence comparison using NGS data. Based on the HMM model for long sequences with random

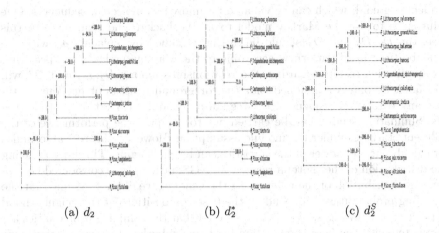

Fig. 1. The clusterings of the 13 tree species with distance measures, from left to right d_2(a), d_2^*(b), and d_2^S(c) with $k = 9$ using all the reads. The number on each internal branch is the fraction of times the branch occurs in 100 random sampling using $\gamma = 5\%$ of the reads.

instances of motif occurrences as in [8,11,12] and a general model for the sampling of NGS reads from the genome, we studied the approximate distributions of D_2, D_2^* and D_2^S. We also studied the power of detecting the relationships between two sequences related through the common motif model by both simulations and theoretical studies, and studied factors affecting the power of these statistics including genome sequence length, coverage of the NGS reads, read length, word length, and the distribution of the reads along the genome sequence. It is shown that D_2^* and D_2^S are more powerful than D_2 for detecting relationships between two sequences related through a common motif model. These results are consistent with those for alignment-free comparison of long sequences found in [8,11]. We also found that D_2^* and D_2^S are generally less powerful when applied to NGS data than when they are applied to complete sequences. Heterogeneity in the sampling of reads along the genome further decreases the power of these statistics. On the other hand, when the sampling of reads is relatively homogeneous across the genome and the coverage is high, the power of D_2^* and D_2^S approaches the power that is achieved when these statistics are applied to complete sequences. Based on these statistics, we defined corresponding distance measures d_2, d_2^* and d_2^S with ranges from 0 to 1. We applied the distance measures with some modifications to cluster five mammalian species and show that they can all cluster them well when the tuple size is 9 or 11. When applied to the real shotgun read data from 13 tree species whose complete genome sequences are unknown, the d_2^* and d_2^S distance measures can correctly separate the two groups of tree species even with 5% of the reads from the shotgun read data sets.

Although we showed the usefulness of D_2^* and D_2^S for detecting the relationships between sequences and for clustering sequences using NGS data without assembly, our study has several limitations. First, we assumed that the background sequences are iid, which can be violated for many real molecular sequences. One solution is to use the Markov model to fit the background sequences. In this case, the D_2, D_2^* and D_2^S should be further modified by replacing $p_{\mathbf{w}}$ with the probability of word pattern $\mathbf{w}$ according to the Markov model. We expect that the qualitative results regarding the relationships among D_2, D_2^* and D_2^S will still hold. Second, we assumed that the foreground consists of just one motif. In many regulatory sequences, the regulatory modules consist of multiple motifs. Simulation studies can be carried out to compare the performance of the different statistics under the module assumption. However, theoretical formulas for calculating the power of the statistics can be challenging. Third, in modeling the distribution of the shotgun reads from NGS, although we considered heterogeneous distribution of the reads along the genome, we did not assume that the sampling probabilities λ_i depend on the base compositions at the neighborhood of position i. Previous studies [14,15] showed that the sampling probabilities are associated with the base composition in the neighborhood of the position. One solution to this problem is to ignore the first 6-10 bases of the reads and only consider the remaining bases of the reads. Without trimming each read, the k-tuple composition vector from the shotgun read data may be significantly different from the k-tuple composition from the original genome which the shotgun read data are sampled from. On the other hand, new sequencing technologies will reduce the dependence of sampling probability on the base composition and the read distributions will be increasingly homogeneous. Despite all these problems, we expect that our study lays the foundations for the study of alignment-free sequence comparison based on NGS shotgun read data.

Acknowledgements. The research is supported by National Natural Science Foundation of China (No.10871009, 10721403, 60928007), National Key Basic Research Project of China (No.2009CB918503), and Graduate Independent Innovation Foundation of Shandong University (GIIFSDU) (ZYZ). FS is partially supported by US NIH P50 HG 002790 and R21AG032743, and NSF DMS-1043075 and OCE 1136818. We sincerely thank Professors Michael S Waterman and Gesine Reinert for collaborations and constant discussion on the development of alignment-free sequence comparison. The paper would have been impossible without their deep insights into the statistics for alignment-free sequence comparison. We also thank Professor Chuck Cannon for his insightful comments on the evolutionary relationships of the 13 tree species, and Mr. Michael Klein for carefully reading the paper and giving valuable suggestions.

References

1. Blaisdell, B.E.: A measure of the similarity of sets of sequences not requiring sequence alignment. Proceedings of the National Academy of Sciences of the United States of America 83(14), 5155–5159 (1986)
2. Domazet-Lošo, M., Haubold, B.: Alignment-free detection of local similarity among viral and bacterial genomes. Bioinformatics 27(11), 1466–1472 (2011)

3. Ivan, A., Halfon, M., Sinha, S.: Computational discovery of cis-regulatory modules in Drosophila without prior knowledge of motifs. Genome Biology 9(1), R22 (2008)
4. Jun, S.R., Sims, G.E., Wu, G.A., Kim, S.H.: Whole-proteome phylogeny of prokaryotes by feature frequency profiles: An alignment-free method with optimal feature resolution. Proceedings of the National Academy of Sciences of the United States of America 107(1), 133–138 (2010)
5. Leung, G., Eisen, M.B.: Identifying cis-regulatory sequences by word profile similarity. PLoS One 4, e6901 (2009)
6. Lippert, R.A., Huang, H.Y., Waterman, M.S.: Distributional regimes for the number of k-word matches between two random sequences. Proceedings of the National Academy of Sciences of the United States of America 100(13), 13980–13989 (2002)
7. Liu, X., Wan, L., Li, J., Reinert, G., Waterman, M.S., Sun, F.: New powerful statistics for alignment-free sequence comparison under a pattern transfer model. Journal of Theoretical Biology 284(1), 106–116 (2011)
8. Reinert, G., Chew, D., Sun, F.Z., Waterman, M.S.: Alignment-free sequence comparison (I): Statistics and power. Journal of Computational Biology 16(12), 1615–1634 (2009)
9. Sims, G.E., Jun, S.R., Wu, G.A., Kim, S.H.: Alignment-free genome comparison with feature frequency profiles (FFP) and optimal resolutions. Proceedings of the National Academy of Sciences of the United States of America 106(8), 2677–2682 (2009)
10. Vinga, S., Almeida, J.: Alignment-free sequence comparison–a review. Bioinformatics 19(4), 513–523 (2003)
11. Wan, L., Reinert, G., Sun, F., Waterman, M.S.: Alignment-free sequence comparison (II): Theoretical power of comparison statistics. Journal of Computational Biology 17(11), 1467–1490 (2010)
12. Zhai, Z.Y., Ku, S.Y., Luan, Y.H., Reinert, G., Waterman, M.S., Sun, F.Z.: The power of detecting enriched patterns: An HMM approach. Journal of Computational Biology 17(4), 581–592 (2010)
13. Zhang, Z.D., Rozowsky, J., Snyder, M., Chang, J., Gerstein, M.: Modeling ChIP sequencing in silico with applications. PLoS Computational Biology 4(8), e1000158 (2008)
14. Hansen, K.D., Brenner, S.E., Dudoit, S.: Biases in Illumina transcriptome sequencing caused by random hexamer priming. Nucleic Acids Research 38(12), e131 (2010)
15. Li, J., Jiang, H., Wong, W.H.: Modeling non-uniformity in short-read rates in RNA-Seq data. Genome Biology 11, R50 (2010)
16. Richter, D.C., Ott, F., Auch, A.F., Schmid, R., Huson, D.H.: MetaSim: a sequencing simulator for genomics and metagenomics. PLoS One 3(10), e3373 (2008)
17. Cannon, C.H., Kua, C.S., Zhang, D., Harting, J.R.: Assembly free comparative genomics of short-read sequence data discovers the needles in the haystack. Molecular Ecology 19(suppl. 1), 146–160 (2010)
18. Miller, W., Rosenbloom, K., Hardison, R.C., Hou, M., Taylor, J., Raney, B., Burhans, R., King, D.C., Baertsch, R., Blankenberg, D., et al.: 28-way vertebrate alignment and conservation track in the UCSC genome browser. Genome Research 17(12), 1797–1808 (2007)

Differential Oestrogen Receptor Binding is Associated with Clinical Outcome in Breast Cancer

Rory Stark

Cambridge Research Institute, Cancer Research UK

This paper, which maps ERα binding via ChIP-seq in tumour tissue from twenty ER+ breast cancer patients, relies on a concurrently developed Bioconductor package, DiffBind, which provides a framework for quantitative differential analysis of protein/DNA binding events. Here we use DiffBind to identify ERα sites significantly differentially bound between those found in tumours from patients with good prognosis vs. those with poor prognosis and metastases. Gene signatures that predict clinical outcome in ER+ disease, validated in publically available breast cancer gene expression datasets, are derived from these sites. These signatures are enriched for genes with relevant proximal cis-regulatory events. Statistical characterization of differentially bound ERα sites enables further downstream analysis, including identification of a differentially enriched motif for the transcription factor FoxA1. Further differential analysis in five ER+ breast cancer cell lines shows how ERα binding is extensively shifted in tamoxifen-resistance, with the FoxA1 motif enriched proximal to ERα binding sites differentially bound in cells resistant to treatment. Analysis of FoxA1 binding at mitogen-induced ERα sites demonstrates that the observed differential ER binding program is not due to the selection of a rare subpopulation of cells, but rather to the FoxA1-mediated reprogramming of ER binding on a rapid time scale. Focusing our analysis on differential binding in primary tumour material allows us to show the plasticity of ERα binding capacity, with distinct combinations of cis-regulatory elements linked with the different clinical outcomes. These techniques are applicable to other cancers (and indeed other diseases) where master transcription factor regulators are known.

http://www.nature.com/nature/journal/vaop/ncurrent/full/
nature10730.html

B. Chor (Ed.): RECOMB 2012, LNBI 7262, p. 286, 2012.
© Springer-Verlag Berlin Heidelberg 2012

Simultaneous Reconstruction of Multiple Signaling Pathways *via* the Prize-Collecting Steiner Forest Problem

Nurcan Tuncbag[1], Alfredo Braunstein[2,3], Andrea Pagnani[3], Shao-Shan Carol Huang[1], Jennifer Chayes[4], Christian Borgs[4], Riccardo Zecchina[2,3], and Ernest Fraenkel[1]

[1] Department of Biological Engineering, Massachusetts Institute of Technology, Cambridge, MA 02139, USA
{ntuncbag,shhuang,fraenkel-admin}@mit.edu
[2] Department of Applied Science, Politecnico di Torino, C.so Duca degli Abruzzi 24, 10129 Torino, Italy
{alfredo.braunstein,riccardo.zecchina}@polito.it
[3] Human Genetics Foundation, Via Nizza 52, 10126 Torino, Italy
andrea.pagnani@hugef-torino.org
[4] Microsoft Research New England, One Memorial Drive, Cambridge, MA 02142, USA
{jchayes,borgs}@microsoft.com

Abstract. Signaling networks are essential for cells to control processes such as growth and response to stimuli. Although many "omic" data sources are available to probe signaling pathways, these data are typically sparse and noisy. Thus, it has been difficult to use these data to discover the cause of the diseases. We overcome these problems and use "omic" data to simultaneously reconstruct multiple pathways that are altered in a particular condition by solving the prize-collecting Steiner forest problem. To evaluate this approach, we use the well-characterized yeast pheromone response. We then apply the method to human glioblastoma data, searching for a forest of trees each of which is rooted in a different cell surface receptor. This approach discovers both overlapping and independent signaling pathways that are enriched in functionally and clinically relevant proteins, which could provide the basis for new therapeutic strategies.

Keywords: Prize-collecting Steiner forest, signaling pathways, multiple network reconstruction.

1 Introduction

High-throughput technologies including mass spectrometry, chromatin immunoprecipitation followed by sequencing (CHIP-Seq), RNA sequencing (RNA-seq), microarray and screening methods have the potential to provide dramatically new insights into biological processes. By providing a relatively comprehensive view of the changes that occur for a specific type of molecule or perturbation, these approaches can uncover previously unrecognized processes in a system of interest. However, interpreting these data types together to provide a coherent view of the biological processes is still a challenging task. In order to discover how changes in

B. Chor (Ed.): RECOMB 2012, LNBI 7262, pp. 287–301, 2012.

different classes of molecules relate to each other, it is possible to map the data onto a network of known or predicted interactions.   In the ideal case, the observed interactions would all lie near each other in a functionally coherent part of the interaction network (the interactome). However, due to false positives and false negatives in both the "omic" data and the interactome, the true situation is much more complex; advanced algorithms are needed to find meaningful connections among the data.   Among the approaches that have been proposed to find these sub-networks from the interactome are network flow optimization [1, 2], network propagation [3], the Steiner tree approach [4-6], network inference from gene expression [7, 8], linear programming [9], maximum-likelihood [10], electric circuits [11-13], network alignment [14] and Bayesian networks [15].

In our previous work, we used the prize-collecting Steiner tree formalism to find an optimum tree composed of nodes detected in experiments (terminals) and nodes that were not detected (Steiner nodes). We assigned costs to each interaction reflecting our confidence that the reported interaction was real and assigned prizes for excluding any of the terminals from the tree based on confidence in the proteomic or transcriptional data. By minimizing the sum of the total cost of all edges in the tree and the total prize of all nodes not contained in the tree, we were able to obtain compact and biologically relevant networks [4, 6]. Despite the power of Steiner tree approach for identifying functionally coherent networks, it is restricted to discovering a connected subgraph, which may be an inadequate representation for many systems. In particular, we often expect there to be many simultaneously acting biological processes in the cell that may not be connected together by interactions in the currently known interactome. These processes may be unconnected either because they may involve essentially independent cell functions, or simply due to our imperfect knowledge of the interactome.

In this work, we formulate a forest (defined as a disjoint union of trees) approach to identify simultaneously acting pathways in biological networks using both proteomic and transcriptional data, We use a generalization of the message-passing algorithm for the Prize-collecting Steiner Tree (PCST) problem [4, 16]. We first demonstrate the forest approach by using it to integrate proteomic and transcriptional data in the yeast pheromone response, showing that the forest consists of trees enriched in specific and distinct biological processes. As an additional feature, directed edges, which are particularly useful for representing the effects of enzymes and transcriptional regulators on their targets, are also incorporated.

We reasoned that the Steiner forest approach could be utilized in modeling mammalian signaling where there are many more cell-surface receptors and downstream pathways than in yeast. In principle, the forest approach could uncover multiple, independent components of the biological response.   Although the interactome data are much less complete for mammals than for yeast, we show that the same methods are applicable. We built prize-collecting Steiner forests derived from proteomic data from a model of glioblastoma multiforme (GBM) in which each tree was rooted in a different cell surface receptor representing independent signaling pathways and potential points of therapeutic intervention.   The solution reveals several known pathways and some unexpected new ones that are altered in the disease and suggests potential therapeutic strategies. The modified algorithm can now be applied to a wide range of complex systems.

2   Methods

2.1   Datasets

Throughout this work, two different biological networks are used: the yeast interactome and the human interactome. We refer to nodes with prize values greater than zero as terminal nodes.

Yeast Dataset. The yeast interactome contains 34,712 protein-protein and transcription factor to target interactions between 5,957 nodes. The terminal node set contains 106 differentially phosphorylated proteins detected by mass spectrometry [17] and 118 differentially expressed genes [18] detected by microarray in response to the mating pheromone alpha factor. The node prizes are computed from the fold changes between treated and non-treated conditions. The edge costs are calculated by taking a negative log of the interaction probability. The details are available in [6]. In this study, we modified the transcription factor–DNA interactions to be directed edges. We also added to the interactome a set of directed edges that represent phosphorylation and dephosphorylation reactions between kinases, phosphatases and their substrates [19]. If these interactions are available in the original interactome, probabilities are retained. If they are not, the probabilities of these interactions are set uniformly to 0.8, based on the distribution of the probabilities in the original interactome. The final interactome contains 35,998 edges between 5,957 nodes. In both cases, the resulting interactomes are comprised of both undirected and directed edges.

Human Dataset. Protein-protein interactions in the STRING database (version 8.3) are used as the data source for the human interactome [20]. Here, the probabilities from experiments and database evidence channels are combined to obtain the final probability of the interactions. Interactions with a combined probability greater than 0.8 are included in the interactome. The receptor molecules are collected from the Human Plasma Membrane Database [21] where 331 receptors are available in the interactome derived from STRING. The phosphoproteomics data in [22] is combined with the interactome in humans for the GBM test case. From this dataset, 72 proteins containing phophorylated tyrosine peptides are present in our human interactome.

2.2   Prize-collecting Steiner Tree Problem

For a given, directed or undirected network $G(V, E, c(e), p(v))$ of node set V and edge set E, where a $p(v) \geq 0$ assigns a prize to each node $v \in V$ and $c(e) \geq 0$ assigns a cost to each edge $e \in E$. The aim is to find a tree $T(V_T, E_T)$, by minimizing the objective function:

$$f(T) = \beta \sum_{v \in V_T} p(v) + \sum_{e \in E_T} c(e) \tag{1}$$

where the first term is β times the sum of the node prizes not included in the tree T and the second part is the sum of the edge costs of T. Note that

$$\sum_{v \in V_T} p(v) = -\sum_{v \in V_T} p(v) + const \tag{2}$$

so that minimizing $f(T)$ amounts to collecting the largest set of high prize vertices while minimizing the set of large cost edges in a trade-off tuned by β. As a starting point, we consider the message-passing algorithm for the PCST problem introduced in [4]. The message-passing algorithm converts the global problem of finding the optimal tree into a set of local problems that can be solved efficiently. These equations are solved iteratively in a computationally efficient way. Here we present a generalization of the message passing algorithm designed to solve the PCST problem on directed networks (*i.e.* where in general c(e{i,j}) might be different from c(e{j,i})). In this variant, the optimization will be done on directed rooted trees, where choice of the root (which will be part of the candidate tree) is an external parameter of the algorithm.

2.3   Prize-collecting Steiner Forest (PCSF) Problem

A type of PCSF has already been considered in [23, 24]. In these works penalties are assigned to each pair of nodes either directly connected in the tree (*i.e.* edges belonging to the forest), or completely disconnected (*i.e.* in different forest components). Here we consider a different PCSF construction for a given, directed or undirected network $G(V, E, c(e), p(v))$ of node set V and edge set E, where a $p(v) \geq 0$ assigns a prize to each node $v \in V$ and $c(e) \geq 0$ assigns a cost to each edge e ∈ F. The aim is to find a forest $F(V_F, E_F)$ that minimizes the objective function:

$$f'(F) = \beta \sum_{v \in V_F} p(v) + \sum_{e \in E_F} c(e) + \omega \cdot \kappa \tag{3}$$

where κ is the number of trees in the forest and ω is new tuning parameter explained below. A practical way of minimizing f' consists in casting the PCSF into a PCST on a slightly modified graph. The idea is to introduce an extra root node v_0 into the network connected to each node $v \in V$ by an edge (v, v_0) with cost ω [25]. The PCST algorithm is employed on the resulting graph $H(V \cup \{v_0,\}, E \cup Vx\{v_0,\})$ and the solution will be called T. We define the forest F as T with all edges that point to the root removed. It is straightforward to see that the tree T is minimal for f if and only if the forest F is minimal for f'. Typically, the algorithm is run for different values of β and ω.

We used the previously published message-passing approach as the underlying implementation for this forest search [4], as many of our networks exceeded the capacity of the linear programming approaches. The message-passing approach is computationally fast and robust to the noise in the network as well. Although this algorithm is not guarranteed to find the optimal solution, in practice the networks it discovers are very similar to the exact solution. Introducing the artificial edges allows the algorithm to identify one or more trees that are only connected to the artificial node and not to each other. Although this modification seems algorithmically straightforward, its biological implications are very important. The concept is illustrated in Figure 1. In that example, two distinct pathways are connected only through spurious edges. The main difference between the tree formalism and the forest formalism is that the former one that connects as many of the experimental data as it can in a single network. As a result, it will either have to exclude some of the data that relate to distinct biological processes or add spurious edges to force these data to connect to the tree while the latter one allows the corresponding nodes to be included in distinct trees. However, the forest formalism is able to locate distinct biological processes into different sub-trees through the artificial node. The artifical node and edges give the flexibility of generating several sub-trees without paying any penalty.

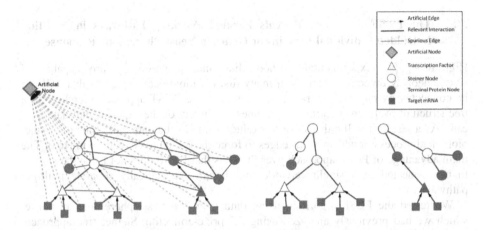

Fig. 1. Conceptual illustration of the PCSF algorithm. The left panel shows an interactome and the right panel shows the Steiner forest constructed from that interactome. The direction of transcription factor to target and kinase/phosphatase to substrate interactions are pointing towards the root node (opposite to the biological direction). In this scenario, there are spurious edges between these two pathways in the interactome. The PCSF algorithm provides the advantage to connect these distinct pathways artificially.

Tuning the Parameters. The parameters to be tuned in this problem are ω and β. The number of components of the solution (κ) depends strongly on the parameter ω, but it also depends on the β value: e.g. for $\beta = 0$ the optimal solution is the empty forest for all values of other parameters. For other values of β, while some sub-trees are composed of a single node, some others are composed of large number of nodes in the resulting forest. A forest with many very small trees (a single node each) would be obtained with very small artificial edges cost. The limiting case in the other direction is a single tree resulting from very large artificial edge cost. Therefore, there is a non-trivial interaction between the two parameters (β and ω). In principle, this two dimensional (β, ω) space of parameters should be explored. In this way, we get many possible types of forest: many small trees, many large trees, few small trees, few large trees. The effect of ω and β intervals highly depends on the distribution of edge costs and node prizes in the targeted interactome, so these parameters will be different for different datasets. For the yeast dataset, ω values are tuned between [0.005, 0.1] and β values are tuned between [1, 20].

Functional Annotation. For functional enrichment analysis, the BINGO plug-in [26] of Cytoscape [27] was used. The p-value significance threshold of 0.05 was used, which is corrected for multiple hypothesis testing, and all yeast proteins were used as the background set for the yeast dataset. For the human dataset, the functional enrichment is performed by using all human proteins as background set. All network visualizations were performed in Cytoscape [27].

3   Results

3.1   The PCSF Approach Reveals Parallel Working Pathways in Addition to Hidden Individual Proteins or Genes in Yeast Pheromone Response

High-throughput experimental methods like mass-spectrometry are capable of simultaneously detecting changes in many distinct biological processes that will not be connected by physical interactions. However, the PCST approach searches for a tree structure in the interactome that connects as many of the experimental data as it can. As a result, it will either have to exclude some of the data that relate to distinct biological process or add spurious edges to force these data to connect to the tree. The main advantage of PCSF approach over PCST is that PCSF does not force the system to be connected in a single network, and it can automatically separate multiple pathways.

We tested the PCSF algorithm using data from the yeast pheromone response, which we had previously analyzed using the prize-collecting Steiner tree approach. The data consist of phosphoproteomic and transcriptional changes induced by mating pheromone, and the network is enriched with directed transcription factor-target and kinase/phosphatase-substrate reactions. The edge costs of the interactome were computed as the negative log of the interaction probabilities, and node prizes were obtained from the scheme detailed in [6]. To explore the space of solutions, we tuned the ω and β parameters between [0.005, 0.1] and [1, 20], respectively. The minimum, maximum and average size and number of trees in the constructed PCSFs are extracted for each (ω, β) pair and the distribution of these values along ω parameter is plotted. We looked for a solution in a region where the number of trees and average size of the trees in the forest are closest to each other. By these criteria, the best solution is found when $\omega = 0.025$ and $\beta = 13$. We note that in order to explore these parameters, we constructed 400 solutions to the PCST problem. This number of calculations is only practical using the message-passing algorithm, but not with the integer linear programming based approaches.

The solution to PCSF problem places distinct functional classes in seperate sub-trees. In this solution, there are six trees, each containing more than 10 nodes. In Figure 2, each tree is labeled with its corresponding pathway. Small sub-trees such as T_{3-6} are enriched in specific biological processes including the PKC pathway, actin organization, protein folding and kinetochore, and DNA and chromatin pathways, while larger trees contain multiple processes. For example, the largest subtree, T_1, contains the pheromone core MAPK pathway with CDC28 related proteins and the second largest one, T_3, contains transcription and transport processes (see Figure 2). There are two different yeast MAPK pathways; the pheromone-induced MAPK and the protein kinase C (PKC) pathways [28, 29]. The PCSF algorithm correctly separates these two pathways into different trees. The largest tree in size is T_1 contains pheromone-induced MAPK pathway but the PKC pathway is located in T_3. While the former one induces cells to differentiate and be prepared for mating, the latter one is involved in cell integrity and new cell wall synthesis.

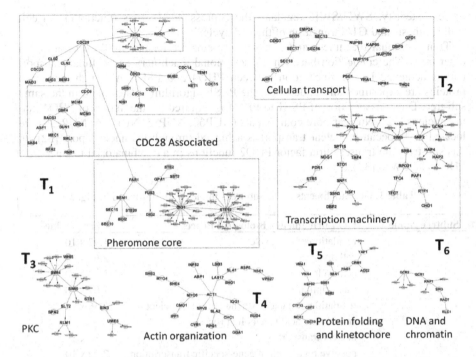

Fig. 2. Prize-collecting Steiner Forest (PCSF) of the Yeast Pheromone Response Network. Functional groups annotated by Gene Ontology (GO) are tagged with red boxes. In this PCSF, the rectangular nodes are DNA, triangular nodes are transcription factors and the circular nodes are proteins. Terminal nodes are colored red.

The core pheromone response pathway component in T_1 includes the STE2 receptor. In this sub-tree, the STE2-GPA1-FUS3 interaction is in the core of pheromone response. In addition, the MAP kinase FUS3 activates several transcription factors such as, STE12, DIG1, DIG2 for the expression of mating related genes. T_1 contains DNA replication proteins and cell cycle proteins associated with CDC28 as well. Here, the connection between the MAP kinase pathway in the pheromone core and the CDC28 associated sub-network is constructed through the interaction between FAR1 and CDC28. FAR1 is a direct inhibitor of CDC28/CLN2 complex and functions in orienting cell polarization. This association blocks the cell cycle progression. FUS3 phosphorylates FAR1, and only phosphorylated FAR1 can associate with CDC28/CLN2 complex. All these interactions and these connected pathways are correctly located into the same sub-tree.

The algorithm correctly identifies SLT2, which was not detected in the phosphoproteomic data, as a key node in regulating new cell wall synthesis. SLT2 is a serine/threonine MAP kinase activated in a cascade starting with PKC. The phosphoproteomic data are not sufficient for the algorithm to pick up the upstream pathway. However, in T_3, the algorithm links SLT2 to several transcription factors that mostly function in cell wall integrity and biosynthesis. SLT2 activates RLM1 [30], SWI4 [31] and SWI6 transcription factors. RLM1 functions in the maintenance of cell integrity. SWI4/SWI6 regulates the expression of genes functioning in cell wall synthesis and G1/S transition of the cell cycle.

Transcriptional machinery and transport proteins are located in T_2, seperate from other trees. The connection between transcriptional machinery and cellular transport part is achieved by the interaction between PHO4 and PSE1. Although these two proteins are experimentally undetected, the PCSF algorithm locates them in the same sub-tree. Direct association of PSE1 to PHO4 is required for the import of PHO4 into the nucleus [32]. Nuclear pore components (NUP60, NUP85, NUP116, NUP159) are located in T_2 because nuclear transport is achieved through the nuclear pore [32]. In this sub-tree, the transcription factor PHO2 functions in a combinatorial manner with PHO4 and SWI5 [33].

Table 1. GO enrichments of the sub-trees in the PCSF illustrated in **Fig. 2**

Subtree Name	GO Enrichment - Biological Process	Corr p-value
T_1	regulation of cell cycle	1.97×10^{-17}
	cell division	2.60×10^{-17}
	cell cycle	3.02×10^{-17}
T_2	transcription	7.07×10^{-13}
	regulation of nucleobase, nucleoside, nucleotide and nucleic acid metabolic process	2.36×10^{-12}
	nuclear transport	7.30×10^{-8}
T_3	positive regulation of gene-specific transcription	2.20×10^{-5}
	regulation of gene-specific transcription	7.75×10^{-5}
	positive regulation of transcription, DNA-dependent	9.11×10^{-5}
T_4	actin filament-based process	1.51×10^{-9}
	endocytosis	4.42×10^{-9}
	actin cytoskeleton organization	9.32×10^{-9}
T_5	protein folding	1.60×10^{-3}
	protein refolding	1.60×10^{-3}
	kinetochore assembly	4.25×10^{-3}
T_6	positive regulation of glycolysis	2.54×10^{-4}
	regulation of glycolysis	2.54×10^{-4}
	positive regulation of transcription	2.54×10^{-4}

In addition to the pathway analysis, we utilized GO biological process annotations to find the specific biological processes enriched in these trees. In **Table 1**, the top three annotations for each tree are tabulated along with their corrected p-values. These results show that this method effectively locates different biological processes into different trees. Instead of forcing all nodes to be connected in a single network, this "forest" representation composed of multiple sub-trees is more useful for distinguishing distinct pathways. The forest solution retains enrichment for the expected biological process, such as response to stress, cell cycle, signaling and transport. Further, by adding directions between transcription factors to targets and enzyme to substrate interactions, we are able to obtain condition-specific transcription factors and compact networks.

3.2   The PCSF Algorithm Reveals Coordinately Acting Receptor Molecules Functioning in Human GBM by Integrating Receptome, Interactome and Proteomics Data

Having demonstrated that the PCSF algorithm can successfully distinguish parallel-working pathways in yeast, we used it to identify cell surface receptors associated with signaling pathways altered in disease. Cell surface receptors are an interesting class of molecules to study, as they may be particularly easy to target with therapeutic agents. There is increasing evidence that some proteins are "undruggable," in other words hard or impossible to target with small molecule-based therapies because their three-dimensional shape does not have any appropriate concave sites to which these proteins can bind. In contrast, cell surface receptors can either be targeted with their natural ligand, modified forms of the natural ligand, small molecules that insert into the naturally occurring binding pocket or antibodies.

We modified our approach to identify cell-surface receptors associated with phosphoproteomic changes that occur in a model of glioblastoma. We use the artificial node to represent external stimuli (including autocrine loops) that potentially activates multiple receptor molecules, by connecting this node only to cell surface receptors, of which 331 are present in our human interactome. After running the prize-collecting Steiner tree algorithm and removing the artificial node, each sub-tree will contain one receptor as the starting node. The receptors selected in the solution of PCSF represent those most closely connected to the measured phosphoproteomic data and are therefore likely to be main contributors of the disease.

We applied this approach to phosphotyrosine data for a model of human GBM [22] representing phosphorylation differences between cells expressing an oncogenic mutation in the EGFR protein and cells with an inactive form of this receptor tyrosine kinase. The result is a set of eleven compact trees each rooted in one of the 331 potential receptors. The selected receptors in order of their tree sizes are EGFR, ERBB2, CD36, IGF1R, PTCH1, A2MR, SDC2, MET, ITGB3, NPR1 and EPHA2 (see Fig. 3). Although the algorithm had no direct knowledge that the data represented the results of mutation in EGFR, it selected this as the root of the largest tree. In fact, each of the four top receptors has a known link to cancer. EGFR and ERBB2 are EGF-family receptors, and it is known that EGFR is mutated in more than 50% GBM cases [34]. IGF1R is overexpressed in many tumors and mediates proliferation and resistance to apoptosis, and it is currently an anti-cancer treatment target [35]. Because IGF1R is also abnormally active in GBM, its inhibition is presented as a potential therapy to arrest the tumor growth [36]. It has been previously shown that the EGF and IGF pathways cross-talk [37], and IGF1R mediates resistance to anti-EGFR therapy in glioma cells [38]. Although CD36 functions in brain specific angiogenic regulation [39] and the interactions between CD36-Fyn-Yes lead to calcium and neurotransmitter release [40], its relation to GBM has not been studied in detail.

Although the algorithm is constrained to identify independent trees, we can observe the potential for cross-talk between different receptors by adding back all the edges among the selected nodes. We noticed two receptors selected by the algorithm, namely MET and ITGB3 (integrin-β3), are also very important, despite the fact that their corresponding sub-trees each contain only two nodes. When all edges are put

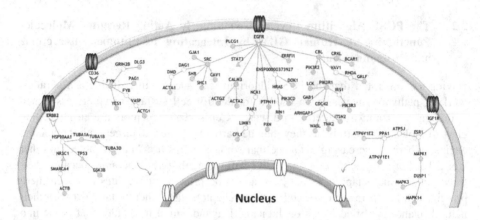

Fig. 3. Network representation of the PCSF for human glioblastoma dataset. Each tree is rooted from a cell surface receptor. The receptor molecules are represented by the arrowheads, transcription factors in triangles and other proteins as circles. Terminal nodes are colored in cyan.

back in the Steiner forest we observe extensive links between these two receptors and the EGFR sub-tree. MET has links to seven proteins out of nine first neighbors of EGFR and to 16 proteins in total of the EGFR rooted sub-tree, and ITGB3 has links to seventeen members of the EGFR sub-tree. By contrast, IGF1R has links only to three proteins and ERBB2 has link only to two proteins in the first neighbors of EGFR. Although this information was not provided to the algorithm, MET is detected as differentially phosphorylated in the original data, and a MET inhibitor synergizes with an EGFR inhibitor [22].

Mammalian signaling systems frequently demonstrate a high degree of cross-talk. If two receptors share many common downstream components, the algorithm need only choose one of these as a root node to explain all the terminal nodes. We, therefore, introduce a perturbation-based approach to improve the sensitivity of the algorithm in identifying receptors that share many downstream components with the selected root nodes. In this analysis, selected receptor molecules (the root of the largest tree in the forest and other receptors in its corresponding sub-family available in the forest) and all their interactions are removed from the interactome and PCSF algorithm is applied to the remaining network. Through this *in silico* knock-out experiment, we can find the other receptor molecules whose role may be masked in the presence of the receptors in the initial forest.

We first knocked-out two EGF/ERBB sub-family receptors, EGFR and ERBB2, from the network and re-generated the PCSF. In the resulting network, 37 out of 42 nodes in the down-stream of EGFR in the original tree are connected to other receptors. PDGFR (Platelet-derived growth factor receptor) is the root of the largest tree, covering 23 nodes linked to EGFR in the original network. This observation suggests that PDGFR may have many overlapping functions with EGFR. In fact, several studies have shown that PDGFR is critical in brain tumorogenesis [41, 42]; mutation of PDGFR causes alteration in the intracellular signaling [43] and it is a therapeutic target in GBM [44]. Another 14 nodes down-stream of EGFR are shared by MET (seven nodes), IGF1R (five nodes) and ITGB3 (two nodes). Although these four receptors (PDGFR, MET, IGF1R and ITGB3) capture many of the nodes that were down-stream of EGFR, five nodes are not captured by any other receptors. These may represent signaling that is uniquely downstream of EGFR.

To further explore the network, we removed PDGFR in addition to EGFR and ERBB2. In the new network, the MET receptor partially replaces PDGFR. It has been shown that the MET receptor is activated in GBM and it might be a therapeutic target [45]. Similar to the MET receptor tree, the sub-tree containing ITGB3 receptor also collects several of the nodes previously associated with EGFR in its corresponding sub-tree. It is interesting to note that integrins function as both upstream and downstream effectors of growth factor receptors, such as EGFR, IGF1R, PDGFR, MET [46]. Integrins and their relation to GBM have not been studied in detail, which may have clinical importance in GBM.

During all these leave-one-receptor-out tests, IGF1R is present in the resulting PCSF, and it retains all proteins in the original network. The downstream network of IGF1R starts with the estrogen receptor (ESR1) interaction and it contains several MAPKs. It has been shown that ESR1 and IGF1R are cross-regulated in the brain and activate the MAPK/ERK pathway. This system of interactions results in some neural functional regulations in the brain; such as, synaptic plasticity, neurotic growth, and neuronal survival [47]. The size of the trees corresponding to the down-stream of MET and ITGB3 receptors increases at each knock-out. Also, the FYN related downstream pathway of CD36 is swapped to be downstream of MET receptor, although CD36 is not knocked-out. This result implies that FYN-related pathway may be activated by several receptors.

To further validate the relevancy of these receptor molecules (EGFR, ERBB2, IGF1R, CD36, PDGFR, MET and ITGB3), we used the TCGA GBM Gene Ranker (http://cbio.mskcc.org/tcga-generanker/). This server combines available literature information and TCGA data for individual genes to score them. All selected receptors are among highly ranked genes (genes having a score greater than 2.0) in GBM (calculated scores are as follows: EGFR: 15.75, MET: 11.75, ERBB2: 9.25, PDGFRB: 7.25, IGF1R: 4.0, ITGB3: 3.75, CD36: 2.0), with EGFR, MET and ERBB2 having the highest rank in the database.

We performed randomization tests to check the reliability of the output of the algorithm. Here, terminal nodes, their prizes and the parameter set are kept same with the original PCSF analysis of GBM. In addition, number of nodes, edges and edge costs are the same as in the original interactome. Only the edges are re-shuffled randomly within the network. The randomization test is repeated ten times on different interactomes. These characteristics show that random PCSFs contain many more sub-trees when compared to the original PCSF and these sub-trees are not structured like the original trees; most of the trees in the random forests are 'stringy', composed of nine proteins at most. Further, random trees are not enriched for a specific biological process and none of the receptors found in the original PCSF are selected in the random PCSFs. The algorithm uses substantially more Steiner nodes to connect terminal nodes in random case. We performed another randomization test by reshuffling the nodes in the original interactome. In this way, the degree distribution is retained. In these randomizations we retain the same terminal nodes, prizes and the parameter set are kept same with the original PCSF analysis, but these proteins have now been randomly mapped to other nodes. The results show that the characteristics of the sub-trees in the random PCSFs are similar with the previous random case; they are 'stringy', not structured and not enriched for functions. However, this time the total number of nodes included in the PCSF is not as large as in the previous random case. These results show that the real PCSF solution is significantly different than the random solutions. It is particularly important that the receptors found by the algorithm run on the GBM data are not selected in the randomizations, supporting the hypothesis that these receptors are biologically relevant.

4   Discussion

We present a method for simultaneously discovery of multiple pathways by searching for "forests" consisting of multiple trees. We are able to solve this problem efficiently, even for large human networks by a simple modification of the previously published message-passing solution for the Steiner tree problem. When applied to the pheromone response data on the directed yeast interactome, the PCSF approach reveals several parallel pathways affected by yeast pheromone. Some of these parallel pathways contain multiple, coherently acting processes, such as pheromone response and the CDC28 associated pathway, or transport and transcriptional activity of PHO4. Others contain only one process, such as actin organization or protein folding.

The advantages of the forest approach are most apparent when used to study mammalian cells, which respond to a large number of hormones, growth factors and cytokines. Applying this approach to proteomic data from a model of GBM results in a forest composed of several sub-trees, each of which is rooted from a receptor molecule. The PCSF algorithm is able to select receptors relevant to GBM from hundreds of molecules in the human receptome. The solution reveals several known pathways and some unexpected new ones. EGFR, ERBB2, IGF1R and CD36 are starting nodes of the largest sub-trees in the PCSF. This set of receptor molecules was selected by the algorithm among hundreds of receptors, and the literature search shows that each of the selected receptors is clinically relevant to GBM. To find additional receptors whose downstream signaling pathways overlap with the selected receptors, we used an iterative approach that can be thought of as an *in silico* knock-out experiment. In this analysis, a selected receptor molecule and all its interactions are removed from the interactome and PCSF algorithm is applied to the remaining network. These calculations revealed the roles of PDGFR, MET and ITGB3 all of which have been previously linked to GBM.

Our method can be efficiently utilized to reconstruct networks that are enriched in functionally and clinically relevant proteins. Further, the algorithm is flexible, and can be modified for other types of data such as protein-small molecule inhibitor interactions and protein-metabolite interactions.

Acknowledgements. We thank Dr. Sara Gosline from MIT and Dr. Oznur Tastan from Microsoft Research for their critical reading and fruitful comments. This work is supported by NIH grants U54CA112967 and R01GM089903 and used computing resources funded by the National Science Foundation under Award No. DB1-0821391. EF receives support from the Eugene Bell Career Development Chair. RZ acknowledges the ERC grant OPTINF 267915. The support from the EC grant STAMINA 265496 is also acknowledged by AB and RZ.

References

1. Lan, A., Smoly, I.Y., Rapaport, G., Lindquist, S., Fraenkel, E., Yeger-Lotem, E.: ResponseNet: Revealing signaling and regulatory networks linking genetic and transcriptomic screening data. Nucleic Acids Res. (2011)
2. Yeger-Lotem, E., Riva, L., Su, L.J., Gitler, A.D., Cashikar, A.G., King, O.D., Auluck, P.K., Geddie, M.L., Valastyan, J.S., Karger, D.R., et al.: Bridging high-throughput genetic and transcriptional data reveals cellular responses to alpha-synuclein toxicity. Nat. Genet. 41(3), 316–323 (2009)
3. Vanunu, O., Magger, O., Ruppin, E., Shlomi, T., Sharan, R.: Associating genes and protein complexes with disease via network propagation. PLoS Comput. Biol. 6(1), e1000641 (2010)
4. Bailly-Bechet, M., Borgs, C., Braunstein, A., Chayes, J., Dagkessamanskaia, A., Francois, J.M., Zecchina, R.: Finding undetected protein associations in cell signaling by belief propagation. Proc. Natl. Acad. Sci. U.S.A. 108(2), 882–887 (2010)
5. Dittrich, M.T., Klau, G.W., Rosenwald, A., Dandekar, T., Muller, T.: Identifying functional modules in protein-protein interaction networks: an integrated exact approach. Bioinformatics 24(13), i223–i231 (2008)
6. Huang, S.S., Fraenkel, E.: Integrating proteomic, transcriptional, and interactome data reveals hidden components of signaling and regulatory networks. Sci. Signal 2(81), ra40 (2009)
7. Friedman, N.: Inferring cellular networks using probabilistic graphical models. Science 303(5659), 799–805 (2004)
8. Bailly-Bechet, M., Braunstein, A., Pagnani, A., Weigt, M., Zecchina, R.: Inference of sparse combinatorial-control networks from gene-expression data: a message passing approach. BMC Bioinformatics 11, 355 (2010)
9. Ourfali, O., Shlomi, T., Ideker, T., Ruppin, E., Sharan, R.: SPINE: a framework for signaling-regulatory pathway inference from cause-effect experiments. Bioinformatics 23(13), i359–i366 (2007)
10. Yeang, C.H., Ideker, T., Jaakkola, T.: Physical network models. J. Comput. Biol. 11(2-3), 243–262 (2004)
11. Kim, Y.A., Wuchty, S., Przytycka, T.M.: Identifying causal genes and dysregulated pathways in complex diseases. PLoS Comput. Biol. 7(3), e1001095 (2011)
12. Missiuro, P.V., Liu, K., Zou, L., Ross, B.C., Zhao, G., Liu, J.S., Ge, H.: Information flow analysis of interactome networks. PLoS Comput. Biol. 5(4), e1000350 (2009)
13. Suthram, S., Beyer, A., Karp, R.M., Eldar, Y., Ideker, T.: eQED: an efficient method for interpreting eQTL associations using protein networks. Mol. Syst. Biol. 4, 162 (2008)
14. Sharan, R., Ideker, T.: Modeling cellular machinery through biological network comparison. Nat. Biotechnol. 24(4), 427–433 (2006)
15. Akavia, U.D., Litvin, O., Kim, J., Sanchez-Garcia, F., Kotliar, D., Causton, H.C., Pochanard, P., Mozes, E., Garraway, L.A., Pe'er, D.: An integrated approach to uncover drivers of cancer. Cell 143(6), 1005–1017 (2010)
16. Bayati, M., Borgs, C., Braunstein, A., Chayes, J., Ramezanpour, A., Zecchina, R.: Statistical mechanics of steiner trees. Phys. Rev. Lett. 101(3), 037208 (2008)
17. Gruhler, A., Olsen, J.V., Mohammed, S., Mortensen, P., Faergeman, N.J., Mann, M., Jensen, O.N.: Quantitative phosphoproteomics applied to the yeast pheromone signaling pathway. Mol. Cell Proteomics 4(3), 310–327 (2005)

18. Issel-Tarver, L., Christie, K.R., Dolinski, K., Andrada, R., Balakrishnan, R., Ball, C.A., Binkley, G., Dong, S., Dwight, S.S., Fisk, D.G., et al.: Saccharomyces Genome Database. Methods Enzymol. 350, 329–346 (2002)

19. Breitkreutz, A., Choi, H., Sharom, J.R., Boucher, L., Neduva, V., Larsen, B., Lin, Z.Y., Breitkreutz, B.J., Stark, C., Liu, G., et al.: A global protein kinase and phosphatase interaction network in yeast. Science 328(5981), 1043–1046 (2010)

20. Jensen, L.J., Kuhn, M., Stark, M., Chaffron, S., Creevey, C., Muller, J., Doerks, T., Julien, P., Roth, A., Simonovic, M., et al.: STRING 8–a global view on proteins and their functional interactions in 630 organisms. Nucleic Acids Res. 37(database issue), 412–416 (2009)

21. Ben-Shlomo, I., Yu Hsu, S., Rauch, R., Kowalski, H.W., Hsueh, A.J.: Signaling receptome: A genomic and evolutionary perspective of plasma membrane receptors involved in signal transduction. Sci. STKE 2003(187), RE9 (2003)

22. Huang, P.H., Mukasa, A., Bonavia, R., Flynn, R.A., Brewer, Z.E., Cavenee, W.K., Furnari, F.B., White, F.M.: Quantitative analysis of EGFRvIII cellular signaling networks reveals a combinatorial therapeutic strategy for glioblastoma. Proc Natl. Acad. Sci. U.S.A. 104(31), 12867–12872 (2007)

23. Chekuri, C., Ene, A., Korula, N.: Prize-Collecting Steiner Tree and Forest in Planar Graphs. Data Structures and Algorithms (2010)

24. Gupta, A., Konemann, J., Leonardi, S., Ravi, R., Schaefer, G.: An efficient cost-sharing mechanism for the prize-collecting Steiner forest problem. In: SODA 2007 Proceedings of the Eighteenth Annual ACM-SIAM Symposium on Discrete Algorithms (2007)

25. Bailly-Bechet, M., Bradde, S., Braunstein, A., Flaxman, A., Foini, F., Zecchina, R.: Clustering with shallow trees. J Stat Mech., 12010 (2009)

26. Maere, S., Heymans, K., Kuiper, M.: BiNGO: A Cytoscape plugin to assess overrepresentation of gene ontology categories in biological networks. Bioinformatics 21(16), 3448–3449 (2005)

27. Shannon, P., Markiel, A., Ozier, O., Baliga, N.S., Wang, J.T., Ramage, D., Amin, N., Schwikowski, B., Ideker, T.: Cytoscape: a software environment for integrated models of biomolecular interaction networks. Genome Res. 13(11), 2498–2504 (2003)

28. Buehrer, B.M., Errede, B.: Coordination of the mating and cell integrity mitogen-activated protein kinase pathways in Saccharomyces cerevisiae. Mol. Cell Biol. 17(11), 6517–6525 (1997)

29. Zarzov, P., Mazzoni, C., Mann, C.: The SLT2(MPK1) MAP kinase is activated during periods of polarized cell growth in yeast. EMBO J. 15(1), 83–91 (1996)

30. Garcia, R., Bermejo, C., Grau, C., Perez, R., Rodriguez-Pena, J.M., Francois, J., Nombela, C., Arroyo, J.: The global transcriptional response to transient cell wall damage in Saccharomyces cerevisiae and its regulation by the cell integrity signaling pathway. J. Biol. Chem. 279(15), 15183–15195 (2004)

31. Baetz, K., Moffat, J., Haynes, J., Chang, M., Andrews, B.: Transcriptional coregulation by the cell integrity mitogen-activated protein kinase Slt2 and the cell cycle regulator Swi4. Mol. Cell Biol. 21(19), 6515–6528 (2001)

32. Kaffman, A., Rank, N.M., O'Shea, E.K.: Phosphorylation regulates association of the transcription factor Pho4 with its import receptor Pse1/Kap121. Genes Dev. 12(17), 2673–2683 (1998)

33. Bhoite, L.T., Allen, J.M., Garcia, E., Thomas, L.R., Gregory, I.D., Voth, W.P., Whelihan, K., Rolfes, R.J., Stillman, D.J.: Mutations in the Pho2 (Bas2) transcription factor that differentially affect activation with its partner proteins Bas1, Pho4, and Swi5. J. Biol. Chem. 277(40), 37612–37618 (2002)

34. Cancer Genome Atlas Research Network: Comprehensive genomic characterization defines human glioblastoma genes and core pathways. Nature 455(7216), 1061–1068 (2008)
35. Macaulay, V.M.: The IGF receptor as anticancer treatment target. In: Novartis Found Symp., 262:235-243; discussion 243-236, 265-238 (2004)
36. Kiaris, H., Schally, A.V., Varga, J.L.: Antagonists of growth hormone-releasing hormone inhibit the growth of U-87MG human glioblastoma in nude mice. Neoplasia 2(3), 242–250 (2000)
37. Adams, T.E., McKern, N.M., Ward, C.W.: Signalling by the type 1 insulin-like growth factor receptor: interplay with the epidermal growth factor receptor. Growth Factors 22(2), 89–95 (2004)
38. Chakravarti, A., Loeffler, J.S., Dyson, N.J.: Insulin-like growth factor receptor I mediates resistance to anti-epidermal growth factor receptor therapy in primary human glioblastoma cells through continued activation of phosphoinositide 3-kinase signaling. Cancer Res. 62(1), 200–207 (2002)
39. Kaur, B., Cork, S.M., Sandberg, E.M., Devi, N.S., Zhang, Z., Klenotic, P.A., Febbraio, M., Shim, H., Mao, H., Tucker-Burden, C., et al.: Vasculostatin inhibits intracranial glioma growth and negatively regulates in vivo angiogenesis through a CD36-dependent mechanism. Cancer Res. 69(3), 1212–1220 (2009)
40. Silverstein, R.L., Febbraio, M.: CD36, a scavenger receptor involved in immunity, metabolism, angiogenesis, and behavior. Sci. Signal 2(72), re3 (2009)
41. Dai, C., Celestino, J.C., Okada, Y., Louis, D.N., Fuller, G.N., Holland, E.C.: PDGF autocrine stimulation dedifferentiates cultured astrocytes and induces oligodendrogliomas and oligoastrocytomas from neural progenitors and astrocytes in vivo. Genes Dev. 15(15), 1913–1925 (2001)
42. Uhrbom, L., Hesselager, G., Nister, M., Westermark, B.: Induction of brain tumors in mice using a recombinant platelet-derived growth factor B-chain retrovirus. Cancer Res. 58(23), 5275–5279 (1998)
43. Clarke, I.D., Dirks, P.B.: A human brain tumor-derived PDGFR-alpha deletion mutant is transforming. Oncogene 22(5), 722–733 (2003)
44. Ziegler, D.S., Wright, R.D., Kesari, S., Lemieux, M.E., Tran, M.A., Jain, M., Zawel, L., Kung, A.L.: Resistance of human glioblastoma multiforme cells to growth factor inhibitors is overcome by blockade of inhibitor of apoptosis proteins. J. Clin. Invest. 118(9), 3109–3122 (2008)
45. Li, Y., Li, A., Glas, M., Lal, B., Ying, M., Sang, Y., Xia, S., Trageser, D., Guerrero-Cazares, H., Eberhart, C.G., et al.: c-Met signaling induces a reprogramming network and supports the glioblastoma stem-like phenotype. Proc. Natl. Acad. Sci. U.S.A. 108(24), 9951–9956 (2011)
46. Alam, N., Goel, H.L., Zarif, M.J., Butterfield, J.E., Perkins, H.M., Sansoucy, B.G., Sawyer, T.K., Languino, L.R.: The integrin-growth factor receptor duet. J. Cell Physiol. 213(3), 649–653 (2007)
47. Cardona-Gomez, G.P., Mendez, P., DonCarlos, L.L., Azcoitia, I., Garcia-Segura, L.M.: Interactions of estrogen and insulin-like growth factor-I in the brain: molecular mechanisms and functional implications. J. Steroid Biochem. Mol. Biol. 83(1-5), 211–217 (2002)

Function-Function Correlated Multi-Label Protein Function Prediction over Interaction Networks

Hua Wang, Heng Huang*, and Chris Ding

Department of Computer Science and Engineering,
University of Texas at Arlington, Arlington, TX 76019
huawangcs@gmail.com, {heng,chqding}@uta.edu

Abstract. Many previous computational methods for protein function prediction make prediction one function at a time, fundamentally, which is equivalent to assume the functional categories of proteins to be isolated. However, biological processes are highly correlated and usually intertwined together to happen at the same time, therefore it would be beneficial to consider protein function prediction as one indivisible task and treat all the functional categories as an integral and correlated prediction target. By leveraging the function-function correlations, it is expected to achieve improved overall predictive accuracy. To this end, we develop a novel network based protein function prediction approach, under the framework of multi-label classification in machine learning, to utilize the function-function correlations. Besides formulating the function-function correlations in the optimization objective explicitly, we also exploit them as part of the pairwise protein-protein similarities implicitly. The algorithm is built upon the Green's function over a graph, which not only employs the global topology of a network but also captures its local structural information. We evaluate the proposed approach on *Saccharomyces cerevisiae* species. The encouraging experimental results demonstrate the effectiveness of the proposed method.

Keywords: Protein Function Prediction, Green's Function, Multi-Label Classification, Function-Function Correlations.

1 Introduction

Many existing *in silico* methods for protein function prediction typically make prediction one function at a time, fundamentally. This turns the problem into a convenient form to use existing machine learning algorithms, which, however, abstracts away the function correlations due to neglecting the fact that most biological functions are interdependent from one another. For example, "Transcription" and "Protein Synthesis" usually appear together [13], one after another, *i.e.*, they tend to appear in the processes of a same protein. Therefore, if a protein

---

* Corresponding author.

B. Chor (Ed.): RECOMB 2012, LNBI 7262, pp. 302–313, 2012.
© Springer-Verlag Berlin Heidelberg 2012

is known to be annotated with function "Transcription", we would have high confidence to annotate the same protein with function "Protein Synthesis" as well. In other words, the *function-function correlations* convey valuable information to understand the biological processes, which provide a potential opportunity to improve the overall accuracy of protein function prediction. However, how to effectively exploit function-function correlations remains a challenge for devising computational methods for protein function prediction. In this study, we tackle this new, yet important, problem by placing protein function prediction under the framework of multi-label classification, an emerging topic in machine learning, to develop a new graph based protein function prediction method to take advantage of function-function correlations.

1.1 Network Based Protein Function Prediction

Recent availability of protein interaction networks for many genomic species has stimulated the development of network based computational methods for protein function prediction. Typically, an interaction network is first modeled as a graph, with the vertices representing proteins and the edges representing the detected protein-protein interactions (PPI), then a graph based statistical learning method is used to infer putative protein functions.

The most straightforward method using network data to predict protein function determines the putative functions of a protein from the known functions of its neighboring proteins on a PPI network [16,9,2], which, though, only leverages the local information of a network. Lately, researchers used the global optimization approaches to improve the protein function predictions by taking into account the full topology of networks [20,11,14]. All these approaches can be summarized as the following common scheme: (1) compute a set of ranking lists, and (2) make prediction using certain thresholds on the ranking lists. In step 1, which is the most critical part of algorithms, these existing methods compute the ranking lists one function at a time, therefore they do not take into account the interrelationships among the functions. Network based approaches using other models for protein function prediction are surveyed in [17].

The aforementioned graph-based protein function prediction methods bank on two assumptions: local consistency and global consistency, which are the exact foundations of label propagation based learning approaches in machine learning. This motivates us to formulate protein function prediction over a PPI network as a label propagation problem on a graph. Among existing label propagation methods, we choose to develop our new method from the Green's function approach [5,23] due to its clear intuition and demonstrated effectiveness in other real-world applications.

1.2 Multi-Label Correlated Protein Function Prediction

Because a protein is usually observed to play several functional roles in different biological processes within an organism, it is natural to annotate it with multiple functions. Thus, protein function prediction is an ideal example of *multi-label classification* [21,22,23,24,25] in machine learning. Multi-label classification is an emerging topic driven by the advances of modern technologies in recent years, in which each object could belong to more than one class. In contrast, in traditional single-label classification, a data object may belong to one and only one class. Placing protein function prediction under the framework of multi-label classification, we use the Green's function approach to integrate the function-function correlations from theory of reproducing kernel Hilbert space (RKHS). Besides incorporating the function-function correlations as a regularizer in the optimization objective explicitly, we also take advantage of them as part of the pairwise protein similarities implicitly.

2   Methods

In this section, we propose a novel Function-function Correlated Multi-Label (FCML) approach using the Green's function on a graph to predict protein functions, which incorporates the function-function correlations by two levels: one from the function perspective to formulate the function-wise similarities explicitly in the optimization objective (Section 2.3), and the other from protein similarity perspective using function assignments to model the function correlations implicitly (Section 2.4).

2.1   Notations and Problem Formalization

In protein function prediction, given K biological functions and n proteins, each protein x_i is associated with a set of labels represented by a function assignment indication vector $\mathbf{y}_i \in \{-1, 0, 1\}^K$ such that $\mathbf{y}_i(k) = 1$ if protein x_i has the k-th function, $\mathbf{y}_i(k) = -1$ if it does not have the k-th function, and $\mathbf{y}_i(k) = 0$ if its function assignment is not yet known in a priori. Given l annotated proteins $\{(x_1, \mathbf{y}_1), \ldots, (x_l, \mathbf{y}_l)\}$ where $l < n$, our task is to predict functions $\{\mathbf{y}_i\}_{i=l+1}^n$ for the unannotated proteins $\{x_i\}_{i=l+1}^n$. We write $Y = [\mathbf{y}_1, \ldots, \mathbf{y}_n]^T$, and $Y_l = [\mathbf{y}_1, \ldots, \mathbf{y}_l]^T = [\mathbf{y}^{(1)}, \ldots, \mathbf{y}^{(K)}]$, where $\mathbf{y}^{(k)} \in \mathbb{R}^l$ is a class-wise function assignment vector. We also define $F = [\mathbf{f}_1, \ldots, \mathbf{f}_n]^T \in \mathbb{R}^{n \times K}$ as the decision matrix for prediction, and $\{\mathbf{f}_i\}_{i=l+1}^n$ include the decision values for prediction.

We formalize a protein interaction network as a graph $\mathcal{G} = (\mathcal{V}, \mathcal{E})$. The vertices $\mathcal{V}$ correspond to the proteins $\{x_1, \ldots, x_n\}$, and the edges $\mathcal{E}$ are weighted by an $n \times n$ similarity matrix W with W_{ij} indicating the similarity between x_i and x_j. In the simplest case, W is the adjacency matrix of the PPI graph where $W_{ij} = 1$ if proteins x_i and x_j interact, and 0 otherwise. In this work, W is computed in Eq. (10) to incorporate more useful information.

2.2   Protein Function Prediction Using the Green's Function over a Graph

In this section, we first briefly review the Green's function approach for label propagation over a graph, from which we will derive the proposed FCML method in later subsections.

Given a graph with edge weights W, its combinatorial Laplacian is defined as $L = D - W$ [3], where $D = \text{diag}(W\mathbf{e})$ and $\mathbf{e} = [1...1]^T$. The Green's function over the graph is defined as the inverse of L with zero-mode discarded, which is computed as following [5]:

$$G = L_+^{-1} = \frac{1}{(D-W)_+} = \sum_{i=2}^{n} \frac{\mathbf{v}_i \mathbf{v}_i^T}{\lambda_i}, \tag{1}$$

where L_+^{-1} and $(D-W)_+$ indicates that the zero eigen-mode of the concerned matrix is discarded, and $0 = \lambda_1 \le \lambda_2 \le \cdots \le \lambda_n$ are the eigenvalues of L and $\mathbf{v}_i$ are the corresponding eigenvectors. Because the Green's function G in Eq. (1) is a kernel [5], from the theory of RKHS, the Green's function approach is to minimize the following objective [5]:

$$J_0(F) = (1 - \mu)\|F - Y\|^2 + \mu \, \text{tr}\left(F^T \mathcal{K}^{-1} F\right), \tag{2}$$

where $\mathcal{K} = G$ is the kernel, $\mu \in (0,1)$ is a constant to control the smoothness regularizer $\text{tr}\left(F^T \mathcal{K}^{-1} F\right)$, and $\text{tr}(\cdot)$ denotes the trace of a matrix.

Thus, we assign functions to unannotated proteins by [5]:

$$\mathbf{y}_i = \text{sign}(\mathbf{f}_i), \; l < j \le n, \quad \text{where} \quad F = GY. \tag{3}$$

We name Eq. (3) as simple Multi-Label Green's Function (MLGF) approach, beyond which we will propose a novel function-function correlated multi-label Green's function approach to leveraging function-function correlations.

2.3 Function-Function Correlated RKHS Approach for Multi-Label Classification

Although the function-function correlations are useful to infer putative functions of unannotated proteins, MLGF approach defined in Eq. (3) does not exploit them because it treats the biological functions of interest as isolated. In multi-label scenarios, however, we concentrate on making use of the function-function correlations, which could be defined as $C \in \mathbb{R}^{K \times K}$ using cosine similarity as following [22,23,24,25]:

$$C_{kl} = \cos\left(\mathbf{y}^{(k)}, \mathbf{y}^{(l)}\right) = \frac{\langle \mathbf{y}^{(k)}, \mathbf{y}^{(l)} \rangle}{\|\mathbf{y}^{(k)}\| \, \|\mathbf{y}^{(l)}\|}. \tag{4}$$

Following [23], we expect to maximize $\text{tr}\left(FCF^T\right)$. In order to make connection with the theory of RKHS, instead of directly using F, we use kernel assisted decision matrix $\mathcal{K}^{-\frac{1}{2}}F$, which leads to the following objective to maximize [23]:

$$J_C(F) = \text{tr}\left(\mathcal{K}^{-\frac{1}{2}} FCF^T \mathcal{K}^{-\frac{1}{2}}\right). \tag{5}$$

Combining Eq. (5) with the original RKHS objective in Eq. (2), we minimize the following objective:

$$J(F) = \beta\|F - Y\|^2 + \text{tr}\left(F^T \mathcal{K}^{-1} F\right) - \alpha \, \text{tr}\left(\mathcal{K}^{-\frac{1}{2}} FCF^T \mathcal{K}^{-\frac{1}{2}}\right), \tag{6}$$

where $\alpha \in (0,1)$ balances the the two objectives, and $\beta = \frac{1-\mu}{\mu}$.

Differentiating J with respect to F, we have:

$$\frac{\partial J}{\partial F} = 2\beta(F - Y) + 2\mathcal{K}^{-1}F - 2\alpha\mathcal{K}^{-1}FC = 0 \implies$$

$$F = \frac{1}{\beta I + \mathcal{K}^{-1}}\beta Y + \alpha\frac{1}{\beta I + \mathcal{K}^{-1}}\mathcal{K}^{-1}FC. \tag{7}$$

Because β is usually very small in typical empirical settings, we have:

$$\frac{F}{\beta} = \mathcal{K}Y + \alpha\frac{F}{\beta}C \implies \tilde{F} = \mathcal{K}Y + \alpha\tilde{F}C = GY + \alpha\tilde{F}C, \tag{8}$$

where $\tilde{F} = \frac{F}{\beta}$. Thus, we have

$$\tilde{F} = GY\left(I - \alpha C\right)^{-1}. \tag{9}$$

We name Eq. (9) as our proposed Function-function Correlated Multi-Label (FCML) approach for protein function prediction.

2.4   Correlation Augmented Interaction Network

Traditional network based protein function prediction approaches only employ biological interaction networks obtained from experimental data such as those from high-throughput technologies. When considering protein function prediction as a multi-label classification problem, we can build a computational interaction network $W_L \in \mathbb{R}^{n \times n}$ from label assignment perspective. As one of our contribution, we make use of this new computational interaction network and propose a *correlation augmented interaction network* as follows:

$$W = W_{\text{Bio}} + \gamma W_L, \tag{10}$$

where W_{Bio} is the biological interaction network, which is same as those in existing approaches. In Eq. (10), γ controls the relative importance of W_L, which is empirically selected as $\gamma = \frac{\sum_{i,j,i \neq j} W_{\text{Bio}}(i,j)}{\sum_{i,j,i \neq j} W_L(i,j)}$.

The true power of the correlation augmented interaction network construction scheme defined in Eq. (10) lies in that, the original biological similarities among proteins are augmented by the function assignment similarities. As a result, label propagation pathways over a graph are reinforced. With this interaction network construction scheme, the correlations among the functional categories are encoded into the graph weights, such that the resulted hybrid graph can be directly used in existing methods to enhance their prediction performance. In this paper, we use W defined in Eq. (10) to compute the Green's function in Eq. (1).

Protein-Protein Similarity from Function Assignments (W_L). Because multiple functions could be assigned to one single protein, the overlap between the function assignments of two proteins can be used to evaluate their similarity. The more functions shared by two proteins, the more similar they are. Consequently, besides the class membership indications, the label assignment vector $\mathbf{y}_i$ is enriched with characteristic meaning, which thereby can be used as an attribute vector to characterize protein x_i. Using cosine similarity, the function assignment similarity between two proteins is computed as:

$$W_L(i,j) = \cos(\mathbf{y}_i, \mathbf{y}_j) = \frac{\langle \mathbf{y}_i, \mathbf{y}_j \rangle}{\|\mathbf{y}_i\| \|\mathbf{y}_j\|}. \tag{11}$$

Note that, the entries of W_L measures *individual protein - individual protein relationships*; whereas the entries of C defined in Eq. (4) assesses the *function category - function category correlations*.

Our task in protein function prediction is to assign functions to unannotated proteins upon annotated ones. In order to compute W_L, however, we need function assignments of all the proteins, including those annotated and unannotated. Therefore, we first initialize unannotated proteins through majority voting [16] approach, which makes prediction using the top three frequent functions appearing one protein's interacting partners.

Biological Protein-Protein Similarity (W_{Bio}) and Multi-Source Integration. W_{Bio} in Eq. (10) computes the protein-protein similarity from biological experimental data, which is same as existing works and could integrate multiple experimental sources. Let $W^{(1)}$ be the graph built from BioGRID PPI data [10,7], $W^{(2)}$ be that from synthetic lethal data [19], $W^{(3)}$ be that from gene co-expression data [6], $W^{(4)}$ be that from gene regulation data [8], *etc.*, W_{Bio} is computed as [15]:

$$W_{Bio}(i,j) = 1 - \prod_k \left[1 - r^{(k)} W^{(k)}(i,j) \right], \tag{12}$$

where $r^{(k)}$ is estimated reliabilities of the corresponding network by Expression Profile Reliability (EPR) index [4]. Eq. (12) reflects the fact that interactions detected in multiple experiments are generally more reliable than those detected by a single experiment [12].

Because in reality the overlap among different biological networks typically is very small, and the BioGRID PPI network data are fairly comprehensive, in this work, we set $W_{Bio} = W^{(1)}$, where $W^{(1)}(i,j) = 1$ if protein x_i and x_j interact, and 0 otherwise.

3   Materials and Data Sets

Two types of data are involved in the experimental evaluations for protein function prediction: function annotation data and PPI data.

The functional catalogue (FunCat) [13] is a project under the Munich Information Center for Protein Sequences (MIPS), which is an annotation scheme for the functional description of proteins from prokaryotes, unicellular eukaryotes, plants and animals. Taking into account the broad and highly diverse spectrum of known protein functions, FunCat of version 2.1 consists of 27 main functional categories, 17 of which are involved in annotating *Saccharomyces cerevisiae*. Although there are still other protein annotation systems such as the Gene Ontology [1], we use the Funcat annotation system due to its clear tree-like hierarchical structure. All 17 level-1 biological functions are listed in Table 1.

Table 1. Function IDs and names by Funcat scheme version 2.1

ID	Name
'01'	Metabolism
'02'	Energy
'10'	Cell Cycle and Dna Processing
'11'	Transcription
'12'	Protein Synthesis
'14'	Protein Fate (Folding, Modification, Destination)
'16'	Protein With Binding Function or Cofactor Requirement (Structural or Catalytic)
'18'	Regulation of Metabolism and Protein Function
'20'	Cellular Transport, Transport Facilitation and Transport Routes
'30'	Cellular Communication/Signal Transduction Mechanism
'32'	Cell Rescue, Defense and Virulence
'34'	Interaction with the Environment
'38'	Transposable Elements, Viral and Plasmid Proteins
'40'	Cell Fate
'41'	Development (Systemic)
'42'	Biogenesis of Cellular Components
'43'	Cell Type Differentiation

The protein-protein interaction data can be downloaded from the BioGRID database [18] and we focus on the *S. cerevisiae*. By using the BioGRDI database of version 2.0.45 and removing the proteins connected by only one PPI, we end up with 4299 proteins annotated by Funcat annotation scheme with 72624 PPIs, as well as 1997 unannotated proteins.

4   Results and Discussions

In this section, using the PPI data from BioGRID database [18] and Funcat annotation scheme [13] on *S. cerevisiae* data, we evaluate our new FCML approach by applying it in protein function prediction, where standard *precision* and *F1 score* are used as statistical metrics.

4.1   Evaluation of the Function-Function Correlations

Because the function-function correlations are one of the most important mechanism to improve the prediction performance in our proposed approach, we first evaluate its correctness. Using the FunCat 2.1 annotation data set for *S. cerevisiae* genome, the function-function correlations defined in Eq. (4) are visualized in Figure 1. The high correlation value between functions "40" (Cell Fate) and "43" (Cell Type Differentiation) depicted in this figure shows that they are highly correlated. In addition, as shown in this figure some other function pairs are also highly correlated, such as functions "11" (Transcription) and "16" (Protein With Binding Function or Cofactor Requirement), "18" (Regulation of Metabolism and Protein Function) and "30" (Cellular Communication/Signal Transduction Mechanism), *etc.* All these observations perfectly comply with the biological facts, which firmly confirm the correctness of our formulation of the function-function correlations defined in Eq. (4) from biological perspective.

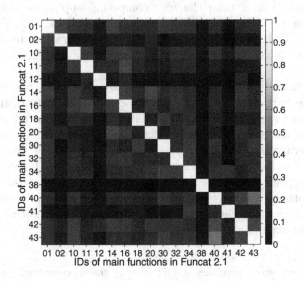

Fig. 1. The visualization of the correlation matrix defined in Eq. (4) among the 17 main functional categories in FunCat 2.1 annotated to *S. cerevisiae* genome

4.2   Improved Function Prediction in Cross-Validation

We compare the performances of the proposed Multi-label Green's Function (MLGF) approach and Function-function Correlated Multi-Label (FCML) approach to several benchmark computational methods for protein function prediction, including Majority Voting (MV) approach [16], Global Majority Voting (GMV) approach [20], χ^2 approach [9], and Functional Flow (FF) approach [14]. The PPI graph is built from BioGRID data of version 2.0.45 with annotation by MIPS Funcat scheme of version 2.1. The standard 10-fold cross-validation is used. For our approach, adaptive decision boundary [23] is used to compute the threshold to make prediction from the ranking list of the decision values for each function. For the four other approaches, we use their respective optimal parameters. In MV approach, we select the 3 most frequently occurring functions in a protein's neighbors. In χ^2 approach, radius $= 1$ gives the best performance. In FF approach, we assign functions according to the proportions of positive and negative training samples as suggested by [14].

The prediction performance of the six compared methods measured by precision and F1 score are shown in Figure 2(a) and Figure 2(b), respectively. The experimental results show that the prediction precision by MLGF approach in 10-fold cross-validation outperforms the other four approaches for most of the level-1 functions of Funcat scheme, which demonstrates the effectiveness of the Green's function method in protein function prediction. Moreover, the precisions by FCML approach are consistently better than those of MLGF, which are clearly better than the four competing approaches for almost all the functions as well. As a balanced performance measurement, the results of F1 score using FCML further demonstrate that incorporating the inherent correlations among biological functions can improve the prediction performance significantly.

In addition to evaluating the prediction performance on each individual function, we also report the overall prediction performance over all functions using the micro averages to address multi-label scenario. The micro average is computed from the sum of per-class contingency table, which can be seen as a weighted average that emphasizes more on the accuracy of classes/functions with more positive samples. The micro average precision and F1 score by the six compared approaches over all 17 level-1 biological functions are listed in Table 2, which give another evidence to support the advantages of the proposed FCML approach. We notice that the performance of FF approach is not as good, which is similar to the experimental results reported in [2].

Table 2. Micro average of precision and F1 score by the six compared approaches over all main functional categories by Funcat Scheme

Approaches	Average Precision	Average F1 score
MV	30.69%	29.04%
GMV	31.13%	22.41%
χ^2	14.8%	7.60%
FF	28.01%	27.05%
MLGF	32.45%	36.36%
FCML	**54.83%**	**43.74%**

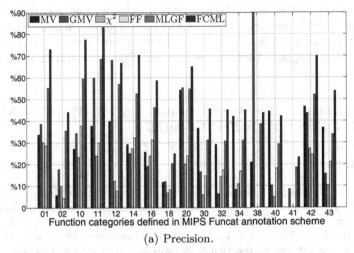

(a) Precision.

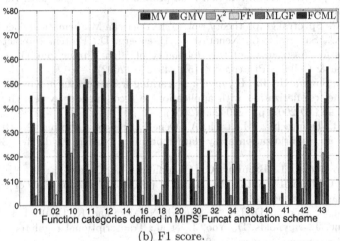

(b) F1 score.

Fig. 2. Performance of 10-fold cross-validation for the main functional categories in Funcat scheme by MV, GMV, χ^2, FF, and proposed MLGF and FCML approaches on BioGRID PPI data

5   Conclusions

We proposed a novel Function-function Correlated Multi-Label (FCML) approach for protein function prediction, and showed its promising performance to outperform other related approaches. Different from most existing approaches that divide protein function prediction into multiple separate tasks and make prediction fundamentally one function at a time, the proposed FCML approach considers all the biological functions as a single correlated prediction target and predict protein functions via an integral procedure. In the proposed approach, correlations among the functional categories are leveraged. By formulating protein function prediction as a multi-label classification problem, we use the Green's function over a graph to efficiently resolve the problem. The Green's function approach takes advantage of both the full topology of the interaction network toward global optimization and the local structures, such that the deficiencies of many existing approaches are gracefully overcome.

Acknowledgments. This research is supported by NSF-IIS 1117965, NSFCCF-0830780, NSF-DMS-0915228, NSFCCF-0917274.

References

1. Ashburner, M., Ball, C., Blake, J., Botstein, D., Butler, H., Cherry, J., Davis, A., Dolinski, K., Dwight, S., Eppig, J., et al.: Gene ontology: Tool for the unification of biology. The Gene Ontology Consortium. Nat. Genet. 25(1), 25 (2000)
2. Chua, H., Sung, W., Wong, L.: Exploiting indirect neighbours and topological weight to predict protein function from protein-protein interactions. Bioinformatics 22(13), 1623–1630 (2006)
3. Chung, F.: Spectral graph theory, vol. (92). American Mathematical Society (1997)
4. Deane, C., Salwinski, L., Xenarios, I., Eisenberg, D.: Protein Interactions Two Methods for Assessment of the Reliability of High Throughput Observations*. Molecular & Cellular Proteomics 1(5), 349–356 (2002)
5. Ding, C., Simon, H., Jin, R., Li, T.: A learning framework using Green's function and kernel regularization with application to recommender system. In: Proc. of ACM SIGKDD 2007, pp. 260–269 (2007)
6. Edgar, R., Domrachev, M., Lash, A.: Gene Expression Omnibus: NCBI gene expression and hybridization array data repository. Nucleic Acids Res. 30(1), 207 (2002)
7. Giot, L., Bader, J., Brouwer, C., Chaudhuri, A., Kuang, B., Li, Y., Hao, Y., Ooi, C., Godwin, B., Vitols, E., et al.: A protein interaction map of Drosophila melanogaster. Science 302(5651), 1727–1736 (2003)
8. Harbison, C., Gordon, D., Lee, T., Rinaldi, N., Macisaac, K., Danford, T., Hannett, N., Tagne, J., Reynolds, D., Yoo, J., et al.: Transcriptional regulatory code of a eukaryotic genome. Nature 431, 99–104 (2004)
9. Hishigaki, H., Nakai, K., Ono, T., Tanigami, A., Takagi, T.: Assessment of prediction accuracy of protein function from protein-protein interaction data. Yeast 18(6), 523–531 (2001)
10. Ho, Y., Gruhler, A., Heilbut, A., Bader, G., Moore, L., Adams, S., Millar, A., Taylor, P., Bennett, K., Boutilier, K., et al.: Systematic identification of protein complexes in Saccharomyces cerevisiae by mass spectrometry. Nature 415(6868), 180–183 (2002)
11. Karaoz, U., Murali, T., Letovsky, S., Zheng, Y., Ding, C., Cantor, C., Kasif, S.: Whole-genome annotation by using evidence integration in functional-linkage networks. Proc. Natl Acad. Sci. U.S.A. 101(9), 2888–2893 (2004)
12. von Mering, C., Krause, R., Snel, B., Cornell, M., Oliver, S.G., Fields, S., Bork, P.: Comparative assessment of large-scale data sets of protein–protein interactions. Nature 417(6887), 399–403 (2002)
13. Mewes, H., Heumann, K., Kaps, A., Mayer, K., Pfeiffer, F., Stocker, S., Frishman, D.: MIPS: a database for genomes and protein sequences. Nucleic Acids Res. 27(1), 44 (1999)

14. Nabieva, E., Jim, K., Agarwal, A., Chazelle, B., Singh, M.: Whole-proteome prediction of protein function via graph-theoretic analysis of interaction maps. Bioinformatics 21, 302–310 (2005)

15. Pei, P., Zhang, A.: A topological measurement for weighted protein interaction network. In: Proceedings of IEEE Computational Systems Bioinformatics Conference, pp. 268–278 (2005)

16. Schwikowski, B., Uetz, P., Fields, S.: A network of protein- protein interactions in yeast. Nat. Biotechnol. 18, 1257–1261 (2000)

17. Sharan, R., Ulitsky, I., Shamir, R.: Network-based prediction of protein function. Mol. System Biol. 3(1) (2007)

18. Stark, C., Breitkreutz, B., Reguly, T., Boucher, L., Breitkreutz, A., Tyers, M.: BioGRID: a general repository for interaction datasets. Nucleic Acids Res. 34(Database Issue), D535 (2006)

19. Tong, A., Lesage, G., Bader, G., Ding, H., Xu, H., Xin, X., Young, J., Berriz, G., Brost, R., Chang, M., et al.: Global mapping of the yeast genetic interaction network. Science 303(5659), 808–813 (2004)

20. Vazquez, A., Flammini, A., Maritan, A., Vespignani, A.: Global protein function prediction from protein-protein interaction networks. Nat. Biotechnol. 21, 697–700 (2003)

21. Wang, H., Ding, C., Huang, H.: Multi-label classification: Inconsistency and class balanced k-nearest neighbor. In: Twenty-Fourth AAAI Conference on Artificial Intelligence (2010)

22. Wang, H., Ding, C., Huang, H.: Multi-label Linear Discriminant Analysis. In: Daniilidis, K., Maragos, P., Paragios, N. (eds.) ECCV 2010, Part VI. LNCS, vol. 6316, pp. 126–139. Springer, Heidelberg (2010)

23. Wang, H., Huang, H., Ding, C.: Image Annotation Using Multi-label Correlated Greens Function. In: Proc. of IEEE ICCV 2009, pp. 2029–2034 (2009)

24. Wang, H., Huang, H., Ding, C.: Multi-label Feature Transform for Image Classifications. In: Daniilidis, K., Maragos, P., Paragios, N. (eds.) ECCV 2010, Part IV. LNCS, vol. 6314, pp. 793–806. Springer, Heidelberg (2010)

25. Wang, H., Huang, H., Ding, C.: Image annotation using bi-relational graph of images and semantic labels. In: Proc. of IEEE CVPR 2011, pp. 793–800. IEEE (2011)

Predicting Protein-Protein Interactions from Multimodal Biological Data Sources via Nonnegative Matrix Tri-Factorization

Hua Wang, Heng Huang*, Chris Ding, and Feiping Nie

Department of Computer Science and Engineering,
University of Texas at Arlington, Arlington, TX 76019
{huawangcs,feipingnie}@gmail.com, {heng,chqding}@uta.edu

Abstract. Due to the high false positive rate in the high-throughput experimental methods to discover protein interactions, computational methods are necessary and crucial to complete the interactome expeditiously. However, when building classification models to identify putative protein interactions, compared to the obvious choice of positive samples from truly interacting protein pairs, it is usually very hard to select negative samples, because non-interacting protein pairs refer to those currently without experimental or computational evidence to support a physical interaction or a functional association, which, though, could interact in reality. To tackle this difficulty, instead of using heuristics as in many existing works, in this paper we solve it in a principled way by formulating the protein interaction prediction problem from a new mathematical perspective of view — sparse matrix completion, and propose a novel Nonnegative Matrix Tri-Factorization (NMTF) based matrix completion approach to predict new protein interactions from existing protein interaction networks. Because matrix completion only requires positive samples but not use negative samples, the challenge in existing classification based methods for protein interaction prediction is circumvented. Through using manifold regularization, we further develop our method to integrate different biological data sources, such as protein sequences, gene expressions, protein structure information, *etc*. Extensive experimental results on *Saccharomyces cerevisiae* genome show that our new methods outperform related state-of-the-art protein interaction prediction methods.

Keywords: Protein-Protein Interaction, Multimodal Biological Data, Nonnegative Matrix Factorization.

1 Introduction

Proteins play an essential role in nearly all cellular functions such as promoting biochemical reactions and composing cellular structures. The multiplicity of functions that proteins execute in most cellular processes and biochemical events

---

* Corresponding author.

B. Chor (Ed.): RECOMB 2012, LNBI 7262, pp. 314–325, 2012.
© Springer-Verlag Berlin Heidelberg 2012

is attributed to their interactions with other proteins. As a result, it is critical to understand protein-protein interactions (PPIs) in both scientific research and practical applications such as new drug development. A variety of techniques are now available to experimental biologists for discovering protein-protein interactions, such as yeast two-hybrid systems [14], mass spectrometry [13], and many others as surveyed in [27]. Although these high-throughput experimental methods have accumulated a large amount of data, interactomes of many organisms are far from complete [27]. The low interaction coverage along with the experimental biases toward certain protein types and cellular localizations reported by most experimental techniques call for the development of computational methods that are able to predict more reliable putative PPI for further experimental screening [28].

Recently, machine learning techniques, such as Bayesian networks [15], decision trees [32,7], random forest [22,6], and support vector machines (SVM) with different kernels [20,2,25,19], have been successfully applied to predict PPIs. These methods used various data sources to train a classifier to distinguish positive examples of truly interacting protein pairs from the negative examples of non-interacting pairs. However, all these methods suffer from a fundamental difficulty — how to choose the negative samples. Compared to the obvious choice of positive samples from truly interacting protein pairs, the selection of negative samples typically is not easy. First, non-interacting protein pairs refer to those currently without experimental or computational evidence to support a physical interaction or functional association. In reality, however, such protein pairs could interact. Second, the number of non-interacting protein pairs is much larger than the number of the interacting ones, therefore unbalanced training data often cause skewed prediction models that lead to unsatisfactory prediction results. In most existing classification based methods, heuristics are often employed to tackle these problems. By recognizing these difficulties, instead of considering PPI prediction as a classification problem, we approach it from a new perspective by using matrix completion, which is an important mathematical topic to address the problem to recover a matrix from what appears to be incomplete, or even corrupted [5]. Because matrix completion only uses truly interacting protein pairs without requiring negative training samples, the difficulty in classification based PPI prediction methods is circumvented.

In this paper, we propose a novel Nonnegative Matrix Tri-Factorization (NMTF) [16,11] based matrix completion approach to predict protein-protein interactions. NMTF focuses on the analysis of data matrices whose elements are nonnegative, such as the adjacency matrix of a PPI graph, and decomposes the input matrix into three nonnegative factor matrices that approximate the input matrix by a low-rank nonnegative representation [9,10]. We first employ NMTF approach to predict putative protein interactions, which only makes use of PPI network data. After that, we extend the standard NMTF framework by adding manifold regularization [12], such that additional biological data, e.g., protein sequences data, protein structures information, and gene expressions, can be incorporated to achieve enhanced PPI prediction performance.

Extensive empirical evaluations on *Saccharomyces cerevisiae* genome have shown encouraging performance, which demonstrate the effectiveness the proposed methods.

2   Methods

We first briefly formalize the problem of PPI prediction. Given a PPI network, we may construct a graph $G = (V, \Omega, \mathbf{X})$, with V corresponding to $n = |V|$ proteins and $\Omega \subseteq V \times V$ corresponding to known PPIs. $\mathbf{X} \in \{0, 1\}^{n \times n}$ is the adjacency matrix, such that $\mathbf{X}_{ij} = 1$ if $(i, j) \in \Omega$, *i.e.*, there exists a PPI between protein i and protein j, and $\mathbf{X}_{ij} = 0$ otherwise. Our task is to identify a subset of non-interacting protein pairs $M \subseteq (V \times V) \setminus \Omega$ which tend to interact and can be served as potential targets for further experimental screening.

Throughout this paper, we denote matrices as boldface upper case characters. Given a matrix $\mathbf{M}$, its Frobenius norm and trace are denoted as $\|\mathbf{M}\|$ and $\mathbf{tr}(\mathbf{M})$ respectively. For convenience, given an index set M of a matrix $\mathbf{X}$, we define $\mathbf{X}_M$ as following:

$$(\mathbf{X}_M)_{ij} = \begin{cases} \mathbf{X}_{ij}, & \forall \ (i, j) \in M, \\ 0 & \text{otherwise.} \end{cases} \tag{1}$$

2.1   Predict New Protein Interactions via PPI Networks

Objective to Predict PPIs. We first predict protein interactions only using PPI network data. Considering the protein interaction prediction as a matrix completion problem, where the input PPI adjacency matrix $\mathbf{X}$ contains missing entries (pairs of proteins whose interactions are yet to be determined), we wish to predict $\mathbf{Y}$ which has full entries, *i.e.*, every elements of $\mathbf{Y}$ is filled with computed values. $\mathbf{Y}$ completes $\mathbf{X}$ in the sense that $\mathbf{Y}_\Omega = \mathbf{X}_\Omega$, or more explicitly, $\mathbf{Y}_{ij} = \mathbf{X}_{ij}, \forall (i, j) \in \Omega$, where Ω denotes the set of edges where the input adjacency matrix $\mathbf{X}$ has known values (the set of interacting edges). Mathematically, the PPI prediction problem can be solved as the following optimization problem:

$$\min_{\mathbf{Y}} \ J_1 = \|\mathbf{X} - \mathbf{Y}\|_\Omega^2 = \sum_{(i,j) \in \Omega} (\mathbf{X} - \mathbf{Y})_{ij}^2 \ . \tag{2}$$

Due to the low-rank nature of the adjacency matrix of an input PPI network as discussed earlier, the completed matrix $\mathbf{Y}$ can be factorized and written as $\mathbf{Y} = \mathbf{HSH}^T$, where $\mathbf{H} \in \mathbb{R}_+^{n \times k}$ and $\mathbf{S} \in \mathbb{R}_+^{k \times k}$ are the factor matrices with nonnegative elements. As a result, $\mathbf{Y} = \mathbf{HSH}^T$ can be seen as a low-rank representation of the input matrix $\mathbf{X}$ with rank of $k \ll n$. Thus we can rewrite Eq. (2) as following:

$$\min_{\mathbf{H} \geq 0, \ \mathbf{S} \geq 0} \ J_2 = \|\mathbf{X} - \mathbf{HSH}^T\|_\Omega^2 \ . \tag{3}$$

Note that, although other low-rank matrix approximation methods, *e.g.*, singular value decomposition (SVD), exist, using NMTF as in Eq. (3) to constrain the factor matrices $\mathbf{H}$ and $\mathbf{S}$ to be nonnegative is a natural choice as all the entries of the adjacency matrix $\mathbf{X}$ of the input PPI network are positive by definition. Moreover, because of the clustering interpretation of NMTF [9,11], other biological data sources can be easily incorporated via manifold regularization as introduced later.

The Solution Algorithm. Different from standard NMTF based objectives as in [16,9,11,31], which are defined over the entire input nonnegative matrix, the objective in Eq. (3) for PPI prediction is defined over a subset of the entries that correspond to known PPIs. Therefore, the solution algorithms to standard NMTF cannot be directly applied to solve Eq. (3). To this end, we present an iterative algorithm in Algorithm 1 to solve Eq. (3). The main computational load of Algorithm 1 is step 4 to solve a symmetric NMF problem, whose numerical solution was just proposed in our recent work in [31].

Algorithm 1. Algorithm to solve Eq. (3).

Input: Input PPI adjacency matrix $\mathbf{X}$;
Index set of known PPIs Ω.
begin
    1. $t = 0$;
    2. Initialize $\mathbf{Z}^{(0)} = \mathbf{X}_\Omega$;
    while *not converge* **do**
        3. $t = t + 1$;
        4. Solve

$$\underset{\mathbf{H} \geq 0\ \mathbf{S} \geq 0}{\arg\min} \left\| \mathbf{Z}^{(t-1)} - \mathbf{H}^{(t)} \mathbf{S}^{(t)} \left(\mathbf{H}^{(t)} \right)^T \right\|^2 \qquad (4)$$

        to obtain $\mathbf{H}^{(t)}$ and $\mathbf{S}^{(t)}$;
        5. Compute $\mathbf{Y}^{(t)} = \mathbf{H}^{(t)} \mathbf{S}^{(t)} \left(\mathbf{H}^{(t)} \right)^T$;
        6. Compute $\mathbf{Z}^{(t)} = \mathbf{X}_\Omega + \mathbf{Y}_M^{(t)}$;
    end
end
Output: Output matrix with filled missing entries $\mathbf{Y}$.

Solving Eq. (3) by Algorithm 1 for matrix completion, our NMTF approach to predict PPIs is proposed.

2.2 Predict New Protein Interactions from Multimodal Biological Data

In last subsection, we infer putative protein interactions only from PPI network data, while in practice we may also have other biological data, such as protein sequence data [20,25,19] and 3D protein structures [2,7,23], and so on. To exploit these useful information, in this subsection, we further develop the proposed NMTF based matrix completion approach.

An important reason of the popularity of NMTF in statistical learning lies in its close connection to k-means clustering [9,11]. Specifically, given a symmetric nonnegative input matrix $\mathbf{W}$, the resulted factor matrix $\mathbf{H}$ can be seen as the clustering indications of the vertices [18]. Therefore, if we have biological data other than PPI networks appearing in form of pairwise similarity, we can incorporate them through manifold regularization [8,26]. Specifically, let $\mathbf{W}_{(k)}$ $(0 \leq k \leq K)$ be a set of pairwise similarities constructed from different biological data, an integrated similarity among proteins can be constructed as $\mathbf{W} = \sum_k \eta_k \mathbf{W}_{(k)}$ $(\eta_k \geq 0, \sum_k \eta_k = 1)$, where η_k are parameters to balance the data from different sources. With $\mathbf{W}$ we further develop the objective in Eq. (3) as following:

$$\min_{\mathbf{H} \geq 0, \, \mathbf{S} \geq 0} J_4 = \|\mathbf{X} - \mathbf{HSH}^T\|_\Omega^2 + 2\lambda \operatorname{tr}\left(\mathbf{H}^T \left(\mathbf{D} - \mathbf{W}\right) \mathbf{H}\right),$$
$$s.t. \ \ \mathbf{H}^T \mathbf{DH} = I, \tag{5}$$

where $\mathbf{D}$ is a diagonal matrix whose diagonal entries $\mathbf{D}_{ii} = \sum_j \mathbf{W}_{ij}$ are the degree of the corresponding data points on $\mathbf{W}$, and λ is a parameter to balance the relative importance of the regularization term which is empirically selected as $\lambda = 0.01$ in all our experimental evaluations. Because $\mathbf{H}$ can be seen as the "soft" clustering labels [9], the second term in Eq. (5) enforces the smoothness over the variation of the clustering labels with respect to the underlying manifold described by $\mathbf{W}$ [4,12], by which additional biological data sources are incorporated.

Equation (5) takes a similar form to Eq. (3), which, again, is not a standard NMTF problem. We use Algorithm 1 to solve it by replacing step 4 to minimize the following objective:

$$J_4 = \|\mathbf{Z} - \mathbf{HSH}^T\|^2 + 2\lambda \operatorname{tr}\left(\mathbf{H}^T \left(\mathbf{D} - \mathbf{W}\right) \mathbf{H}\right),$$
$$s.t. \ \mathbf{H} \geq 0, \mathbf{S} \geq 0, \mathbf{H}^T \mathbf{DH} = I. \tag{6}$$

Solving Eq. (5) for matrix completion, our Regularized Non-negative Matrix Tri-Factorization (R-NMTF) approach for PPI prediction is proposed, which is able to utilize both PPI network data as well as other biological data.

3   Experimental Results and Discussions

3.1   Materials and Data Sources

Protein Interaction Networks. We construct PPI graphs using the protein interaction networks compiled by BioGRID database [29]. We evaluate our methods on the *Saccharomyces cerevisiae* genome, for which an undirected graph is constructed, with vertices representing proteins and edges representing observed physical interactions. When constructing the graph, we only consider the largest connected component of the physical interaction map from BioGRID database of version 2.0.56. The details of the PPI graph of *S. cerevisiae* genome are listed in Table 1, where "coverage" stands for the percentage of known PPIs against the total number of protein pairs $(n \times (n-1)/2)$.

Table 1. PPI graphs of the *S. cerevisiae* genome constructed using BioGRID database of version 2.0.56.

Number of proteins	5056
Edges (number/coverage)	9439/0.738%

Protein Sequence Data. We downloaded protein sequence data from Gen-Bank [3] and computed the sequence based similarity using the mismatch kernel [17]. A protein sequence s_i is first mapped to a feature vector $\Phi_{k,m}(s_i) = \{\phi_\beta(\alpha)\}_{\beta \in \mathcal{A}^k}$, where $\mathcal{A}$ is the alphabet of 20 amino acids. The neighborhood $\mathcal{N}_{k,m}(\alpha)$ of a k-mer α is the set of k-mers that differs in at most m positions. The feature vector encodes all the k-mers in the neighborhood for $\phi_\beta(\alpha) = 1$ if $\alpha \in \mathcal{N}_{k,m}(\beta)$, and 0 otherwise. Then the mismatch kernel, thereby the induced pairwise similarity between two protein sequences s_i and s_j, is computed as $W_{ij}^{(1)} = \mathcal{K}(s_i, s_j) = \langle \Phi_{k,m}(s_i), \Phi_{k,m}(s_j) \rangle$. In our empirical studies, we set $k = 6$ and $m = 1$, which is the same as in [19]. We use protein sequence data as the additional biological data source, *i.e.*, $W = W^{(1)}$.

Protein Annotation Data. We use the functional annotations defined by Gene Ontology (GO) Consortium [1], which is a set of structured vocabularies organized in a rooted directed acyclic graph (DAG), describing attributes of gene products (proteins or RNA) in three categories of "cellular component", "molecular function" and "biological process".

3.2   Improved Prediction Capability in Cross-Validation

We first evaluate the proposed methods and compare their prediction capabilities against three most recent PPI prediction methods:

(1) Tensor product pairwise kernel (TPPK) method [20]: This method builds a kernel for pairwise objects. In order for a fair comparison, protein sequence and protein interaction network topology are used for kernel construction. PPI prediction is then carried out by the ranking scores for non-interacting protein pairs yielded by a SVM on the resulted score.

(2) Metric learning pairwise kernel (MLPK) method [23]: This method represents a pair of objects as the difference between its members, such that the resulted kernel is invariant with respect to the order of the proteins. Again, SVM is used to compute the ranking score for putative PPIs.

(3) Nearest neighbor (NN) [19] method: NN is the simplest classification method in machine learning. In [19], a ranking score for each non-interacting protein pair is computed as

$$f_{NN}(x_i) = \sum_{x_j \in (\mathcal{N}_k(x_i) \cap E)} d(x_i, x_j) - \sum_{x_j \in (\mathcal{N}_k(x_i) \cap ((V \times V) \setminus E))} d(x_i, x_j), \qquad (7)$$

where $\mathcal{N}_k(x_i)$ is the set of k-nearest neighbors of x_i, and $d(\cdot, \cdot)$ is distance function built from a kernel by $d(x_i, x_j) = \sqrt{\mathcal{K}(x_i, x_i) - 2\mathcal{K}(x_i, x_j) + \mathcal{K}(x_j, x_j)}$. In our evaluations, we use the mismatch kernel for protein sequence data.

Experimental Procedures. For each method, we perform 20-fold cross-validation as following. For each trail, we remove 5% known edges (PPIs) from the input graph and try to recover them using the remained graph, which is repeated by 10 times. The average results over the 10 trials on the *S. cerevisiae* genome are reported in Fig. 1. During each trial, an internal 5-fold cross validation is performed for parameter selection. For our NMTF and R-NMTF methods, the parameter is the rank k of the factor matrices H and S. For TPPK and MLPK methods, we use the Gaussian kernel, therefore the parameters are the two regularization parameters. For NN method, we select the k of NN from $\{1, 2, 3, 5, 10, 15\}$, which is the same as in [19]. We fine tune the parameters for best prediction precision for all the compared methods.

Results. Because all compared methods produce a list of ranking scores for non-interacting protein pairs, we employ precision-recall curves to measure the prediction performance. We compute the precisions and recalls when picking up a range of top k non-interacting protein pairs as predictions, and average them over the 10 trials. The resulted precision-recall curves on the *S. cerevisiae* genome are reported in Fig. 1. From the results, we can see that the both proposed methods, NMTF and R-NMTF, consistently outperform the compared methods, sometimes very significantly. In addition, the prediction performances of R-NMTF method are always better than those of NMTF method, which is consistent with our previous theoretical analysis in that multimodal biological data, *i.e.*, protein interaction network plus protein sequence data, offer enhanced prediction performance.

A more careful analysis on the prediction results shows that the non-interacting protein pairs (including the non-interacting protein paris in the original PPI graphs and those removed due to cross-validation) with high ranking scores identified by the proposed methods typically exhibit high similarities in their functional roles. In Table 2, we list the predicted protein pairs with top 5 highest ranking scores by R-NMTF method on *S. cerevisiae* species, in which the biological functions of all protein pairs are very similar to each other. For example, "PHO91" works as "Low-affinity phosphate transporter of the vacuolar membrane", which is also the main functional role of its putative interacting partner "PHO90". Moreover, both of them have functionalities of "transcription independent of Pi and Pho4p activity" and "overexpression results in vigorous

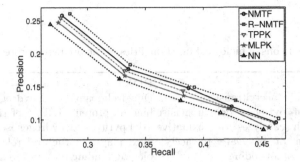

Fig. 1. Precision-recall curves by 20-fold cross-validation on *S. cerevisiae* genome by the compared PPI prediction methods

growth". These observations clearly demonstrate that these two proteins are functionally related and tend to interact with each other, which provide a concrete evidence to support that the predicted protein interactions by the proposed R-NMTF are biologically meaningful.

Note that, in our empirical studies we only use one additional data source, *i.e.*, protein sequence data, for the purpose of demonstration. In practice, more biological data could be also incorporated through a proper kernel construction under the R-NMTF prediction framework to achieve better prediction results.

3.3 Improved Protein Function Prediction Using Predicted PPI Networks

Protein interaction networks are broadly used in various biological applications, whose performances are inevitably determined by the quality input PPI graphs. Therefore, we assess the quality of predicted PPI networks in protein function prediction on *S. cerevisiae* species.

We predict protein functions on the original PPI graph constructed from the BioGRID database, the PPI graph filled by the top 1000 putative interacting protein pairs predicted by NMTF method, and that by R-NMTF method. We make predictions using the following three benchmark graph-based protein function prediction methods:

(1) Majority voting (MV) [24] method: This method assigns functions to a protein via its connecting neighbors in certain ranges.

(2) Iterative majority voting (IMV) [30] method: This method is same as MV method, but iteratively repeats the function assignment process until certain conditions are satisfied.

(3) Function Flow (FF) [21] method: This method formulates the annotation problem as a minimum multiway-cut problem, where the goal is to assign a unique function to all un-annotated proteins so as to minimize the cost of edges connecting proteins with different assignments.

Table 2. Comparisons of biological functionalities of the predicted protein interactions

LAT1	PDX1
Dihydrolipoamide acetyltransferase component (E2) of pyruvate dehydrogenase complex, which catalyzes the oxidative decarboxylation of pyruvate to acetyl-CoA—Lat1p; protein coding	Dihydrolipoamide dehydrogenase (E3)-binding protein (E3BP) of the mitochondrial pyruvate dehydrogenase (PDH) complex, plays a structural role in the complex by binding and positioning E3 to the dihydrolipoamide acetyltransferase (E2) core—Pdx1p; protein coding
PHO91	**PHO90**
Low-affinity phosphate transporter of the vacuolar membrane; deletion of pho84, pho87, pho89, pho90, and pho91 causes synthetic lethality; transcription independent of Pi and Pho4p activity; overexpression results in vigorous growth—Pho91p; protein coding	Low-affinity phosphate transporter; deletion of pho84, pho87, pho89, pho90, and pho91 causes synthetic lethality; transcription independent of Pi and Pho4p activity; overexpression results in vigorous growth—Pho90p; protein coding
PHO91	**PHO87**
Low-affinity phosphate transporter of the vacuolar membrane; deletion of pho84, pho87, pho89, pho90, and pho91 causes synthetic lethality; transcription independent of Pi and Pho4p activity; overexpression results in vigorous growth—Pho91p; protein coding	Low-affinity inorganic phosphate (Pi) transporter, involved in activation of PHO pathway; expression is independent of Pi concentration and Pho4p activity; contains 12 membrane-spanning segments—Pho87p; protein coding
PHO87	**PHO89**
Low-affinity inorganic phosphate (Pi) transporter, involved in activation of PHO pathway; expression is independent of Pi concentration and Pho4p activity; contains 12 membrane-spanning segments—Pho87p; protein coding	Na+/Pi cotransporter, active in early growth phase; similar to phosphate transporters of Neurospora crassa; transcription regulated by inorganic phosphate concentrations and Pho4p—Pho89p; protein coding
COY1	**SVP26**
Coy1p—Golgi membrane protein with similarity to mammalian CASP; genetic interactions with GOS1 (encoding a Golgi snare protein) suggest a role in Golgi function; protein coding	Integral membrane protein of the early Golgi apparatus and endoplasmic reticulum, involved in COP II vesicle transport; may also function to promote retention of proteins in the early Golgi compartment—Svp26p; protein coding

Table 3. Performance of protein function prediction by involved method on compared PPI graphs

	MV	IMV	FF
Original PPI graph	30.12%	30.92%	32.99%
Predicted PPI graph by NMTF method	34.85%	35.21%	36.02%
Original PPI graph by R-NMTF method	35.98%	36.33%	38.19%

We implement these methods following the details in the original literatures. Because FF method produces a ranking list of predicted protein functions, we select a threshold such that the prediction precision is maximized. 5-fold cross-validation is performed to predict the functions in "biological process" of GO. The average prediction precision over all test functions and 5 trials of cross-validation of the involved methods on different PPI graphs are reported in Table 3.

The results in Table 3 show that the function prediction performance for all three methods are improved when the predicted PPI graphs are used. Such results experimentally prove that the predicted PPI graphs have higher quality than the original one, which demonstrates that the filled putative protein interactions by the proposed methods are largely biological meaningful. Thus, we can tentatively conclude that the proposed NMTF and R-NMTF indeed can improve the protein interaction networks. Again, multimodal biological data sources based R-NMTF method is better than single data source based NMTF method.

4   Conclusions

In this paper, instead of considering protein-protein interaction prediction as a binary classification problem as in many existing works, we formulated it as a matrix completion problem. Taking this different perspective, the difficulty of selecting negative training samples required by classification based methods is averted. Moreover, because the number of protein interaction types is small, the recovery of missing PPIs from an incomplete observed protein interaction network can be suitably solved under the framework of matrix completion. We first proposed to use NMF approach to predict PPIs only from protein interaction network data, and then extended it through manifold regularization to incorporate multimodal biological data sources. We have conducted extensive empirical studies to evaluate different aspects of the proposed methods on *S. cerevisiae* genome. Promising results in the experiments validate our methods that are consistent with our theoretical analysis.

Acknowledgments. This research is supported by NSF-IIS 1117965, NSFCCF-0830780, NSF-DMS-0915228, NSFCCF-0917274.

References

1. Ashburner, M., Ball, C., Blake, J., Botstein, D., Butler, H., Cherry, J., Davis, A., Dolinski, K., Dwight, S., Eppig, J., et al.: Gene ontology: tool for the unification of biology. Nature Genetics 25(1), 25–29 (2000)
2. Ben-Hur, A., Noble, W.: Kernel methods for predicting protein–protein interactions. Bioinformatics 21(suppl. 1), i38 (2005)
3. Benson, D., Karsch-Mizrachi, I., Lipman, D.: GenBank. Nucleic Acids Res. 34, D16–D20 (2006)
4. Cai, D., He, X., Wu, X., Han, J.: Non-negative matrix factorization on manifold. In: Proceedings of the 2008 Eighth IEEE International Conference on Data Mining, pp. 63–72. IEEE (2008)
5. Candès, E., Plan, Y.: Matrix completion with noise. Proceedings of the IEEE (2009)
6. Chen, X., Jeong, J.: Sequence-based prediction of protein interaction sites with an integrative method. Bioinformatics 25(5), 585 (2009)
7. Chen, X., Liu, M.: Prediction of protein–protein interactions using random decision forest framework. Bioinformatics 21(24), 4394 (2005)
8. Chung, F.: Spectral Graph Theory. Amer. Math. Society (1997)
9. Ding, C., He, X., Simon, H.: On the equivalence of nonnegative matrix factorization and spectral clustering. In: Proc. SIAM Data Mining Conf., Citeseer, pp. 606–610 (2005)
10. Ding, C., Li, T., Jordan, M.: Convex and semi-nonnegative matrix factorizations for clustering and low-dimension representation. Lawrence Berkeley National Laboratory, Tech. Rep. LBNL-60428 (2006)
11. Ding, C., Li, T., Peng, W., Park, H.: Orthogonal nonnegative matrix t-factorizations for clustering. In: Proceedings of the 12th ACM SIGKDD International Conference on Knowledge Discovery and Data Mining, pp. 126–135. ACM (2006)
12. Gu, Q., Zhou, J.: Co-clustering on manifolds. In: Proceedings of the 15th ACM SIGKDD International Conference on Knowledge Discovery and Data Mining, pp. 359–368. ACM (2009)
13. Ho, Y., Gruhler, A., Heilbut, A., Bader, G., Moore, L., Adams, S., Millar, A., Taylor, P., Bennett, K., Boutilier, K., et al.: Systematic identification of protein complexes in Saccharomyces cerevisiae by mass spectrometry. Nature 415(6868), 180–183 (2002)
14. Ito, T., Tashiro, K., Muta, S., Ozawa, R., Chiba, T., Nishizawa, M., Yamamoto, K., Kuhara, S., Sakaki, Y.: Toward a protein–protein interaction map of the budding yeast: a comprehensive system to examine two-hybrid interactions in all possible combinations between the yeast proteins. Proceedings of the National Academy of Sciences of the United States of America 97(3), 1143 (2000)
15. Jansen, R., Yu, H., Greenbaum, D., Kluger, Y., Krogan, N., Chung, S., Emili, A., Snyder, M., Greenblatt, J., Gerstein, M.: A Bayesian networks approach for predicting protein-protein interactions from genomic data. Science 302(5644), 449 (2003)
16. Lee, D., Seung, H.: Learning the parts of objects by non-negative matrix factorization. Nature 401(6755), 788–791 (1999)
17. Leslie, C., Eskin, E., Weston, J., Noble, W.: Mismatch string kernels for SVM protein classification. In: Advances in Neural Information Processing Systems, pp. 1441–1448 (2003)

18. Luo, D., Ding, C., Huang, H., Li, T.: Non-negative laplacian embedding. In: 2009 Ninth IEEE International Conference on Data Mining, pp. 337–346. IEEE (2009)

19. Martial, H., Michael, R., Jean-Philippe, V., William, N.: Large-scale prediction of protein-protein interactions from structures. BMC Bioinformatics 11(114) (2010)

20. Martin, S., Roe, D., Faulon, J.L.: Predicting protein–protein interactions using signature products. Bioinformatics 221(2), 218 (2005)

21. Nabieva, E., Jim, K., Agarwal, A., Chazelle, B., Singh, M.: Whole-proteome prediction of protein function via graph-theoretic analysis of interaction maps. Bioinformatics 21(suppl. 1), i302 (2005)

22. Qi, Y., Klein-Seetharaman, J., Bar-Joseph, Z.: Random forest similarity for protein-protein interaction prediction from multiple sources. In: Pac. Symp. Biocomput., vol. 10, pp. 531–542 (2005)

23. Qiu, J., Hue, M., Ben-Hur, A., Vert, J., Noble, W.: A structural alignment kernel for protein structures. Bioinformatics 23(9), 1090 (2007)

24. Schwikowski, B., Uetz, P., Fields, S.: A network of protein–protein interactions in yeast. Nature Biotechnology 18(12), 1257–1261 (2000)

25. Shen, J., Zhang, J., Luo, X., Zhu, W., Yu, K., Chen, K., Li, Y., Jiang, H.: Predicting protein–protein interactions based only on sequences information. Proceedings of the National Academy of Sciences 104(11), 4337 (2007)

26. Shi, J., Malik, J.: Normalized cuts and image segmentation. IEEE. Trans. on Pattern Analysis and Machine Intelligence 22, 888–905 (2000)

27. Shoemaker, B.A., Panchenko, A.R.: Deciphering protein-protein interactions. Part I. experimental techniques and databases. PLoS Computational Biology 3(3), 334–337 (2007)

28. Shoemaker, B., Panchenko, A.: Deciphering protein-protein interactions. Part II. Computational methods to predict protein and domain interaction partners. PLoS Computational Biology 3(4), 595–601 (2007)

29. Stark, C., Breitkreutz, B.J., Reguly, T., Boucher, L., Breitkreutz, A., Tyers, M.: BioGRID: a general repository for interaction datasets. Nucleic Acids Res. 34(database issue), D535 (2006)

30. Vazquez, A., Flammini, A., Maritan, A., Vespignani, A.: Global protein function prediction from protein-protein interaction networks. Nature Biotechnology 21(6), 697–700 (2003)

31. Wang, H., Huang, H., Ding, C.: Simultaneous Clustering of Multi-Type Relational Data via Symmetric Nonnegative Matrix Tri-factorization. In: The 20th ACM Conference on Information and Knowledge Management. ACM (2011)

32. Zhang, L., Wong, S., King, O., Roth, F.: Predicting co-complexed protein pairs using genomic and proteomic data integration. BMC Bioinformatics 5(1), 38 (2004)

CNVeM: Copy Number Variation Detection Using Uncertainty of Read Mapping

Zhanyong Wang[1,*], Farhad Hormozdiari[1,*], Wen-Yun Yang[1], Eran Halperin[2,3], and Eleazar Eskin[1,**]

[1] Computer Science Department, University of California, Los Angeles
[2] Blavatnik School of Computer Science,
The Department of Molecular Microbiology and Biotechnology, Tel-Aviv University
[3] International Computer Science Institute, Berkeley
eeskin@cs.ucla.edu

Abstract. Copy number variations (CNVs) are widely known to be an important mediator for diseases and traits. The development of high-throughput sequencing (HTS) technologies has provided great opportunities to identify CNV regions in mammalian genomes. In a typical experiment, millions of short reads obtained from a genome of interest are mapped to a reference genome. The mapping information can be used to identify CNV regions. One important challenge in analyzing the mapping information is the large fraction of reads that can be mapped to multiple positions. Most existing methods either only consider reads that can be uniquely mapped to the reference genome, or randomly place a read to one of its mapping positions. Therefore, these methods have low power to detect CNVs located within repeated sequences. In this study, we propose a probabilistic model, CNVeM, that utilizes the inherent uncertainty of read mapping. We use maximum likelihood to estimate locations and copy numbers of copied regions, and implement an expectation-maximization (EM) algorithm. One important contribution of our model is that we can distinguish between regions in the reference genome that differ from each other by as little as 0.1%. As our model aims to predict the copy number of each nucleotide, we can predict the CNV boundaries with high resolution. We apply our method to simulated datasets and achieve higher accuracy compared to CNVnator. Moreover, we apply our method to real data from which we detected known CNVs. To our knowledge, this is the first attempt to predict CNVs at nucleotide resolution, and to utilize uncertainty of read mapping.

1 Introduction

Genetic variation between individuals can range from single nucleotide differences to differences in large segments of DNA. Variations on the nucleotide level are referred to as Single Nucleotide Polymorphisms (SNPs) and on the segment level are referred to as Structural Variations (SVs), including insertions,

---

B. Chor (Ed.): RECOMB 2012, LNBI 7262, pp. 326–340, 2012.
© Springer-Verlag Berlin Heidelberg 2012

deletions, and copy number variations (CNVs). SVs and in particular CNVs, in which a large region of genome is deleted or duplicated, play an important role in the genetics of complex diseases and traits [14,21]. Many recent studies have shown a correlation between CNVs and different genomic disorders, ranging from brain related diseases (such as autism, schizophrenia and idiopathic learning disability [17]) to cancers (e.g. non-small cell lung cancer [3]).

Common methods to detect CNVs were until recently based on whole genome array comparative genome hybridization (ArrayCGH). In ArrayCGH, both a genome of interest (donor genome) and a reference genome are hybridized to a tiling array and the intensity ratio of the two genomes (donor/reference) provides an estimate of the copy number gain or loss [4,5]. Although a powerful method to detect the presence of CNVs and estimate copy numbers, the ArrayCGH approach is unable to identify the boundaries of CNVs with high resolution.

The development of high-throughput sequencing (HTS) technologies provides great opportunities to detect CNV regions. With HTS technologies, whole genome shotgun sequencing of one or more individuals becomes possible. Methods to detect the CNVs from short reads generated by HTS technologies can be categorized by two main ideas. The first category of methods divides the genome into small windows and the number of reads mapped to each specific window (read depth) is used as a proxy for the copy number of that window [2,20,18,6,22,8]. Alkan et. al [2] used a set of fixed regions which are unique among all primates as control windows and calculated the average read depth for those regions. Then they scaled the results to predict the copy number of other windows. Simpson et al. [18] used the same idea of splitting the genome into windows while incorporating read depth and heterozygous SNPs information (in inbred mouse) into a Hidden Markov Model (HMM). Adjacent windows with same copy number state are combined into one CNV region. Abyzov et al. [1] developed a method for CNV discovery from statistical analysis of read depth. The method is based on the established mean-shift approach [7], which is a popular method in computer vision. This approach is able to detect the presence of large CNVs and the copy numbers. However, the resolution of this approach is limited by the size of the windows, which is typically at least one kilo-base.

In the second strategy, "paired-end" reads, where "paired-end" refers to the two ends of the same segment of a DNA molecule, are used to detect CNVs. A short gap appears between the two paired-end reads and the distance of this gap is roughly fixed and known. The second class of approaches utilizes the discordant paired-end reads, which are the reads mapped to the reference genome in an unexpected way [16,11,13]. Discordant reads may indicate the presence of CNVs. Read depth information is then used to compute the copy number for each candidate CNV region [20,2]. Medvedev et al. [16] introduced the idea of using both the read depth as well as the discordant reads to detect CNVs. This method first clusters the discordant reads to identify the CNVs boundary, after which they build a "donor graph" representing the genome as segments of sequence connected by edges. Moreover, they use the maximum flow to estimate the most likely copy numbers for the donor genome. One limitation of this strategy is that it only detects CNVs in regions which are not repeat-rich. This may reduce the applicability of this method given the existence of many repeat-rich regions in the genome. Also, the CNVs may have complex structure. For example, there exist multiple copies of CNVs in the reference genome. This method can not detect the variation within different copies.

Another important challenge for CNV detection lies in the uncertainty of read mapping. All of the mentioned methods use read depth information. The read depth is obtained by mapping the short reads to the reference genome and then calculating number of reads within a region. However, a read can be mapped to multiple locations while the read originated from one specific locus in the donor genome. This mapping uncertainty can be due to short read length, sequencing errors, and the presence of repetitive regions. With few exceptions [12], most studies either consider all possible locations or randomly pick one mapping location, or even discard all such reads. These methods have difficulty in detecting CNVs with high accuracy, especially CNVs in repeat-rich regions.

In this study, we show that handling the uncertainty of read mapping can help us in predicting the copy number of CNVs, especially in repeat-rich regions. We propose a probabilistic model, CNVeM, that utilizes the uncertainty of read mapping. We use maximum likelihood to estimate locations and copy number of copied regions, and implement an expectation-maximization (EM) algorithm. One important contribution of our model is that we distinguish between similar copies of a region in the reference genome. We can predict exactly which copy of a region is duplicated or deleted utilizing the differences between copies and handling uncertainty of read mapping.

In our model, we predict the copy number for each nucleotide and adjacent nucleotides with same copy number are then combined to form a full CNV region. In this way, we can detect the boundaries precisely and are able to predict small CNVs. To our knowledge, this is the first attempt to detect CNVs at nucleotide resolution and to distinguish between similar sequences in the reference genome.

2   Methods

2.1   A Motivating Example

One important contribution of our method is that we distinguish between regions in the reference genome that differ from each other by a single nucleotide. Figure 1 illustrates an example. The reference genome has two nearly identical copies of a CNV region, represented as A and B. They only differ by one nucleotide as indicated in the figure, where the nucleotide is 'C' in region A and 'T' in region B. In the donor genome, region B is copied twice as B1 and B2. Reads $\{r1, r2, \ldots, r6\}$ are obtained from the donor genome as shown in the lower part of Figure 1 and then mapped to the reference genome as shown in the upper part of Figure 1. As shown in the figure, reads $\{r1, r3, r5\}$ can be mapped to both region A and B in the reference. However, read $\{r2\}$ can only be mapped to region A and reads $\{r4, r6\}$ can only be mapped to region B. If we assign a read to one of multiple mapping positions randomly following the traditional

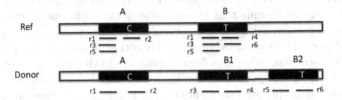

Fig. 1. Similar copies of a CNV region exist in the reference genome. 'C' and 'T' are the only different nucleotides between region A and B. Reads $\{r1, r2, \ldots, r6\}$ are obtained from the donor genome as shown in the lower part of the figure. Furthermore, these reads can be mapped to the reference genome as shown in the upper part of the figure.

strategy, we would determine the copy number of both region A and B to be 1.5. However, in CNVeM, we use the EM algorithm to find the optimal solution. In each iteration, we assign a read to different mapping positions according to the distribution of copy numbers of those positions, and update the copy number of each position. Upon convergence, the EM algorithm assigns reads $\{r1, r3, r5\}$ to region A with probability $1/3$ and to region B with probability $2/3$. We correctly predict the copy number of region A to be 1 and copy number of region B to be 2.

2.2   The Generative Model

We use short read information from HTS technologies to detect copy number variants. Let $\mathcal{G} = (g_1, g_2, \ldots, g_K)$ be K continuous nucleotides in the reference genome, where g_i is the ith nucleotide. We assign the copy number of each nucleotide in the reference genome to be 1. The donor genome is also composed of these nucleotides. However, large regions of the genome can be either deleted or duplicated and thus the copy number is changed. For each nucleotide g_i, we denote the copy number to be C_i in the donor genome. If $C_i < 1$, we call it a copy loss. If $C_i > 1$, we call it a copy gain. $\mathcal{C} = (C_1, C_2, \ldots, C_K)$ can be interpreted as the copy number vector of the donor genome. For most nucleotides, the copy numbers are the same in the donor genome and in the reference genome. So one can assume that the length of donor genome is the same as the length of the reference genome, i.e. $\sum_{i=1}^{K} C_i = K$. We define vector $(\frac{C_1}{K}, \frac{C_2}{K}, \ldots, \frac{C_K}{K})$ to be the normalized copy number vector of the donor genome.

Using HTS technology, millions of short reads are sampled from the donor genome. We assume that a read r_j of length l is generated by randomly picking a position i from $\mathcal{G}$ according to distribution $\mathcal{C}/K$, and then copying l consecutive positions starting from position i. The copying process is error-prone, with known probability ϵ for a sequencing error rate at any position of the read. This process is repeated until we have a set of N reads $\mathcal{R} = \{r_1, r_2, \ldots, r_N\}$. The objective is to infer $\mathcal{C} = (C_1, C_2, \ldots, C_K)$ from $\mathcal{R}$. Since the reads are mapped to the reference genome, mapping information is utilized to infer CNVs.

In our model, each read r_j is sequenced starting from one position in the donor genome. As we assume that the donor genome is obtained from the reference genome by alternating the copy number of some regions, each position in the donor genome "originates" from a nucleotide in the reference genome. Consequently, each read originates from a position in the reference genome. If a region in the reference genome is duplicated in the donor genome, any read generated from the duplicated segments of the donor genome originates from a unique position in the reference genome. $\mathcal{Z} = (Z_1, Z_2, \ldots, Z_N)$ is the origin for each read in the reference genome, where $Z_j \in \{1, 2, \ldots, K\}$. We then define the following likelihood model of all reads given copy number $\mathcal{C}$ and reference genome $\mathcal{G}$

$$P(\mathcal{R}|\mathcal{C}, \mathcal{G}) = \prod_{j=1}^{N} P(r_j|\mathcal{C}, \mathcal{G}) = \prod_{j=1}^{N} \sum_{i=1}^{K} P(r_j, Z_j = i|\mathcal{C}, \mathcal{G}) \tag{1}$$

where the first equality follows from that the probability that read set $\mathcal{R}$ is composed of independent probabilities of all the reads, and the second equality follows from the fact that the read probability is equal to the marginalization of read mapping uncertainty, i.e., $P(r) = \sum_i P(r, Z = i)$.

The interpretation of the above probability definition $P(r_j, Z_j = i|\mathcal{C}, \mathcal{G})$ is straightforward: the probability of j-th read originating from i-th position of the reference genome, given the copy numbers and reference genome. We can further expand this probability as follows

$$P(r_j, Z_j = i|\mathcal{C}, \mathcal{G}) = P(Z_j = i|\mathcal{C})P(r_j|Z_j = i, \mathcal{G}) \tag{2}$$

where the equality follows from the fact that the read origin Z is independent of reference genome $\mathcal{G}$ and the sequence of read r is independent of copy number $\mathcal{C}$. We define the first term $P(Z_j = i|\mathcal{C}) = C_i/K$ to be the probability for read r_j originating from position i. For each position i and read r_j, we have a probability $P(r_j|Z_j = i, \mathcal{G})$, which stands for the probability of observing read sequence r_j given that the origin of read r_j is position i. We can write $P(r_j|Z_j = i, \mathcal{G})$ as

$$P(r_j|Z_j = i, \mathcal{G}) = \prod_{x=1}^{l} \gamma(g_{i+x-1}, r_j^x)$$

and

$$\gamma(g_{i+x-1}, r_j^x) = \begin{cases} \epsilon/3 & \text{if } r_j^x \neq g_{i+x-1} \\ 1 - \epsilon & \text{otherwise} \end{cases}$$

where r_j^x stands for the x-th nucleotide of read r_j, and the l consecutive nucleotides starting from position i in the reference genome are $g_i, g_{i+1}, \ldots, g_{i+l-1}$. In practice, for each read r_j, the probability $P(r_j|Z_j = i, \mathcal{G})$ will be close to zero for all but a few positions, which are reported by the mapping methods.

We also take the prior probability of the donor genome into consideration. As we assume the donor genome sequence can be obtained by either deleting or duplicating large regions of nucleotides from the reference genome, adjacent positions will have similar copy numbers in the donor genome. Then, in our probabilistic model, it is natural to assume that the copy number of the current nucleotide is only dependent on the previous nucleotide. We have $P(\mathcal{C}) = P(C_1, C_2, \ldots, C_K) = P(C_1) \prod_{i=2}^{K} P(C_i|C_{i-1})$.

Using Bayes rule, we can get the posterior probability of $\mathcal{C}$ given the read set $\mathcal{R}$ and reference genome $\mathcal{G}$:

$$P(\mathcal{C}|\mathcal{R}, \mathcal{G}) \propto P(\mathcal{R}|\mathcal{C}, \mathcal{G})P(\mathcal{C})$$

$$\propto \left(\prod_{j=1}^{N} \sum_{i=1}^{K} \frac{C_i}{K} P(r_j|Z_j = i) \right) \times \left(P(C_1) \prod_{i=2}^{K} P(C_i|C_{i-1}) \right) \quad (3)$$

2.3 Optimization

Maximizing the posterior of copy number $\mathcal{C}$ in (3) is equal to maximizing the following log probability with respect to $\mathcal{C}$:

$$\sum_{j=1}^{N} \left(\log \sum_{i=1}^{K} \frac{C_i}{K} P(r_j|Z_j = i) \right) + \log \left(P(C_1) \prod_{i=2}^{K} P(C_i|C_{i-1}) \right)$$

In this section, we illustrate a lower level description of our method. In order to make the above objective function simpler we eliminate the constraint $\sum_{i=1}^{K} C_i = K$ by introducing a penalty function $g(\mathcal{C}) = K - \sum_{i=1}^{K} C_i$, which prevents the $C_i's$ from growing unbounded (the above objective function will have a higher value if the $C_i's$ grows larger). Incorporating the penalty function, our objective function now becomes

$$\sum_{j=1}^{N} \left(\log \sum_{i=1}^{K} \frac{C_i}{K} P(r_j|Z_j = i) \right) + \log \left(P(C_1) \prod_{i=2}^{K} P(C_i|C_{i-1}) \right) + \delta \left(K - \sum_{i=1}^{K} C_i \right)$$

$$(4)$$

where δ is a penalty function coefficient (We set $\delta = \frac{N}{K}$ in our experiments, from which we achieve best results). We optimize the objective function (4) through an expectation-maximization (EM) algorithm. The algorithm iteratively applies the following two steps until convergence.

Expectation-step:

$$Q(\mathcal{C}|\mathcal{C}^{(t)}) = \sum_{j=1}^{N} \sum_{i=1}^{K} \log \left(\frac{C_i}{K} \right)^{P(Z_j=i|r_j)}$$

$$+ \sum_{j=1}^{N} \sum_{i=1}^{K} \log P(r_j|Z_j = i)^{P(Z_j=i|r_j)} + \log P(\mathcal{C}) + \delta \left(K - \sum_{i=1}^{K} C_i \right)$$

Maximization-step:

$$C^{(t+1)} =$$

$$\arg\max_{\mathcal{C}} \log\left[P(C_1) \left(\frac{C_1}{K}\right)^{d_1} \times e^{-\delta C_1} \times \prod_{i=2}^{K} \left(P(C_i|C_{i-1}) \left(\frac{C_i}{K}\right)^{d_i} \times e^{-\delta C_i}\right)\right] \tag{5}$$

where $d_i = \sum_{j=1}^{N} P(Z_j = i|r_j)$.

We solve the M-step using dynamic programming. Denote the objective function in the M-step to be

$$f = \log\left[P(\mathcal{C}) \times \prod_{i=1}^{K} \left(\left(\frac{C_i}{K}\right)^{d_i} \times e^{-\delta C_i}\right)\right] \tag{6}$$

Then we define $f(k,x)$ to be the maximum function value for the first k positions when the copy number of kth position is $C_k = x$. Now we design the dynamic programming indicated in Equation (7).

$$f(k,x) = \begin{cases} \log[P(C_k = x) \times (\frac{C_k}{K})^{d_k} \times e^{-\delta C_k}] & \text{if } k = 1 \\ \max_{C_{k-1}} \{f(k-1, C_{k-1}) + \log[P(C_k|C_{k-1})]\} & \text{otherwise} \\ \quad + \log[(\frac{C_k}{K})^{d_k} \times e^{-\delta C_k}] \end{cases} \tag{7}$$

We prove that the above dynamic programming returns the global optimal solution for the objective function in (6).

Lemma 1. *The objective function in (6) is solved optimally using the dynamic programming mentioned in (7) .*
Proof. For the sake of space we describe the correctness of the dynamic programming in the extended version of the paper.

The maximum value of the objective function in the M-step is then $\max_x f(K, x)$. Using a backtracking process, we find the vector $C = (C_1, C_2, \ldots, C_K)$ that maximizes function f in the M-step. By iteratively running E-step and M-step, we achieve local optima.

2.4   Implementation

This optimization process requires an initial input of copy numbers. Different initial inputs will affect the convergence time. To achieve better performance, it is important to start with a "*good*" initial guess. In order to obtain a good initial input, we split the genome into non-overlapping bins of 300 bp. All nucleotides within one bin share the same copy number. Using a similar model as in (1), we get a initial guess of copy numbers by optimizing the objective function (8).

$$P(\mathcal{R}|\mathcal{C}, \mathcal{G}) = \prod_{j=1}^{N} P(r_j|\mathcal{C}, \mathcal{G}) = \prod_{j=1}^{N} \sum_{i=1}^{\lceil K/300 \rceil} \frac{C_i \times 300}{K} P(r_j|Z_j \in i\text{-th bin}) \tag{8}$$

Algorithm 1. The complete algorithm of CNVeM

Input: Read mapping information, allowing reads map to multiple locations.
Output: Copy number variations compared to reference genome.
Initialization: Choose an initial configuration of copy numbers $\mathcal{C}^{(0)}$.
STAGE ONE:
      Optimize the function in (8) using a standard EM algorithm based on bins.
We get an initial solution of copy numbers for each bin.

STAGE TWO:
    2.1 Use the output from STAGE ONE as an initial guess.
    2.2 For each read r_j with $j \in \{1, 2, \ldots, N\}$, consider all mapping positions, calculate the posterior probability of each position according to the joint probability in formula (2). Then map the read to multiple locations fractionally according to the posterior probability.
    2.3 Calculate the total number of reads mapped to each position.
    2.4 Update the copy numbers of all nucleotides using the dynamic programing in formula (7).
    2.5 Repeat Steps 2.3-2.4 until it converges.

where $P(r_j | Z_j \in i\text{-th bin}) = \frac{1}{300} \sum_{s=1}^{300} \prod_{x=1}^{l} \gamma(g_{i \times 300 + s + x - 1}, r_j^x)$. Similarly, we can optimize function (8) by the EM algorithm. As proved in [10], the likelihood function (8) is concave. The EM algorithm will converge to global optimal solution and it will be a good initial guess for the objective function in formula (4).

After obtaining a solution using a standard EM algorithm, we conduct our extended EM algorithm introduced in section 2.3. We summarize our method in Algorithm 1.

2.5 GC-Bias Correction

One of the short comings of the HTS technologies is the existence of different biases in the sequencing process. Some biases are due to the environment while others are due to chemical reactions (DNA amplifications, GC content). Studies show that both Sanger and HTS sequencing have bias toward high GC regions. GC-bias can influence the number of reads generated from a position and thus the reads are no longer uniformly generated. There have been a number of papers [2,1,20,22] which deal with GC-bias in CNV calling. In this work, we adapted the idea mentioned in [1,22] to correct for GC-bias. In equation (5), d_i is the number of reads mapped to position i. We correct this bias by updating the definition of d_i to be $d_i^c = d_i \times \frac{\overline{DOC}_{global}}{\overline{DOC}_{gc}}$, where d_i^c is the corrected number of reads mapped to position i, d_i is the original number of reads mapped to position i, $\overline{DOC}_{global}$ is the average depth of coverage (DOC) over all positions, and $\overline{DOC}_{gc}$ is the average DOC over all positions where the reads have the same GC content as in the reads mapped to position i.

Table 1. The results on the simulated mouse chromosome 17 under different sequencing depth and mutation rates between duplicated segments. No. of predicated CNVs are the number of regions CNVeM reports as CNVs. False discovery rate is the ratio between number of false positives and number of predicted CNVs, while false negative rate is the ratio between number of false negatives and number of true CNVs. It is obvious that CNVeM reports false positive regions due to that fact that it calls more CNVs than implanted in the donor genome.

mutation rate between duplicated segments	Depth of Coverage	No. of Predicted CNVs	No. of Correct CNVs	False Discovery Rate	False Negative Rate
1%	30X	102	100	2.0%	0
	15X	102	100	2.0%	0
	5X	105	100	4.8%	0
0.5%	30X	102	100	2.0%	0
	15X	105	100	4.8%	0
	5X	109	100	8.3%	0
0.1%	30X	101	97	4.0%	3.0%
	15X	107	98	8.4%	2.0%
	5X	116	96	17.2%	4.0%

3   Results

3.1   Simulation Results

In order to assess our method, we carried out experiments on simulation datasets. We developed a simulation framework, in which a donor genome is obtained by altering the copy number of some regions in the reference genome.

Experiment on a Simulated Mouse Chromosome. We first tested CNVeM on a simulated mouse genome. We obtained the masked reference chromosome 17 of Mus Musculus. After pruning all the 'N's, the length of the chromosome 17 reduced to 58Mb. This can be used as the "template sequence". We then duplicate segments of the sequence to generate a reference genome. The lengths of the duplicated segments are chosen from the range [1000, 10000]. We allow nucleotides to mutate with probability 1% in the duplication process. The copy numbers of these segments are then altered to generate the donor genome. The copy numbers are chosen from the set $\{0, 1, 2, 3, 4, 5\}$. In each experiment, we simulated 100 copy number variations between the reference genome and donor genome. To generate a read, we randomly picked a position from the donor genome and copied 36 consecutive bases starting from this position. The copying process is repeated until we have the desired coverage. All reads are then mapped to the reference genome using mrsFast [9], allowing reads to map with two mismatches. In addition to detecting the existence of copy number variants, CNVeM especially aims to distinguish which copy is duplicated or deleted in

Table 2. Measuring the accuracy of CNV break points by base pairs under different sequencing depth and mutation rates between duplicated segments. False discovery rate is the ratio between length of false positive regions and total length of predicted CNVs, while false negative rate is the ratio between length of false negative regions and total length of true CNVs.

Mutation Rate between Duplicated Segments	Depth of Coverage	Length of Predicted CNVs(bp)	Length of overlap(bp)	False Discovery Rate	False Negative Rate
1% (504000bp)	30X	506755	502183	0.9%	0.3%
	15X	506162	501291	1.0%	0.5%
	5X	507703	495074	1.8%	2.5%
0.5% (493000bp)	30X	492271	488114	0.9%	1.0%
	15X	500460	488387	2.4%	0.9%
	5X	501139	483830	3.5%	1.9%
0.1% (492000bp)	30X	469821	452120	3.8%	9.1%
	15X	465518	433495	6.9%	11.9%
	5X	462193	417340	9.7%	15.2%

the donor genome, while others have the same number of copy occurrences compared to the reference genome. Simulations are performed using various depth of coverage settings. A CNV is considered to be detected correctly when it overlaps with the true CNV region, meanwhile the predicted copy numbers should be the same as the true copy numbers. The results are shown in the first row of Table 1.

We also compared our reported CNVs to true CNVs by base pairs. The overlap is calculated by intersecting the coordinates of predicted CNVs with those of true CNVs. The results in the first row of Table 2 indicate high accuracy of CNVeM in predicting the break points.

Furthermore, we simulated the duplicated segments under different mutation rates to assess the power of our method in locating the copy variation origin. All results are summarized in Table 1 and Table 2. We see that both the mutation rate between duplicated segments and sequencing depth can affect the accuracy of our program. The smaller the mutation rate, the more similar the duplicated sequence, and the more difficult to distinguish which segment has copy number variation in the donor sequence. We have higher false discovery rate when the read depth is lower and the difference between duplicated copies is smaller, but we manage to recall almost all copy number variations.

The key observation in comparing the two tables (Table 1 and Table 2) is that the false negative rate in predicting the correct quantitive copy number is always lower than the false negative rate in calling the breakpoints of CNVs, moreover the false discovery rate of quantitative value for CNV is always higher than the false discovery rate in breakpoint calling. This illustrates that CNVeM is robust in detecting the existence of CNVs and determining the break points of CNVs. To achieve high sensitivity in CNV calling, CNVeM inevitably reports false positive regions. However, most of these false positive regions are short and thus we have low false discovery rate in break points calling.

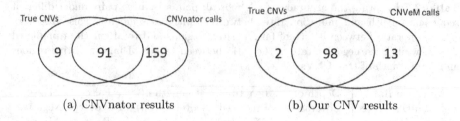

(a) CNVnator results   (b) Our CNV results

Fig. 2. Intersection of two CNV detection results with true CNVs. (a) We illustrate the venn diagram of the CNVnator calling with the true CNV regions. (b) We illustrate the intersection between the CNVeM calls and the true CNV regions. This Figure indicates that we have less false positives and false negatives compared to the CNVnator.

Comparing CNVeM with CNVnator on GC-Biased Data. In this section we compare CNVeM with the CNVnator [1], which is freely available online. Using a similar framework, we generated a reference genome and donor genome from chromosome 17 of Mus Musculus. We set the mutation rate between duplicated segments to be 0.1%. Reads are then simulated from the donor genome, allowing GC-bias [1,22]. In order to make the comparison fair for CNVnator, we used Bowtie [15] to do the mapping with option '-best -M 1'. With this option, Bowtie returns the best mapping for each read and in the case of tie it will randomly pick one mapping location for a read. This step is due to the fact CNVnator assumes there exists one mapping location for each read. However, for CNVeM, we use mrsFAST [9] to return all possible mapping positions for each read. Figure 2 illustrates the intersection of CNVs found by CNVeM and CNVnator on the simulated dataset, where 100 CNVs are implanted to the donor genome. CNVeM finds 111 CNVs which includes 98 of the true CNVs. This indicates that CNVeM has 13 false positives and 2 false negatives. However, CNVnator finds 250 CNV regions among which 91 regions are true CNVs. CNVnator fails to find 9 regions which are CNVs. Moreover, CNVnator reports 159 false positives. This results from the fact that CNVnator randomly places a read to one of its multiple mapping positions, and thus affects the read depth (RD) information, from which CNVnator determines the copy variation status. All the results indicates CNVeM has lower false discovery rate and false negative rate compared to CNVnator. We successfully locate the copy variation origins while CNVnator reports all possible CNV regions. Another disadvantage of CNVnator is that it can only determine the CNV to be a copy gain or copy loss, instead of recalling the exact quantitive copy number as in CNVeM.

Comparison between Different Strategies Dealing with Read Mapping Uncertainty. When handling reads that can be mapped to multiple positions, existing methods either discard those reads, or randomly place the read to one of the multiple mapping positions. CNVeM considers all possible mapping positions, and a read can be placed to one of the positions with a probability.

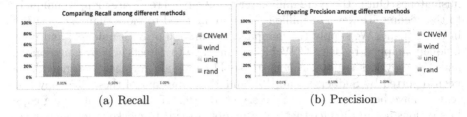

(a) Recall   (b) Precision

Fig. 3. Comparison between several strategies dealing with read mapping uncertainty. The x-axis represents the mutation rate between duplicated segments. The shorthands *CNVeM*, *wind*, *uniq* and *rand* represent the results from *CNVeM*, the results from *wind* which divides the genome into bins, the results from only considering reads mapped to unique positions, and results from placing a read to one of multiple mapping positions randomly, respectively.

We compared the performance of these different strategies. Furthermore, we consider the popular strategy which divides the genome into bins. All nucleotides within one bin have the same copy number. We develop a method 'wind' using the same EM framework as in section 2.3 for the bin strategy.

We run these methods on the same simulated datasets. Following the same process as mentioned above, we generated the reference genome and donor genome from chromosome 17 of Mus Musculus, with mutation rate between duplicated segments set to be 0.1%, 0.5%, and 1%, respectively. Reads are simulated at 30X coverage. The results plotted in Figure 3 illustrate that CNVeM has highest recall and precision at different mutation rates.

Time and Memory Usage. When dealing with HTS technology which generates tens of gigabytes of data per day, not only the accuracy of the method becomes important, but memory and time usage become important factors. The time and memory usage is estimated for the CNV calling process, we assume the mapping is done in a separate step. Our program takes 30 minutes to detect all the CNVs in the simulation dataset on masked chromosome 17 of the mouse genome, where we had 30X coverage (having around 50 million reads). All the experiments run on 64-bit AMD Opteron processor, furthermore, our program used 2Gb of memory at the peak of usage. In order to run CNVeM on the whole genome sequencing data, the memory usage increases linearly with the size of the genome.

3.2   Results on Real Data

We used the data published by Sanger Institute [19], where chromosome 17 of mouse strain A/J is deeply sequenced using Illumina technology to test our method on real data. The data contains 112 million (56 million pair-end) reads and the length of each read is 36bp. This results in a 42X coverage. We aligned the reads to the masked chromosome 17 using mrsFast [9], allowing up to 2 mismatches. Out of these 112 million reads, 39 million reads are mapped uniquely to the genome. However, 4 million reads are mapped to more than one position in the genome. We supply the mapping information of both uniquely and non-uniquely mapped reads to CNVeM, and manage to detect 44 copy gain regions and 355 copy loss regions. Among those 44 copy gain regions, 28 regions have been reported by Sudbery et al. [19], and 15 regions out of these 355 copy loss regions have been reported by Sudbery et al. [19]. The difficulty of calling copy loss regions lies in the fact that when mapping reads to masked mouse genome, large regions are 'N's thus no reads can be mapped to these masked regions. We investigate the coordinates of those remaining 340 novel copy loss regions and found all of them have an overlap with those masked regions. Furthermore, we apply CNVnator on this real data where it manages to detect 42 copy gain regions and 264 copy loss regions. Comparing the CNVeM calls with those of CNVnator, we see 26 copy gain regions overlap, and 86 copy loss regions are found by both methods.

4   Conclusion and Discussions

CNV regions have been shown to be correlated with many diseases ranging from cancers to learning disabilities [3,17]. Two main strategies exist to improve the CNV detection, either to improve the technology from which we gather data from individuals, or to design better algorithms. The shift from ArrayCGH to HTS is a great indication of improving the data gathering process, as current studies suggest that the use of HTS results in higher power in detecting CNV breakpoints and quantifying the true copy number for each region.

It has been shown previously that we can use both the depth of coverage (DOC) and paired-end information to detect CNVs accurately [16]. We illustrate the correct usage of DOC improves the accuracy of CNV detection greatly. In this work we present a probabilistic model for detecting CNVs, based on an expectation-maximization (EM) method. Our method incorporates all possible mapping information in the CNV prediction. It not only has higher accuracy in detecting the CNVs but also can detect which of the paralog regions in the genome is copied or deleted. All previous methods fail to distinguish paralog regions as they either discard all multiple mapping reads (reads mapped to multiple positions) or randomly place a read to one of the mapping positions.

Another main contribution of this work is that we can predict the CNV breakpoints in base-pair resolution. Unlike previous methods which define CNV for each bin (segment of fixed or variable length), our objective function is defined for each base-pair. In other words we are predicting the CNV for each base-pair. This helps us to detect the breakpoint of each CNV with high accuracy.

Although we mention that using DOC can improve the accuracy of CNV detection, we do not deny the fact that paired-end mapping has valuable information. Our future work is to incorporate paired-end reads information into our probabilistic model.

Acknowledgement. Z.W., F.H., W.Y. and E.E. are supported by National Science Foundation grants 0513612, 0731455, 0729049 and 0916676, and NIH grants K25-HL080079 and U01-DA024417. This research was supported in part by the University of California, Los Angeles subcontract of contract N01-ES-45530 from the National Toxicology Program and National Institute of Environmental Health Sciences to Perlegen Sciences. E.H. is a faculty fellow of the Edmond J. Safra Bioinformatics program at Tel-Aviv University. E.H. was supported by the Israel Science Foundation grant 04514831.

References

1. Abyzov, A., Urban, A.E., Snyder, M., Gerstein, M.: CNVnator: an approach to discover, genotype, and characterize typical and atypical CNVs from family and population genome sequencing. Genome Research 21(6), 974–984 (2011)
2. Alkan, C., Kidd, J.M., Marques-Bonet, T., Aksay, G., Antonacci, F., Hormozdiari, F., Kitzman, J.O., Baker, C., Malig, M., Mutlu, O., Cenk Sahinalp, S., Gibbs, R.A., Eichler, E.E.: Personalized copy number and segmental duplication maps using next-generation sequencing. Nature Genetics 41(10), 1061–1067 (2009)
3. Cappuzzo, F., Hirsch, F.R., Rossi, E., Bartolini, S., Ceresoli, G.L., Bemis, L., Haney, J., Witta, S., Danenberg, K., Domenichini, I., Ludovini, V., Magrini, E., Gregorc, V., Doglioni, C., Sidoni, A., Tonato, M., Franklin, W.A., Crino, L., Bunn Jr., P.A., Varella-Garcia, M.: Epidermal growth factor receptor gene and protein and gefitinib sensitivity in non-small-cell lung cancer. Journal of National Cancer Institute 97(9), 643–655 (2005)
4. Carter, N.P.: Methods and strategies for analyzing copy number variation using dna microarrays. Nature Genetics 39(suppl. 7), 16–21 (2007)
5. Chen, P.-A., Liu, H.-F., Chao, K.-M.: CNVDetector: locating copy number variations using array CGH data. Bioinformatics 24(23), 2773–2775 (2008)
6. Chiang, D.Y., Getz, G., Jaffe, D.B., O'Kelly, M.J.T., Zhao, X., Carter, S.L., Russ, C., Nusbaum, C., Meyerson, M., Lander, E.S.: High-resolution mapping of copy-number alterations with massively parallel sequencing. Nature Methods 6(1), 99–103 (2009)
7. Comaniciu, D., Meer, P.: Mean shift: A robust approach toward feature space analysis. IEEE Transactions on Pattern Analysis and Machine Intelligence 24, 603–619 (2002)
8. 1000 Genomes Project Consortium: A map of human genome variation from population-scale sequencing. Nature 467(7319), 1061–1073 (2010)
9. Hach, F., Hormozdiari, F., Alkan, C., Hormozdiari, F., Birol, I., Eichler, E.E., Cenk Sahinalp, S.: mrsfast: a cache-oblivious algorithm for short-read mapping. Nature Methods 7(8), 576–577 (2010)
10. Halperin, E., Hazan, E.: HAPLOFREQ-estimating haplotype frequencies efficiently. Journal of Computational Biology 13(2), 481–500 (2006)
11. He, D., Furlotte, N., Eskin, E.: Detection and reconstruction of tandemly organized de novo copy number variations. BMC Bioinformatics 11(suppl. 11), S12 (2010)
12. He, D., Hormozdiari, F., Furlotte, N., Eskin, E.: Efficient algorithms for tandem copy number variation reconstruction in repeat-rich regions. Bioinformatics 27(11), 1513–1520 (2011)

13. Hormozdiari, F., Alkan, C., Eichler, E.E., Cenk Sahinalp, S.: Combinatorial algorithms for structural variation detection in high-throughput sequenced genomes. Genome Research 19(7), 1270–1278 (2009)

14. John Iafrate, A., Feuk, L., Rivera, M.N., Listewnik, M.L., Donahoe, P.K., Qi, Y., Scherer, S.W., Lee, C.: Detection of large-scale variation in the human genome. Nature Genetics 36(9), 949–951 (2004)

15. Langmead, B., Trapnell, C., Pop, M., Salzberg, S.L.: Ultrafast and memory-efficient alignment of short dna sequences to the human genome. Genome Biology 10(3), R25 (2009)

16. Medvedev, P., Fiume, M., Dzamba, M., Smith, T., Brudno, M.: Detecting copy number variation with mated short reads. Genome Research 20(11), 1613–1622 (2010)

17. Sebat, J., Lakshmi, B., Malhotra, D., Troge, J., Lese-Martin, C., Walsh, T., Yamrom, B., Yoon, S., Krasnitz, A., Kendall, J., Leotta, A., Pai, D., Zhang, R., Lee, Y.-H., Hicks, J., Spence, S.J., Lee, A.T., Puura, K., Lehtimki, T., Ledbetter, D., Gregersen, P.K., Bregman, J., Sutcliffe, J.S., Jobanputra, V., Chung, W., Warburton, D., King, M.-C., Skuse, D., Geschwind, D.H., Conrad Gilliam, T., Ye, K., Wigler, M.: Strong association of de novo copy number mutations with autism. Science 316(5823), 445–449 (2007)

18. Simpson, J.T., McIntyre, R.E., Adams, D.J., Durbin, R.: Copy number variant detection in inbred strains from short read sequence data. Bioinformatics 26(4), 565–567 (2010)

19. Sudbery, I., Stalker, J., Simpson, J.T., Keane, T., Rust, A.G., Hurles, M.E., Walter, K., Lynch, D., Teboul, L., Brown, S.D., Li, H., Ning, Z., Nadeau, J.H., Croniger, C.M., Durbin, R., Adams, D.J.: Deep short-read sequencing of chromosome 17 from the mouse strains A/J and CAST/Ei identifies significant germline variation and candidate genes that regulate liver triglyceride levels. Genome Biology (October 2009)

20. Sudmant, P.H., Kitzman, J.O., Antonacci, F., Alkan, C., Malig, M., Tsalenko, A., Sampas, N., Bruhn, L., Shendure, J., 1000 Genomes Project, Eichler, E.E.: Diversity of human copy number variation and multicopy genes. Science 330(6004), 641–646 (2010)

21. Tuzun, E., Sharp, A.J., Bailey, J.A., Kaul, R., Anne Morrison, V., Pertz, L.M., Haugen, E., Hayden, H., Albertson, D., Pinkel, D., Olson, M.V., Eichler, E.E.: Fine-scale structural variation of the human genome. Nature Genetics 37(7), 727–732 (2005)

22. Yoon, S., Xuan, Z., Makarov, V., Ye, K., Sebat, J.: Sensitive and accurate detection of copy number variants using read depth of coverage. Genome Research 19(9), 1586–1592 (2009)

Structure-Based Whole Genome Realignment Reveals Many Novel Non-coding RNAs

Sebastian Will[1,*], Michael Yu[1,*], and Bonnie Berger[1,**]

Math. Department and CSAIL, MIT, 77 Massachusetts Ave, Cambridge, MA, USA
bab@mit.edu

Recent genome-wide computational screens that search for conservation of RNA secondary structure in whole genome alignments (WGAs) have predicted thousands of structural non-coding RNAs (ncRNAs). The sensitivity of such approaches, however, is limited due to their reliance on sequence-based whole-genome aligners, which regularly misalign structural ncRNAs. This suggests that many more structural ncRNAs may remain undetected. Structure-based alignment, which could increase the sensitivity, has been prohibitive for genome-wide screens due to its extreme computational costs. Breaking this barrier, we present the pipeline REAPR (RE-Alignment for *de novo* Prediction of structural ncRNA) that realigns whole genomes based on RNA sequence *and* structure and then evaluates the realignments for potential structural ncRNAs with a ncRNA predictor such as RNAz 2.0. For efficiency of the pipeline, we develop a novel banding realignment algorithm for the RNA multiple alignment tool LocARNA. This allows us to perform very fast structure-based realignment within limited deviation of the original multiple alignment from the WGA. We apply REAPR to the complete twelve Drosophila WGAs to predict ncRNAs across all these Drosophila species. Compared to direct prediction from the original WGA at the same False Discovery Rate (FDR), we predict twice as many high-confidence ncRNA candidates in *D.melanogaster* while less than doubling the run-time. As a novelty in ncRNA prediction, we control the FDR, going beyond the usual *a posteriori* FDR estimation. Applying the sequence-based alignment tool MUSCLE for realignment, we demonstrate that structure-based methods are necessary for effective prediction of originally misaligned ncRNAs. Comparing to recent screens of Drosophila and ENCODE we show that REAPR outperforms the widely-used *de novo* predictors RNAz, EvoFold, and CMFINDER. Finally, we reveal, with high confidence, a putative structural motif in the long ncRNA roX1 of *D.melanogaster*, known to regulate X chromosome dosage compensation in male flies. Interestingly, we recapitulate the Drosophila phylogeny, based on co-predicted ncRNAs across all fly genomes.

* Joint first authors.
** Corresponding author.

B. Chor (Ed.): RECOMB 2012, LNBI 7262, p. 341, 2012.
© Springer-Verlag Berlin Heidelberg 2012

Probabilistic Inference of Viral Quasispecies Subject to Recombination

Osvaldo Zagordi[1,2,*,**], Armin Töpfer[1,2,*], Sandhya Prabhakaran[3], Volker Roth[3], Eran Halperin[4,5], and Niko Beerenwinkel[1,2,***]

[1] Department of Biosystems Science and Engineering, ETH Zurich, Basel, Switzerland
[2] SIB Swiss Institute of Bioinformatics, Switzerland
[3] Computer Science Department, University of Basel, Switzerland
[4] Department of Molecular Microbiology and Biotechnology, Tel-Aviv University, Israel
[5] International Computer Science Institute, Berkeley, California, USA
niko.beerenwinkel@bsse.ethz.ch

Abstract. RNA viruses are present in a single host as a population of different but related strains. This population, shaped by the combination of genetic change and selection, is called quasispecies. Genetic change is due to both point mutations and recombination events. We present a jumping hidden Markov model that describes the generation of the viral quasispecies and a method to infer its parameters by analysing next generation sequencing data. The model introduces position-specific probability tables over the sequence alphabet to explain the diversity that can be found in the population at each site. Recombination events are indicated by a change of state, allowing a single observed read to originate from multiple sequences. We present an implementation of the EM algorithm to find maximum likelihood estimates of the model parameters and a method to estimate the distribution of viral strains in the quasispecies. The model is validated on simulated data, showing the advantage of explicitly taking the recombination process into account, and applied to reads obtained from two experimental HIV samples.

Keywords: Molecular sequence analysis, Sequencing and genotyping technologies, Next-generation sequencing, Viral quasispecies, Hidden Markov model.

1 Introduction

Next-generation sequencing (NGS) technologies have transformed experiments previously considered too labour intensive into routine tasks [11]. One application of NGS is the sequencing of genetically heterogeneous populations to quantify their genetic diversity. The genetic diversity is of primary clinical relevance, for example, in infection by RNA viruses, such as HIV and HCV. In these

---

* These authors contributed equally.
** Current affiliation: Institute of Medical Virology, University of Zurich, Switzerland.
*** Corresponding author.

B. Chor (Ed.): RECOMB 2012, LNBI 7262, pp. 342–354, 2012.
© Springer-Verlag Berlin Heidelberg 2012

systems, the high mutation rate of the pathogen, together with recombination, which can occur when different viral particles infect a single cell, give rise to a population of different but related individuals, referred to as viral quasispecies. We denote the different viral strains in this population as haplotypes. Studying the features of the viral quasispecies can shed light on the mechanisms of pathogen evolution in the host and it is of direct clinical relevance. In fact, the diversity of the quasispecies is known to affect virulence [19], immune escape [12], and drug resistance [10].

The quasispecies equation is a mathematical model for RNA virus populations evolving according to a mutation-selection process [6]. The dynamics of the model are described by a mutation term accounting for transformation of one viral haplotype (or strain) into another at the time of replication and a selection term that accounts for varying replication rates of different strains. The mutation process is generally considered as the result of point mutations only, although recombination is known to be frequent in many clinically relevant viruses, including HIV and HCV. For example, the recombination rate of HIV is estimated to be about ten fold higher than its point mutation rate. Therefore, the quasispecies model has been extended to account for both mutation and recombination [5]. At equilibrium, the model predicts the viral population to be dominated by one or a few haplotypes, which are surrounded by a cloud of mutants they constantly generate.

Recent NGS technologies allow for observing viral quasispecies at an unprecedented level of detail by producing millions of DNA reads in a single experiment. However, this high yield comes at a cost, because reads are usually short (up to 700bp with the latest technology and much shorter than the smallest viral genomes) and error-prone [8]. As a result, since the data obtained are incomplete and noisy, a meaningful characterization of viral populations by means of NGS requires careful analysis of the sequencing data [4].

In this manuscript, we aim at making inference of the viral quasispecies based on NGS data by explicitly modeling mutation and recombination. To this end, we use a hidden Markov model (HMM) to generate viral populations, i.e., haplotype distributions, and their probing by means of NGS. In our model, the haplotypes are originating from a small number of generating sequences via recombination (described as change of state in the HMM that selects from which sequence the haplotype derives) and mutation (described by position-specific probability tables for the generating sequences). The sequencing reads are obtained from the haplotypes subject to observation error. HMMs allowing for a switch between generating sequences, termed jumping HMMs, have been applied, for example, to sequence alignment of protein domains [18] and to detecting inter-host HIV circulating recombinant forms [17].

In order to reliably identify the haplotypes shaping intra-host quasispecies, including variants of low frequencies, sequencing errors must be corrected, as they will confound the true variation present in the sample. This has been approached, for example, by clustering reads or flowgrams and removing the within-cluster variation [15,21,7,22]. Rather than addressing error correction, in the present manuscript, we present a novel generative probabilistic model for making inference of viral quasispecies, i.e., for estimating the intra-patient viral haplotype distribution. Specifically, we assume that the true genetic diversity is generated by a few sequences, called generators, through mutation and recombination, and that the observed diversity results from additional sequencing errors. Throughout, the sequencing error rate is assumed to be known (either from control experiments or from complementary analyses) and fixed.

We present the model for local haplotype inference, meaning that we aim at inferring the population structure in a genomic region of a size that can be covered by individual reads. Extending this model to global haplotype inference, i.e., to longer genomic regions, is straightforward and will be discussed briefly. Nevertheless, local inference will generally be more reliable and sufficient for many applications. For example, the HIV protease gene, an important target of antiretroviral therapy, is 297 bp long, and it is now standard to obtain reads of 400 bp and longer with the Roche/454 GS Junior platform, a common pyrosequencing platform for clinical diagnostics. Local haplotype reconstructions can also be used as a starting point for global reconstruction [20,7,14,2,13].

We show that our model is able to estimate the true distribution of the haplotypes with high reliability, by applying it to simulated data, where we have access to the ground truth. We also present an application to experimental data obtained from sequencing mixtures of HIV clones.

2   Methods

2.1   Hidden Markov Model

During infection a viral strain can change either by point mutation, when a single base is copied with error, or by recombination, when a cell is infected by more than one viral particle, and viruses in subsequent generations produce a sequence that is a mosaic of those of the progenitors. The model we present here does not aim at representing these evolutionary processes mechanistically. Rather, it is a descriptive probabilistic model, in which the quasispecies is generated by switching among K different generating sequences, each of length L. These generators are defined as sequence profiles (μ_{jkv}) indicating the probability over the alphabet $\mathcal{A} = \{\mathtt{A}, \mathtt{C}, \mathtt{G}, \mathtt{T}, \mathtt{-}\}$ of base $v \in \mathcal{A}$ at position j of the k-th generating sequence. The set of sequences generates viral haplotypes $H \in \mathcal{A}^L$ by mutation (modeled by the probability tables (μ_{jkv})) and by recombination (modeled by switching to a different profile) as follows.

Let Z_j be the hidden random variable with state space $[K] = \{1, \ldots, K\}$ indicating the parental sequence generating H_j, the haplotype character at position j. We denote by ρ_j the probability of recombination, i.e., of switching the generating sequence right before position j. Each observed read R with bases R_j is obtained from a haplotype subject to noise (sequencing errors) at rate ε. The model is depicted in Figure 1 and defined as

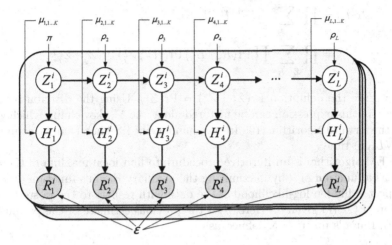

Fig. 1. Graphical representation. of the model. Only one observation i is depicted; for the full model, the graph is replicated for $i = 1, \ldots, N$.

$$\Pr(Z_1 = k) = \pi_k \tag{1a}$$

$$\Pr(Z_j = k \mid Z_{j-1} = l) = \begin{cases} \rho_j, & \text{if } k \neq l \\ 1 - (K-1)\rho_j, & \text{otherwise} \end{cases} \tag{1b}$$

$$\Pr(H_j = v \mid Z_j = k) = \mu_{jkv} \tag{1c}$$

$$\Pr(R_j = b \mid H_j = v) = \begin{cases} \varepsilon, & \text{if } b \neq v \\ 1 - (n-1)\varepsilon, & \text{otherwise} \end{cases} \tag{1d}$$

where $n = |\mathcal{A}|$ is the size of the alphabet.

The full model consists, for each observation $i = 1, \ldots, N$, of the hidden random variables Z_j^i (indicating generator sequences) and H_j^i (the haplotypes of the quasispecies), and the observed reads R_j^i, for all sequence positions $j = 1, \ldots, L$. The model parameters are summarized as $\theta = (\pi, \rho, \mu)$, and ε is a fixed constant.

2.2   EM Algorithm (Baum-Welch)

In the following we compute the likelihood $\Pr(R \mid \theta)$ and develop an expectation-maximization (EM) algorithm for finding the maximum likelihood estimates (MLE) of the parameters θ. The likelihood factorizes into the product over independent reads and, for each read, it can be computed efficiently using the Markov property. We have

$$\Pr(R) = \prod_i \sum_{Z^i, H^i} \Pr(Z^i, H^i, R^i)$$

$$= \prod_i \sum_{Z^i, H^i} \prod_j \Pr(R_j^i \mid H_j^i) \Pr(H_j^i \mid Z_j^i) \Pr(Z_j^i \mid Z_{j-1}^i),$$

where we use the definition $\Pr(Z_1^i \mid Z_0^i) = \Pr(Z_1^i)$. Using the distributive law, each sum in this expression can be factored along the Markov chain, which gives rise to the forward algorithm [16]. In this manner, the likelihood can be computed in $O(NLK^2)$ time.

The EM algorithm is an iterative procedure to find local maxima of the likelihood as a function of θ by maximizing the auxiliary Baum's function $Q(\theta, \theta')$, the expected hidden log-likelihood of the data with respect to the posterior distribution of (Z, H) given θ'. Here, θ' is the previous estimate of the parameters (π, ρ, μ). Baum's function is defined as

$$Q(\theta, \theta') = E_{Z, H \mid \theta'} [\log \Pr(R, Z, H \mid \theta)].$$

It bounds the log-likelihood from below, and repeated iterations of the maximization step with respect to θ (M-step) alternated with estimations of the distributions $\Pr(Z, H \mid R, \theta)$ (E-step) are guaranteed to find a local maximum of the likelihood function.

For the E-step, we compute

$$Q(\theta, \theta') = \sum_{j,k,v} N_j(k, v) \log \mu_{jkv} + \sum_{j=2}^{L} \left[N_j^{\neq} \log \rho_j + N_j^{=} \log(1 - (K - 1)\rho_j) \right] +$$

$$+ N^{\neq} \log \varepsilon + N^{=} \log(1 - (n - 1)\varepsilon) + \sum_k N_1(k) \log \pi_k,$$

where $N_1(k)$ is the expected number of times a Markov chain starts in state k at position 1, $N_j^{\neq}$ is the expected number of times that a Markov chain switches from a state k to a state $l \neq k$ at position j, $N_j^{=}$ is the expected number of times it does not switch, and $N_j(k, v)$ is the expected number of times the Markov chain is in state k and emits haplotype character v at position j. These expected counts are estimated for all reads by computing posterior probabilities of the hidden variables H, Z given the data and the current estimate of θ, using the forward and backward algorithm [16].

In the M-step, the parameters θ are updated by solving

$$\theta^{\text{new}} = \arg\max_\theta Q(\theta, \theta').$$

This is achieved by setting

$$\pi_k = \frac{N_1(k)}{\sum_k N_1(k)}, \qquad \rho_j = \frac{1}{K-1} \frac{N_j^{\neq}}{N}, \qquad \mu_{jkv} = \frac{N_j(k, v)}{\sum_v N_j(k, v)}.$$

The two steps are iterated until convergence (defined here as a relative change of the log-likelihood smaller than 10^{-6}). Since the EM algorithm is only guaranteed to find a local maximum, we performed 50 random restarts and chose the solution with the largest likelihood. The initial parameter values were drawn at random from the following distributions:

$$\pi \sim \mathrm{Dir}(2, \ldots, 2),$$
$$\mu_{jk} \sim \mathrm{Dir}(2, 2, 2, 2),$$
$$\rho_j \sim \frac{1}{K - 1 + M} \mathrm{Beta}(2, 2),$$

where Dir and Beta are, respectively, the Dirichlet and Beta distributions, and M the number of positions where ρ_j is non-zero (see below). Reads are hashed at the beginning in order to identify identical ones and to avoid unnecessary computations. The sequencing error rate ε was fixed to 0.01%

2.3   Sparse Recombination Rates

The model (1a–d) allows for recombination at each site $j = 1, \ldots, L$ with rate ρ_j, but it is not possible to estimate, for each sequence position of a read separately, from which generating sequence it has originated. However, despite high recombination rates, real RNA virus populations always display genomic regions that are conserved or nearly conserved (and that define the virus). In these regions, the different generating sequences cannot be distinguished because there is no or little diversity. Therefore, recombination among different sequences is expected to occur only at a small fraction of genomic sites, i.e., most recombination rates ρ_j are expected to be zero.

Sparse estimators of ρ can be found by considering the regularized maximum likelihood problem

$$\max_{\theta}\ \log \Pr(R \mid \theta) - f(\theta).$$

A natural choice for the regularization term is $f(\theta) = \lambda||\theta||_p$, where λ controls the degree of regularization. Values of p between 0 and 1 result in the desired sparsity of the parameters. The choice $p = 1$ is the lasso regularization, and approximation algorithms for solving it have been proposed [9].

Here, we consider the combinatorial model selection ($p = 0$) and select an optimal small subset of non-zero transition parameters ρ_j. We only allow ρ_j to be different from zero at M out of the $L-1$ positions. In practice, this is achieved by choosing $\rho_j = 0$ as initial value for $L - 1 - M$ positions, as they will remain zero throughout the EM procedure. In each re-start, the M positions for which we allow $\rho_j > 0$ are sampled uniformly at random. We choose $M = 3$ because more than three recombination sites are unlikely to be detectable for the size and the diversity of the genomic regions considered here.

2.4   Estimating the Haplotype Distribution

The quantity of greatest interest we derive from the model is the haplotype distribution $P(H)$, i.e., the structure of the viral quasispecies. For given model parameters θ, we can compute the probability of each haplotype efficiently using the forward algorithm. The distribution $P(H)$ might be estimated by computing the probability of each haplotype. However, since there are 4^L possible haplotypes, enumeration is infeasible. We estimate $P(H)$ by sampling from the model using equations 1a–1c. We sampled 10,000 haplotypes at the MLE of θ obtained in the previous step. This procedure is efficient because almost all model parameters μ_{jkv} and ρ_j are very close to either zero or one and as a result the probability mass will be centered on a few haplotypes.

3 Results

3.1 Simulated Datasets

We assessed the performance of our model on four different datasets, corresponding to different distributions of haplotypes of 300 bp length. In the first dataset, the haplotypes have frequencies 80%, 10%, 5% and 5% respectively. The two most frequent ones have a mutual sequence identity of 94%. They recombine to produce the latter two at recombination breakpoint at position 198. The second population serves as a negative control. It consists of three haplotypes with pairwise similarities 94%, 91%, and 91%, and we sampled reads at frequencies 80%, 10% and 10% without recombination. The third dataset comprises the same three haplotypes at frequencies 28%, 28% and 26%, and six others at frequency 3% each obtained by recombining the dominating ones at position 198. The fourth dataset consists again of the same three dominating haplotypes at frequencies 22%, 21% and 21%, and 24 recombinants at frequency 1.5% each obtained by recombining the three at two breakpoints at positions 198 and 280.

We performed model selection by evaluating the BIC score, defined as

$$\log \Pr(R \mid \theta) - \frac{\nu \log N}{2},$$

where ν is the number of free parameters of the model. For each distribution of haplotypes, we sampled 50 datasets of 2000 reads with point mutations at an error rate of 0.03% per base and evaluated the BIC score. Figure 2 reports the BIC scores for the three datasets. In all cases, the BIC score is maximum at the correct number of generators (two for the first dataset, three for the others). For comparison, we additionally learned a model in which recombination is not possible, i.e., where $\rho_j = 0$ for all j. For datasets 1 and 3, the BIC score is maximized at $K = 2$ and $K = 3$, respectively, but without recombination the correct number of generators (and of haplotypes) is $4 = 2 + 2$ and $9 = 3 + 6$, respectively. Thus, the recombination-free model fails to reconstruct the quasispecies structure in these cases.

In order to study the impact of the sample size, we sampled instances of the first dataset of different sizes and repeated the model selection procedure. We analyzed datasets of 500, 750, 1000 and 2000 reads. The results are reported in Figure 3. For the smallest sample, the BIC score erroneously selects $K = 1$,

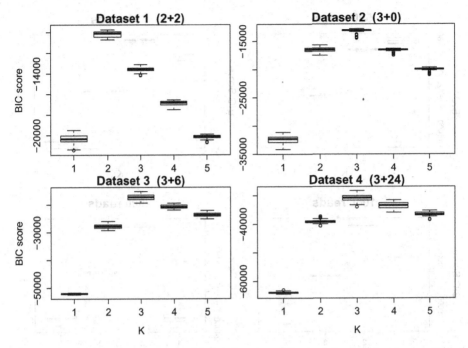

Fig. 2. BIC score for the four simulated datasets. The model correctly chooses $K = 2$ for the first dataset and $K = 3$ for the others. The boxplots summarize results of 50 independent datasets. The numbers in parentheses report the number of original generators plus the number of recombinants.

whereas for 750 and more reads, the procedure correctly selects $K = 2$ generating sequences indicating that sufficient coverage is an important prerequisite.

For parameter estimation, we sampled an additional set of reads from the haplotype distribution, and ran the EM algorithm with the value of K inferred before. Then, we inspected the MLE of the parameters μ, π and ρ. For datasets 1 and 2, where $K = 2$ and $K = 3$, respectively, were chosen, the estimates of μ_{jkv} are very close to either zero or one. The recombination parameter ρ is different from zero before the recombination hotspot and close to it.

In the second dataset, where no recombinants are present, the three generating sequence profiles corresponded exactly to the original haplotypes (i.e., all μ_{jkv} were either close to zero or to one) and no recombination was found (i.e., $\rho_j = 0$ for all j). In this case, π represents the frequency of the original haplotypes. Its estimate was very close to the original distribution and the remaining discrepancy can be explained by the sampling variance of the reads alone.

We also inspected the estimated parameters when all ρ_j are constrained to zero (no recombination), for dataset 1 with $K = 2$ and dataset 3 with $K = 3$. The estimates of μ_{jkv} reflected the distribution of bases at each individual position, but, as expected, the generating sequences cannot predict the original haplotypes. This is a consequence of the poor performance of the model selection in this case.

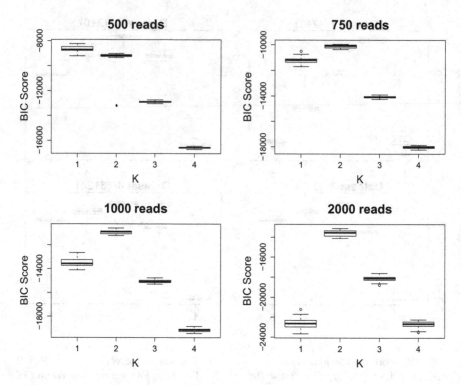

Fig. 3. BIC score for simulated dataset 1 at different sample sizes between 500 and 2000 reads. The model selection correctly selects $K = 2$ already with 750 reads. The boxplots summarize results on 50 independent datasets.

We assessed the reconstruction of the population by comparing it to the original set of sequences using the proportion close measure, φ_q, defined as the fraction of inferred haplotypes that match an original one with at most q mismatches [7]. Figure 4 reports the proportion close, as a function of the number of allowed mismatches, for the models with and without recombination estimated from datasets 1 and 3. The advantage of modeling recombinants is evident as the fraction of the population reconstructed is always higher than in the recombinant-free case. The proportion close is at least 99% for $q \geq 3$.

3.2   Real HIV Dataset

Using our model, we analyzed two sets of experimental NGS reads downloaded from the Sequence Read Archive (SRA). The first one (SRA run SRR069887) was obtained by sequencing a clinical sample from an HIV-infected patient in the context of a study of viral tropism [1]. We selected 1517 reads overlapping a 179 bp long region of the *env* gene (positions 6321–6499 in the HXB2 reference strain). We ran the EM algorithm on 50 datasets, generated by bootstrapping 1517 reads each, and selected the model with $K = 2$ generators (Figure 5, left).

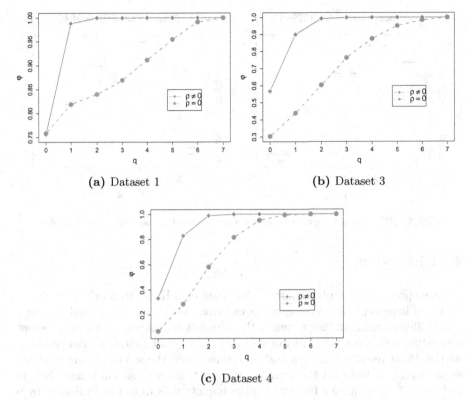

(a) Dataset 1   (b) Dataset 3

(c) Dataset 4

Fig. 4. Proportion close, φ_q, as a function of q. The fraction of the population reconstructed with q mismatches is higher than 99% already for $q = 3$ if one allows recombination, and, depending on the dataset, for $q = 6$ or 7 if one does not allow recombination.

By visual inspection of the estimated parameters, we found two positions where ρ is significantly different from zero. This translates into 8 possible different reconstructed recombination patterns and additional haplotypes of low frequencies generated by mutation.

The second sample (SRA run SRR002680) consists of 3954 reads overlapping a region of length 155 bp of the *env* gene (HXB2 positions 7085–7239). Under the same bootstrapping procedure used for the other dataset, model selection resulted in $K = 3$ generators (Figure 5, right).

In order to appreciate the compactness of the model inferred with the jumping HMM, we compared its solutions with those of another tool to reconstruct haplotypes, implemented in the software ShoRAH [20]. This method does not take recombination into account and it identified 15 haplotypes in the dataset SRR069887, which can be further reduced to 10 if one excludes those with frequencies lower than 1% and those which harbor a frameshift due to a deletion. Similarly, more than 50 haplotypes were found in the dataset SRR002680, which can be reduced to 18 by the same analysis.

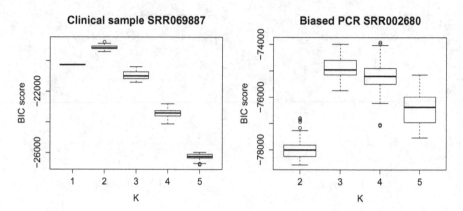

Fig. 5. BIC scores for experimental datasets based on 50 bootstrap samples

4   Discussion

We have presented a probabilistic model based on a HMM that infers the distribution of haplotypes in a viral quasispecies from NGS data. The model describes these different viral strains present in the population as originating from different generating sequences by means of two processes: point mutation and recombination. Point mutation is captured by the fact that the sequences are modeled as probability tables over the sequence alphabet. Recombination is modeled via a change of the sequence from which the haplotype is drawn as indicated by a change of state, or a jump, in the hidden Markov chain. Due to the possibility to switch between sequences, the number of tables necessary to describe the population structure remains small, whilst offering an excellent fit to the data. This results in a more compact and structured description of the viral population.

Using the EM algorithm, we find the MLE of the model parameters, namely the emission and transition probabilities. We have also introduced sparse recombination rates accounting for the fact that only few sites exist where generators recombine. Our results on simulated data demonstrate the usefulness of incorporating such sparsity while inferring haplotypes from recombinant reads.

There are several ways to extend the methodology presented here. For example, a different strategy for the regularization of the transition parameters ρ_j could be explored. We have observed in some cases a slow optimization of the likelihood, a behavior that might be related to the unidentifiability of the model, which implies that there are regions in the parameter space where the likelihood is flat. A Bayesian approach in which one could run a modified EM algorithm to maximize the posterior distribution of the parameters might solve this issue. In this framework, an appropriate prior might also provide efficient regularization.

Additionally, the transition matrix might be generalized allowing a different parameter for each pair of different states. This would account better for recombinant sequences present in the population at different frequencies. In this case, an efficient regularization of the ρ_j would be even more necessary.

Previous work on the analysis of NGS data to estimate genetic diversity have approached model selection in a non-parametric way by using the Dirichlet process mixture [21]. Extension of the HMM in this direction have been proposed and might be explored in this context as well [3].

Currently, we have presented our results in a local reconstruction setting, but our method generalizes to global inference. In this scenario, the population structure inferred locally is extended to longer genomic regions (longer than the typical read length). This can be achieved, for example, by allowing for longer generating sequences, along with two additional silent states to describe the unobserved regions before and after each read in the same fashion as the pair-HMM can be used for semi-global sequence alignment.

Although no current sequencing technology produces reads longer than 1000 bp, new platforms are foreseen to push this limit further. With such long reads, the probability to observe recombinations on a single read will be higher, and the necessity to keep the number of generators small will be even more compelling.

References

1. Archer, J., Rambaut, A., Taillon, B.E., Harrigan, P.R., Lewis, M., Robertson, D.L.: The evolutionary analysis of emerging low frequency HIV-1 CXCR4 using variants through timean ultra-deep approach. PLoS Comput. Biol. 6(12), e1001022 (2010), http://dx.doi.org/10.1371/journal.pcbi.1001022

2. Astrovskaya, I., Tork, B., Mangul, S., Westbrooks, K., Mandoiu, I., Balfe, P., Zelikovsky, A.: Inferring viral quasispecies spectra from 454 pyrosequencing reads. BMC Bioinformatics 12(suppl. 6) (2011), doi:10.1186/1471-2105-12-S6-S1

3. Beal, M., Ghahramani, Z., Rasmussen, C.: The infinite hidden Markov model. Advances in Neural Information 14, 577–584 (2002), http://citeseerx.ist.psu.edu/viewdoc/download?doi=10.1.1.62.8840 &rep=rep1&type=pdf

4. Beerenwinkel, N., Zagordi, O.: Ultra-deep sequencing for the analysis of viral populations. Current Opinion in Virology (January 2011) (in press), http://dx.doi.org/10.1016/j.coviro.2011.07.008

5. Boerlijst, M., Bonhoeffer, S., Nowak, M.: Viral quasi-species and recombination. Proceedings: Biological Sciences 263(1376), 1577–1584 (1996), http://www.jstor.org/stable/50405

6. Eigen, M.: Selforganization of matter and the evolution of biological macromolecules. Naturwissenschaften (January 1971), http://www.springerlink.com/index/Q47866457218X543.pdf

7. Eriksson, N., Pachter, L., Mitsuya, Y., Rhee, S.Y., Wang, C., Gharizadeh, B., Ronaghi, M., Shafer, R.W., Beerenwinkel, N.: Viral population estimation using pyrosequencing. PLoS Computational Biology 4(4), e1000074 (2008), http://www.ploscompbiol.org/article/info:doi/10.1371/journal.pcbi.1000074

8. Gilles, A., Meglécz, E., Pech, N., Ferreira, S., Malausa, T., Martin, J.F.: Accuracy and quality assessment of 454 GS-FLX Titanium pyrosequencing. BMC Genomics 12, 245 (2011), http://www.biomedcentral.com/1471-2164/12/245

9. Graça, J., Ganchev, K., Taskar, B., Pereira, F.: Posterior vs. parameter sparsity in latent variable models. In: NIPS 2009 (2009)

10. Johnson, J.A., Li, J.F., Wei, X., Lipscomb, J., Irlbeck, D., Craig, C., Smith, A., Bennett, D.E., Monsour, M., Sandstrom, P., Lanier, E.R., Heneine, W.: Minority HIV-1 drug resistance mutations are present in antiretroviral treatment-naïve populations and associate with reduced treatment efficacy. Plos Med. 5(7), 158 (2008)

11. Metzker, M.L.: Sequencing technologies - the next generation. Nat. Rev. Genet. 11(1), 31–46 (2010)

12. Nowak, M.A., Anderson, R.M., McLean, A.R., Wolfs, T.F., Goudsmit, J., May, R.M.: Antigenic diversity thresholds and the development of AIDS. Science 254(5034), 963–969 (1991), http://www.sciencemag.org/cgi/reprint/254/5034/963

13. Prabhakaran, S., Rey, M., Zagordi, O., Beerenwinkel, N., Roth, V.: HIV-haplotype inference using a constraint-based dirichlet process mixture model. In: Machine Learning in Computational Biology (MLCB) NIPS Workshop 2010, pp. 1–4 (October 2010)

14. Prosperi, M.C., Prosperi, L., Bruselles, A., Abbate, I., Rozera, G., Vincenti, D., Solmone, M.C., Capobianchi, M.R., Ulivi, G.: Combinatorial analysis and algorithms for quasispecies reconstruction using next-generation sequencing. BMC Bioinformatics 12(1), 5 (2011), http://www.biomedcentral.com/1471-2105/12/5

15. Quince, C., Lanzen, A., Davenport, R.J., Turnbaugh, P.J.: Removing noise from pyrosequenced amplicons. BMC Bioinformatics 12, 38 (2011)

16. Rabiner, L.: A tutorial on hidden Markov models and selected applications in speech recognition (with erratum). Proceedings of the IEEE 77(2), 257–286 (1989), doi:10.1109/5.18626

17. Schultz, A.K., Zhang, M., Leitner, T., Kuiken, C., Korber, B., Morgenstern, B., Stanke, M.: A jumping profile hidden Markov model and applications to recombination sites in HIV and HCV genomes. BMC Bioinformatics 7, 265 (2006)

18. Spang, R., Rehmsmeier, M., Stoye, J.: A novel approach to remote homology detection: jumping alignments. J. Comput. Biol. 9(5), 747–760 (2002)

19. Vignuzzi, M., Stone, J., Arnold, J., Cameron, C., Andino, R.: Quasispecies diversity determines pathogenesis through cooperative interactions in a viral population. Nature 439(7074), 344–348 (2006), http://www.nature.com/doifinder/10.1038/nature04388

20. Zagordi, O., Bhattacharya, A., Eriksson, N., Beerenwinkel, N.: Shorah: estimating the genetic diversity of a mixed sample from next-generation sequencing data. BMC Bioinformatics 12, 119 (2011), http://www.biomedcentral.com/1471-2105/12/119

21. Zagordi, O., Geyrhofer, L., Roth, V., Beerenwinkel, N.: Deep sequencing of a genetically heterogeneous sample: local haplotype reconstruction and read error correction. J. Comput. Biol. 17(3), 417–428 (2010), http://www.liebertonline.com/doi/abs/10.1089/cmb.2009.0164

22. Zagordi, O., Klein, R., Däumer, M., Beerenwinkel, N.: Error correction of next-generation sequencing data and reliable estimation of HIV quasispecies. Nucleic Acids Res. 38(21), 7400–7409 (2010), http://nar.oxfordjournals.org/lookup/pmid?view=long&pmid=

Simultaneously Learning DNA Motif along with Its Position and Sequence Rank Preferences through EM Algorithm

ZhiZhuo Zhang[1], Cheng Wei Chang[2], Willy Hugo[1], Edwin Cheung[2], and Wing-Kin Sung[1,2,*]

[1] National University of Singapore
{zhizhuo,hugowill,ksung}@comp.nus.edu.sg
[2] Genome Institute of Singapore
{changcw99,cheungcwe}@gis.a-star.edu.sg

Abstract. Although de novo motifs can be discovered through mining over-represented sequence patterns, this approach misses some real motifs and generates many false positives. To improve accuracy, one solution is to consider some additional binding features (i.e. position preference and sequence rank preference). This information is usually required from the user. This paper presents a *de novo* motif discovery algorithm called SEME which uses pure probabilistic mixture model to model the motif's binding features and uses expectation maximization (EM) algorithms to simultaneously learn the sequence motif, position and sequence rank preferences without asking for any prior knowledge from the user. SEME is both efficient and accurate thanks to two important techniques: the variable motif length extension and importance sampling. Using 75 large scale synthetic datasets, 32 metazoan compendium benchmark datasets and 164 ChIP-Seq libraries, we demonstrated the superior performance of SEME over existing programs in finding transcription factor (TF) binding sites. SEME is further applied to a more difficult problem of finding the co-regulated TF (co-TF) motifs in 15 ChIP-Seq libraries. It identified significantly more correct co-TF motifs and, at the same time, predicted co-TF motifs with better matching to the known motifs. Finally, we show that the learned position and sequence rank preferences of each co-TF reveals potential interaction mechanisms between the primary TF and the co-TF within these sites. Some of these findings were further validated by the ChIP-Seq experiments of the co-TFs.

Keywords: Motif Finding, Expectation Maximization, Importance Sampling, Binding Preference.

Program and Supplementary Online: http://biogpu.ddns.comp.nus.edu.sg/~chipseq/SEME/

---

* This work was supported in part by the MOEs AcRF Tier 2 funding R-252-000-444-112.

1 Introduction

Motif finding is an important classical bioinformatics problem. Given a set of biopolymer sequences (DNA or proteins), the motif finding problem aims to identify the recurring patterns (motifs) in them. Motif finders can generally be classified into two approaches: combinatorial searching and probabilistic modeling. The former approach enumerates the consensus patterns which are over-represented in the set of sequences. Using indexing data structures (e.g suffix tree [22], suffix array [16], and hash table [23]), it can efficiently identify short consensus motifs. Weeder [22], Trawler [7], YMF [29], DREME [2] are a few examples representing this line of approach. On the other hand, (most) probabilistic modeling approaches represent motifs using position weighted matrices (PWM) [28]. A PWM represents a length-w DNA motif as a $4 \times w$ matrix. It is more informative than a consensus pattern but it is also more difficult to compute. The high computational complexity of probabilistic modeling approach is a formidable bottleneck for its practical use. Expectation maximization [3] and Gibbs sampling [25] are the two most common approaches to find a PWM but they require long running time. Recently, some hybrid algorithms combined both approaches to get a good balance between accuracy and efficiency (e.g. [27], [17] and [15]).

By only examining the over-representation of sequence patterns, the previous generation motif finders often miss some real motifs and generate many false positives. On the other hand, additional information for the input sequences are found to be helpful to improve motif finding. For example, some transcription factor (TF) binding motifs (e.g. TATA-box) are localized to certain intervals with respect to the transcription start site(s) (TSS) of the gene. In this case, the position information can help to filter spurious sites. In protein binding microarray (PBM) [4] data, the *de Bruijn* sequences are ranked by their binding affinities and we expect the correct motif occurs in the high ranking sequences; such data has a rank preference. In the ChIP-Seq data [30], the ChIPed TF's motif (ChIPed TF is the TF pulled down in the ChIP experiment) prefers to occur in sequences with high ChIP intensity and also near the ChIP peak summits (thus having both position and rank preference). Hence, if we know the position preference and the sequence rank preference of the TF motifs in the input sequences, we can improve motif finding. In fact, many existing motif finders already utilize such additional information. MDscan [18] only considers high ranking sequences to generate its initial candidate motifs. Other programs allow users to specify the prior distribution of position preference or sequence rank preference [3,22,2,15,12] or add such preferences as a prior knowledge component in their scoring functions [5,21,17,13,9]. However, the users may not know the correct prior(s) to begin with. Even worse, different motifs may have different preferences. For example, in ChIP-Seq experiments, some motifs prefer to occur in high ranking sequences and at the center of the ChIP peak summit while others do not.

To resolve such problem, we propose a novel motif finding algorithm called SEME (**S**ampling with **E**xpectation maximization for **M**otif **E**licitation).

SEME assumes the set of input sequences is a mixture of two models: a motif model and a background model. It uses EM-based algorithm to learn the motif pattern (PWM), position preference and sequence rank preference at the same time; instead of asking users to provide them as inputs. SEME does not assume the presence of both preferences but automatically detect them during the motif refinement process through statistical significance testing. We also observe that EM algorithms are generally slow in analyzing large scale high throughput data. Speeding up EM using suffix tree was recently proposed [24] but the technique cannot be applied when one wants to also learn the position and sequence rank preferences. To improve the efficiency, SEME developed two EM procedures. The two EM procedures are based on the observations that the correct motifs usually have a short conserved pattern in it and majority of the sites in the input sequences are non-motif sites. For the first EM procedure, called extending EM (EEM), starts by finding all over-represented short l-mers and then attempts to include and refine the flanking positions around the l-mers within the EM iterations. This way, SEME recovers the proper motif length within a single run thus saving a substantial amount of time by avoiding multiple runs with different motif length (as done in many existing motif finders [3,22,17,15,12]). The second EM procedure, called the re-sampling EM (REM), tries to further refine the motif produced by EEM. It is based on a theorem similar to importance sampling [11], which stated that the motif parameters can be learned unbiasedly using a biased subsampling. By this principle, we can sample more sites which are similar to the EEM's motif and less sites from the background. This way, REM is able to learn the correct motifs using significantly less background sites. In our implementation, REM is capable to produce the correct TF motifs using approximately 1% of the sites normally considered in a normal EM procedure.

Using 75 large scale synthetic datasets, we show that SEME is better both in terms of accuracy and running time when compared to MEME, a popular EM-based motif finding program [3]. We found that MEME is unable to find motifs with gap regions while SEME's EEM procedure can successfully extend the motifs to include them. In the real experimental datasets, we perform comparison using 32 metazoan compendium datasets and 164 ChIP-Seq libraries. SEME consistently outperformed seven existing motif finding programs that we compare with. In general, we found that SEME not only finds more TF motifs but also gives more accurate results (as evaluated using either PWM divergence, AUC score or STAMP's p-value [20]). When we compare the programs to find co-regulated TF (co-TF) motifs[1] from 15 ChIP-Seq datasets, the superior performance of SEME is more pronounced. We propose that SEME's ability to learn the underlying motif binding preference is crucial in its performance. We further confirmed the correctness of the position and sequence rank preference of the coTF motifs learned by SEME on three ChIP-Seq datasets. The actual ChIP-Seq data of the predicted co-TFs clearly shows that SEME managed to infer the correct preferences. We also show that such preferences provide biological insights on the mechanism of the ChIPed TF–coTF interactions.

---

[1] Other TFs which bind nearby and function together with the ChIPed TF.

2 SEME Algorithm

SEME uses a probabilistic framework known as the two component mixture model (TCM) which is first proposed by MEME [3]. It assumes that the observed data is generated by two independent components: a motif model and a background model. Given an ordered list X of equal length DNA sequences, each site X_i in X is associated with a DNA sequence $X_i^{(seq)}$ and two integers: the rank of the sequence containing X_i ($X_i^{(rank)}$) and the position of the site X_i in the sequence ($X_i^{(pos)}$). We use an indicator variable Z_i to indicate if X_i is from the motif model or the background model, i.e., denote $Z_i = 1$ if X_i is from the motif model and 0 otherwise. The likelihood of an observed site X_i is written as:

$$Pr(X_i) = Pr(X_i|Z_i = 1)Pr(Z_i = 1) + Pr(X_i|Z_i = 0)Pr(Z_i = 0) \quad (1)$$

We use a naïve bayesian approach to combine three types of preferences (sequence, position, rank):

$$Pr(X_i|Z_i) = Pr(X_i^{(seq)}|Z_i)Pr(X_i^{(pos)}|Z_i)Pr(X_i^{(rank)}|Z_i) \quad (2)$$

For sequence preference, we model the motif site sequence with a position weight matrix (PWM) Θ, and the background sequence with a 0-order markov model θ_0. Θ is a $4 \times w$ matrix where $\Theta_{j,a}$ is the probability that the nucleotide a occurs at position j. For any length-w sequence X_i, the probability that X_i is generated from the motif model and the background model are as follows.

$$Pr(X_i^{(seq)}|Z_i = 1) = Pr(X_i^{(seq)}|\Theta) = \prod_{j=1}^{w} \Theta_{j,X_{i,j}^{(seq)}} \quad (3)$$

$$Pr(X_i^{(seq)}|Z_i = 0) = Pr(X_i^{(seq)}|\overrightarrow{\theta_0}) = \prod_{j=1}^{w} \theta_{0,X_{i,j}^{(seq)}} \quad (4)$$

where $X_{i,j}^{(seq)}$ is the nucleotide in the j-th position of the site X_i.

The position and sequence rank preferences are modeled using multinomial distributions. The position preference models the preference of the motif site to certain positions. Similarly, the sequence rank preference tries to model if the motif site prefers the sequences with certain range of ranks assuming input sequences are ordered by some criteria. To this end, we discretize both the positions and sequence ranks into K bins. The probability a binding site occurs at the k-th position bin is denoted as α_k, for $k = 1, \ldots, K$, while the background distribution is assumed to be uniform. Precisely, for every X_i, we have

$$Pr(X_i^{(pos)} = k|Z_i = 1) = \alpha_k; Pr(X_i^{(pos)} = k|Z_i = 0) = \frac{1}{K}$$

Similarly for sequence rank preferences, the probability a motif site occurs at the k-th sequence rank bin is denoted as β_k and,

$$Pr(X_i^{(rank)} = k|Z_i = 1) = \beta_k; Pr(X_i^{(rank)} = k|Z_i = 0) = \frac{1}{K}$$

Let $Pr(Z_i = 1)$ be λ. The parameters of the mixture model in SEME are $\Phi = (\lambda, \Theta, \theta_0, \{\alpha_1, ..., \alpha_K\}, \{\beta_1, ..., \beta_K\})$. We estimated these parameters by maximizing the log likelihood $\sum_{i=1}^{n} \log Pr(X_i | \Phi)$ using expectation maximization (EM) procedure. Given a set of sequences X, the classical EM algorithm is as follows. It first gives an initial guess of the parameter $\Phi^{(0)}$. Then, it iteratively performs two steps: E-step and M-step. Given $\Phi^{(t-1)}$, the t-th iteration of the E-step estimates $Z_i^{(t)} = Pr(Z_i | \Phi^{(t-1)}, X)$. Then, given $Z_i^{(t)}$, the t-th iteration of the M-step computes $\Phi^{(t)} = \arg\max_{\Phi} \sum_{i=1}^{n} \log Pr(X_i, Z_i^{(t)} | \Phi)$. The E-step and M-step are iterated until $\Phi^{(t)}$ is converged.

In this work, we developed four phases in the SEME pipeline (see Figure 1). To search for a good starting point, SEME first enumerates a set of over-represented short l-mers (phase 1) and extends each short l-mer to a proper length PWM motif by the extending EM (EEM) procedure (phase 2). The PWM reported by the extending EM procedure will approximate the true motif when its starting l-mer captures the conserved region of the motif. To further refine EEM's PWM motif, SEME applies the re-sampling EM (REM) procedure (phase 3). It is an importance sampling version of the classical EM algorithm which greatly speed up the EM iterations. Finally, the refined PWM motifs are scored and filtered for redundancies (phase 4). Below, we briefly describe these four phases (see the Supplementary section 3 for details).

SEME Pipeline

Require: A set of input DNA sequences (fasta format)
Ensure: Return a set of non-redundant motifs M
1: Identify a set of over-represented short l-mers Q in X ;
2: **for** every $q \in Q$ **do**
3:     Extend q to full length PWM motif Θ^{EEM} using extending EM procedure;
4:     Refine Θ^{EEM} to a more accurate PWM Θ^{REM} using re-sampling EM procedure;
5:     Add Θ^{REM} to candidate motif set M;
6: **end for**
7: Compute empirical scores (AUC or Z-score) for all the PWMs in M;
8: Sort all the PWMs in M and filter lower scoring redundant PWMs in M;

Fig. 1. Algorithm description for SEME Pipeline

Identifying Over-represented l-mers. In the first phase, SEME computes the frequencies of all short l-mers ($l = 5$ by default) in the input sequences and the background. If no background sequences are provided, a 1st-order markov model will be learned from the input sequences as the background model. We output all l-mers whose frequencies in the input sequences are higher than in the background to the next phase.

Extending EM. For each l-mer q obtained from the first phase, the aim of the extending EM (EEM) procedure is to extend the l-mer to a longer motif which maximizes the likelihood of observing the sites with q. In this phase, the EEM procedure only needs to study the sites containing the l-mer q, i.e., $Y = \{X_i \in X \mid X_i^{(seq)}$ matches $(N)^{w-|q|}q(N)^{w-|q|}\}$ ("N" is a wild char), and w is the maximum length of a motif. For example, if l-mer is "GGTCA" and the predefined longest possible motif length is 10, then EEM considers only those sites in X matching the string pattern "NNNNNGGTCANNNNN". The EEM procedure first initializes the parameters $\Phi^{(0)} = (\lambda^{(0)}, \Theta^{(0)}, \theta_0^{(0)}, \{\alpha_1^{(0)}, \ldots, \alpha_K^{(0)}\}, \{\beta_1^{(0)}, \ldots, \beta_K^{(0)}\})$ where $\lambda^{(0)}$ is the estimated percentage of Y not from background, $\Theta^{(0)}$ is PWM representing q, $\theta_0^{(0)}$ is the frequency of A,C,G,T in Y excluding the conserved l-mer q, $\alpha_i^{(0)} = \beta_i^{(0)} = 1/K$ for $i = 1, \ldots, K$ (uniform distribution). Then, it performs E-step (expectation) and M-step (maximization) iteratively.

Extending EM

Require: l-mer q, maximum allowed motif length w , input sequences X
Ensure: final extended PWM $\Theta^{(t)}$
1: $Y := \{X_i \in X \mid X_i^{(seq)}$ matches $(N)^{w-|q|}q(N)^{w-|q|}\}$;
2: Initialize the parameter set $\Phi^{(0)}$ for the mixture model;
3: $t := 1$;
4: **repeat**
5:     E-step: $\forall X_i \in Y$, compute the expectation of $Z_i^{(t)}$ using the parameter set $\Phi^{(t-1)}$ in the last iteration;
6:     M-step: update the parameter set $\Phi^{(t)}$ by maximizing log likelihood $Pr(Y, Z^{(t)}|\Phi)$;
7:     **if** length of $\Theta^{(t)} < w$; **then**
8:         Find a position j which maximizes the log likelihood increment in Equation5 and denote J to be the corresponding nucleotide distribution of position j;
9:         **if** J is significantly different from the background distribution $\overrightarrow{\theta_0}^{(t)}$ using Chi-square test; **then**
10:            Use J as the distribution in position j of PWM $\Theta^{(t)}$;
11:        **end if**
12:    **end if**
13:    $t := t+1$;
14: **until** PWM $\Theta^{(t)}$ converges;
15: The columns representing the l-mer q in $\Theta^{(t)}$ are diluted;

Fig. 2. Pseudocode for Extending EM procedure

In each iteration of the M-step, the EEM procedure will also try to include one additional column into $\Theta^{(t)}$ if such extension improves the likelihood. Precisely, for each position $j = 1, \ldots, 2w - |q|$ not in $\Theta^{(t)}$, we show that the maximum increment of the log likelihood before and after including the position j is $G(j)$ where

$$G(j) = \sup_J \sum_{X_i \in Y} Z_i^{(t)} \log(\frac{Pr(X_{i,j}^{(seq)}|J)}{Pr(X_{i,j}^{(seq)}|\theta_0^{(t)})}) \tag{5}$$

where J is any probability distribution over the nucleotides {A,C,G,T}.

Re-sampling EM

Require: the extended PWM $\Theta^{(EEM)}$, sampling rate μ, input sequences X

Ensure: Final refined PWM $\Theta^{(t)}$

1: Initialize the parameter set $\Phi^{(0)}$ for the mixture model;

2: $X_Q := \{X_i \in X \mid Q(X_i) = 1\}$ according to the probability $Pr(Q(X_i) = 1) = \min\{4^w \mu Pr(X_i|\Theta^{(EEM)}), 1\}$;

3: $t := 1$;

4: **repeat**

5:     E-step: $\forall X_i \in X_Q$, compute $Z_i^{(t)}$ using the parameter set $\Phi^{(t-1)}$ in the last iteration;

6:     M-step: update $\Phi^{(t)}$ by maximizing the weighted log likelihood $\sum_{X_i \in X_Q} \frac{\log Pr(X_i, Z_i|\Phi)}{Pr(Q(X_i)=1)}$;

7:     **if** the position distribution of $\{Z_i^{(t)}\}$ is significantly different from uniform distribution **then**

8:         include position preference in the model;

9:     **end if**

10:    **if** sequence rank distribution of $\{Z_i^{(t)}\}$ is significantly different from uniform distribution **then**

11:        include sequence rank preference in the model;

12:    **end if**

13:    $t := t+1$;

14: **until** $\Theta^{(t)}$ converge;

Fig. 3. Pseudocode for Re-sampling EM procedure

While the length of $\Theta^{(t)}$ is less than w, we extend the PWM $\Theta^{(t)}$ to include position j which brings the largest $G(j)$. To avoid over-fitting, the selected column also has to be tested (Chi-square) significantly different from the background frequency θ_0. The EEM procedure ends when PWM Θ converges. Finally, the columns in Θ representing the l-mer q will be further diluted (by setting all $[1.0, 0.0, 0.0, 0.0]$ columns representing "A" to $[0.5, \frac{0.5}{3}, \frac{0.5}{3}, \frac{0.5}{3}]$—other nucleotides are handled similarly) before Θ is returned as the output of the EEM procedure. In Supplementary section 1.1, we confirmed that EEM consistently recovers the correct motif length.

Re-sampling EM. The EEM procedure identifies an approximate motif model $\Theta^{(EEM)}$ with a proper motif length. This motif can be further refined using the classical EM algorithm to improve accuracy. However, when the input data X is big, this step will be slow. Using the idea of importance sampling, we proposed the re-sampling EM (REM) procedure which reduces the running time by running EM algorithm on a subsample of the original data.

Let $Q(\cdot)$ be the sampler function, where $Q(X_i) = 1$ if the sequence X_i is sampled; and 0 otherwise. In the Supplementary Theorem 1 (Supplementary section 3.4), we show that the log likelihood function $\log Pr(X, Z|\Phi)$ can be unbiasedly approximated by

$$\sum_{X_i \in X} \log Pr(X_i, Z_i | \Phi) = E_{X_Q} [\sum_{X_i \in X_Q} \frac{\log Pr(X_i, Z_i | \Phi)}{Pr(Q(X_i) = 1)}] \tag{6}$$

where each sampled site is weighted by a factor $\frac{1}{Pr(Q(X_i)=1)}$. The theorem implies that we need only run the EM algorithm on X_Q. Moreover, in the M-step of the original EM, instead of maximizing $\log Pr(X, Z | \Phi)$, we maximize $\sum_{X_i \in X_Q} \frac{\log Pr(X_i, Z_i | \Phi)}{Pr(Q(X_i)=1)}$.

Although Equation 6 is true for any arbitrary sampler function $Q(\cdot)$, running EM using different $Q(.)$ yields different sampling efficiencies. For example, we can use a uniform random sampler, i.e., $Pr(Q(X_i) = 1) = \mu$ for every $X_i \in X$, where $\mu \in [0, 1]$ is the sampling ratio. This function is expected to only cover $100\mu\%$ of the correct motif sites from X which prohibits the use of small μ. In our work, we employ the idea of importance sampling. Our sampling function $Q(.)$ satisfies $Pr(Q(X_i) = 1) = min\{4^w \mu Pr(X_i | \Theta^{(EEM)}), 1\}$, where w is motif length. This sampling function gives higher probabilities for the sites that are more consistent to $\Theta^{(EEM)}$ thus it is expected to sample more from the correct motif sites and less from the background. (assuming $\Theta^{(EEM)}$ models more of the correct motif site signal than the background signal). This strategy is useful since we avoid most of the background sites in X. In fact, our simulation reveals that the REM procedure can achieve nearly 60% recall rate (of the correct motif sites) at the sampling ratio as small as $2^{-10}(\approx 0.001)$ and 90% recall rate at the sampling ratio of $2^{-5}(\approx 0.031)$ (see Supplementary section 1.2). We choose a default sampling ratio of 0.01 in all experiments of this paper.

The position and sequence rank preferences are assumed to be non-existent at the beginning of the REM iterations (i.e., $Pr(X_i | Z_i) = Pr(X_i^{(seq)} | Z_i)$). The position and/or sequence rank preferences are considered only when the position and/or sequence rank distributions of $\{Z_i^{(t)}\}$ are significantly different from the uniform distribution (by Chi-square test). This strategy allows SEME to tell users which preference is really important for the predicted motif. Figure 3 is the pseudocode for this procedure.

Sorting and Redundancy Filtering. The PWMs output by REM are evaluated and sorted by empirical ROC-AUC (the area under the receiver-operator characteristic curve) or over-representation Z-score (representing the motif abundance) with the input data (details on each scoring are in the Supplementary section 4). The first score is preferred for the case when input sequences are short and most of sequences contain at least one motif site (e.g., ChIPed TF motif finding); for the other cases, we suggest to use the Z-score. We eliminate redundant PWMs from the sorted list as follows. When the sites of a PWM motif overlap with those of another PWM motif by more than 10%, we will treat the PWM motif with the lower score as redundant and remove it.

3   Result

3.1   Profiling Two Novel EM Procedures

SEME Significantly Outperforms MEME in Recovering the Planted PWM. To analyze SEME's performance, we extract all seventy-five motifs of lengths > 9 in JASPAR[31] vertebrate core database. For each such motif, we generated a training dataset of 1000 random sequences of length 400bp where 500 of them have a motif instance. The instances are planted uniformly across all positions and sequences.

For each dataset, we run SEME (EEM only), SEME (EEM + REM), and MEME (the classical EM-based motif finder) and obtain the top 5 predicted PWMs from each program. To test the goodness of the predicted PWMs, we compare the PWM divergence between the predicted PWMs and the actual planted PWMs. We also generate independent testing sequences with length 400bp (1000 positive sequences with one implanted motif site, 1000 negative sequences without motif site), and compute the ROC-AUC value for each predicted PWM. Figure 4(a) shows the comparison result. As expected, the random PWMs have the worst AUC values while the actual planted PWMs have the best AUC values. EEM's predicted PWMs have significantly better discriminative capability (AUC) and similarity (less PWM divergence) to the actual planted PWM as compared to random PWMs. This indicates that EEM's PWMs are good starting points for the subsequent REM procedure. REM's predicted PWMs further

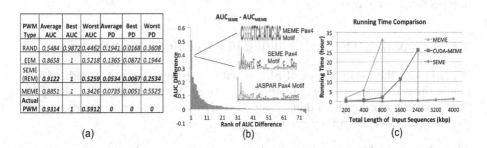

PWM Type	Average AUC	Best AUC	Worst AUC	Average PD	Best PD	Worst PD
RAND	0.5484	0.9872	0.4462	0.1941	0.0168	0.3608
SEME (EEM)	0.8658	1	0.5218	0.1365	0.0872	0.1944
SEME (REM)	0.9122	1	0.5259	0.0534	0.0067	0.2534
MEME	0.8851	1	0.3426	0.0735	0.0051	0.5525
Actual PWM	0.9314	1	0.5912	0	0	0

(a)   (b)   (c)

Fig. 4. The empirical performance of SEME on synthetic datasets. (a) The accuracy of SEME's PWM (both EEM step's (unrefined) PWM and the REM's (final) PWM are listed). We quantify accuracy using the commonly used Area-Under-ROC Curve (AUC) score and PWM divergence (PD) . We show that EEM's predicted PWM is already significantly stronger than random; indicating the goodness of EEM's PWM as starting point for the subsequent REM step. The scores also show that SEME's PWMs are significantly better when compared to MEME's. (b) Based on the performances of SEME and MEME on the Pax4 motif dataset, we observe that MEME has serious difficulties in mining PWMs with long gap region within them. (c) The running time of SEME is shown against increasing input size. We observe that CUDA-MEME, the GPU enabled version of MEME, still runs slower than SEME running on normal CPU (it takes 1 day to handle ≈ 6000 sequences while SEME takes around 1 hour for 10000 sequences).

improve the AUC score and are similar to the actual planted PWM (as indicated by the small PWM divergence).

Figure 4(a) also shows that SEME outperforms MEME. In fact, SEME is better than MEME in 42 out of 75 experiments (the cases with positive AUC differences in Figure 4(b)). The cases where SEME performed worse have relatively small AUC score differences (less than 0.04). We examined the Pax4 dataset in which SEME gains the highest improvement against MEME. The implanted JASPAR Pax4 motif is a diverged PWM of length 30. SEME successfully extended and recovered the full Pax4 motif; thanks to the ability of its EEM procedure to handle long gaps in its extension step. In contrast, MEME failed to model the long gaps due to their starting point finding procedure which assumes that all of the PWM positions are equally important.

SEME Is More Suitable in Handling Large Scale Data. We further generated 7 large datasets to observe the capability of SEME in handling large scale data. Each dataset consists of different number of sequences (from 500 to 10000, each of length 400bp). Figure 4(c) showed that the original MEME program cannot process more than 2000 sequences within one day, hence we also used the GPU-accelerated version of MEME, CUDA-MEME[19] (run on two Intel X5670 CPUs and two Fermi M2050 GPUs with 48GB RAM). SEME was run as normal CPU program. SEME is still around 60 times faster than CUDA-MEME which runs on the highly parallelized GPU system. In addition, SEME can process up to 10000 sequences (a typical dataset size for ChIP-Seq experiments) in 1 hour while the CUDA-MEME took more than one day to process 6000 sequences.

3.2   Comparing TF Motif Finding in Large Scale Real Datasets

We compare the performance of SEME with other existing motif-finding programs on two large scale TF binding site data. We also study the ability of SEME in uncovering the hidden position and/or sequence rank preferences in the input dataset when they are present.

The Metazoan Compendium Datasets. The first benchmark is a metazoan compendium dataset published by Linhart et.al[17]; consisting of 32 datasets based on experimental data from microarray, ChIP-chip, ChIP-DSL, and DamID as well as Gene Ontology data[1]. A list of the promoter sequences of many target genes (1000bp upstream and 200bp downstream the Transcription Start Site (TSS)) are used as the positive input for each motif-finding program and promoter sequences of other non-target genes are used as background sequences. The performance of six existing motif-finding programs, namely AlignACE [25], MEME [3], YMF [29], Trawler [7], Weeder [22], and Amadeus [17], were compared in the original benchmark study [17]. Each program's predicted PWMs are evaluated by the PWM divergence. Only PWMs with medium and strong matching with the known motifs (PWM divergence < 0.18) are considered to be successfully detected [17].

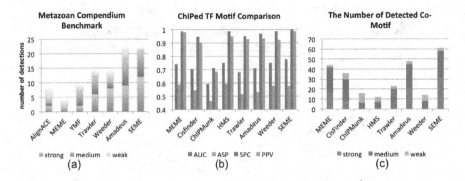

Fig. 5. The performance of SEME compared to existing motif finding programs from large scale real data (a) Comparison result on the metazoan compendium datasets. Four PWM motifs returned by each motif finding program are then compared to the known Transfac motifs using PWM divergence (PD) (as in [17]) and further classified into three matching categories (strong, medium, weak) corresponding to different PD cut-offs (0.12,0.18,0.24). (b) Comparison result on 164 ChIP-Seq libraries over four different measurements? AUC, PPV (Positive Predictive Value), ASP (Average Site Performance) and SPC (Specificity). The result shows that most motif finders perform similarly well in detecting ChIPed TF (but SEME is consistently better than all of them). (c) Comparison result for Co-TF motif finding on 15 ChIP-Seq libraries. The quality of reported PWMs is classified into three categories (strong, medium, weak) corresponding to different STAMP p-value cut-offs (0.0001, 0.01, 0.05). SEME reported the most number of co-TF's motif which match the known PWM with STAMP p-value ≤ 0.0001 (strong match, blue bar). Overall, SEME also found the most number of co-TF motif (61) as compared to the second best program, Amadeus (48).

The result of this comparison is shown in Figure 5(a). We find that SEME successfully detected the correct motifs in 21 datasets whereas the second best program, Amadeus, succeeded in 18. Weeder and Trawler found correct PWMs in 11 and 12 datasets, respectively. SEME also found more accurate motifs than the rest; it found 12 motifs with PWM divergence < 0.12. SEME further detected a significant position preference for the correct motifs for many datasets in this benchmark: most of them tend to bind nearer to the TSS position (see the Supplementary section 1.4 for details).

ChIP-Seq Experimental Datasets: Discovery of the ChIPed TF Motif from ChIP-Seq Data. The second benchmark is a collection of large scale ChIP-Seq experimental data which consists of 164 published ChIP-Seq libraries from the ENCODE project[8] and our lab over different cell-lines and TFs[6,33,14]. ChIP-Seq usually reports more than 10000 target sequences with narrower target regions (100bp). We compute the Area Under ROC Curve , Positive Predictive Value, Average Site Performance and Specificity scores of each program's predicted PWM. The formula for the above scores are given in the Supplementary section 4. From each library, the 100bp sequences around the top 10000 ChIP-Seq peaks were selected (sorted by ChIP intensity) as our

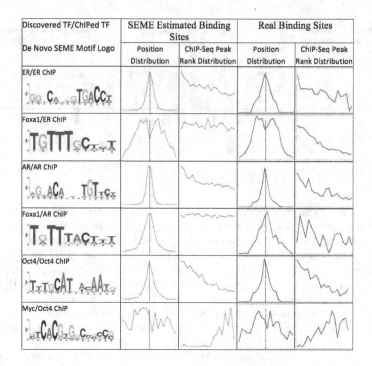

Fig. 6. Automatic learning of the position and sequence rank preference from the input data. Instead of requiring the user to input the expected co-TF motif preference distribution (position and/or sequence rank distribution), SEME learns such distributions directly from the input data. We show that most of the time, SEME can learn the correct distributions of each TF (as compared to real binding sites distribution in the rightmost column, defined by the ChIP-Seq and the known PWM of the TF). For position distribution, the x-axis is +/-200bp from ChIP-seq peak summit (the black dash line), and the y-axis is the fraction of binding sites in a given position. For rank distribution, the x-axis is the rank of ChIP-seq peak (left : high ChIP intensity, right : low ChIP intensity), and the y-axis is the fraction of binding sites in a given rank. The ChIP-seq peak rank distributions (MCF7 ER ChIP, LNCaP AR ChIP) of FoxA1 and the position distribution of Myc are tested to be insignificant by SEME.

input data. MEME and Weeder only use the top 2000 peaks due to their long running time. Peaks with odd numbered ranks were used for training while the even numbered peaks were used as positive testing data. The negative dataset is generated a 1st-order Markov model trained using the same number of 100bp random sequences extracted from the regions 1000bp away from the ChIP-Seq peaks.

We compared SEME with 7 popular *de novo* motif finding programs for ChIP data: MEME, Weeder, Cisfinder, Trawler, Amadeus, ChIPMunk and HMS. Each program's top 5 motifs are evaluated using the four statistics measurements on the test data. For each scoring, the best of the 5 motifs will be used to represent the performance of a program. Figure 5(b) shows the average performances of the motif finders. Again, we find that SEME is consistently better than all other programs (1st rank in Area Under ROC Curve , Positive Predictive Value and Specificity, and 3rd rank in Average Site Performance).

Discovery of co-TF Motifs from ChIP-Seq Data. We note that most motif finders show good performance in finding the ChIPed TF motifs. This is expected since the ChIPed TFs are highly enriched[33]. Compared to finding ChIPed TF motifs in ChIP-Seq datasets, the problem of finding co-TF motifs in the ChIP-Seq datasets is much more challenging. The co-TF motif instances are less abundant and most are not located exactly at the ChIP-Seq peaks. Nevertheless, finding the co-TF(s) could potentially uncover previously unknown TF-TF interaction.

For co-TF motif comparison, we used 15 ChIP-Seq libraries whose co-TFs have been characterized (the list of co-TFs for each ChIP-Seq is in Supplementary section 2.5). We extracted 400bp sequences around the ChIP-Seq peaks and compared the top 20 *de novo* motifs of each program to the known co-TF motifs in the JASPAR and Transfac database; we cannot use the previous statistical measurements since co-TFs may not occur in all ChIP-Seq peaks. Furthermore, the ChIPed TF binding sites need to be masked before we start the co-TF motif finding. SEME and ChIPMunk can do this automatically and, for other programs without auto-masking mode, the input sequences were masked by the top 2 motifs reported from their ChIPed motif finding results.

STAMP program[20] was used to compute the p-value of the match between a predicted co-TF motif against the known co-TF motif. STAMP p-value provides a better match measurement compared to PWM divergence since it removes the motif length bias [20]. We separated the p-value of the PWM matching into three significance levels: (1) weak match ($0.05 \geq$ p-value > 0.01), (2) medium match ($0.01 \geq$ p-value > 0.0001) and (3) strong match (p-value ≤ 0.0001). Figure 5(c) shows the performances for different motif-finding programs in finding the co-TF motifs from the 15 datasets. SEME recovered 61 known co-TF motifs; compared to Amadeus and MEME which find 48 and 44 co-TF motifs, respectively. 31 out of the 61 co-TF motifs of SEME belong to the strong match category (Amadeus only found 20) and another 27 are in the medium match category. This indicates that SEME's predicted co-TF PWMs are highly accurate.

To study the biological significance of the learnt preferences, we further study the output of three datasets, involving the ER, AR, FoxA1, Oct4 and c-Myc TFs, in details (see Figure 6). The real binding site of each TF is defined to be the site around +/-100bp around the TF's ChIP-Seq peak whose known PWM score is better than a cutoff that yields FDR= 0.01. If multiple matches occur, only the best scoring site is chosen. Comparison between SEME's learnt distributions (Figure 6, middle columns) and the real binding site distributions (Figure 6, rightmost columns) indicates that SEME is able to learn the correct co-TF position and sequence rank preferences. We also found that the motif positions of FoxA1, a known co-TF of ER, is not enriched exactly at the ER ChIP-Seq peak in the MCF7 data; instead it is found in the flanking regions near the ER peaks. Interestingly, in the LnCAP AR ChIP-Seq dataset (FoxA1 is also a known co-TF of AR), we found that FoxA1 binds very closely to AR—it is enriched at the AR ChIP-Seq peak summits. This observation is consistent with the previous report that FoxA1 can physically interact with AR [10]. This observation also indicates the different roles FoxA1 assumes when working with AR and ER[26]. In the ChIP-Seq data of Oct4 from mouse's ES cell, SEME found the motif of c-Myc enriched within Oct4's low intensity peaks regions. We conjectured that, in these regions, Oct4 indirectly bind the DNA through c-Myc (hence explaining the ChIP-Seq's low intensity). An earlier report showed that Oct4, along with Sox2, Nanog, and Stat3 form an enhancer module while c-Myc along with n-Myc, E2F1 and Zfx form a promoter module in the ES cell[6]. In fact, interaction between these enhancer and promotor modules had also been reported previously[32].

These examples indicate that the position and sequence rank distribution learnt by SEME are reasonably accurate and users could use them to infer the nature of the interaction between the ChIPed TF and the co-TF(s). In this manner, SEME can be used to generate biological hypothesis for further experimental validations. Moreover, the highly diverse preferences that we observe highlight the difficulty for users to provide the correct prior in the first place.

4   Conclusion

This paper developed a novel algorithm called SEME for mining motifs using mixture model and EM algorithm. We presented three important contributions: (1) automatic detection and learning of the position and sequence rank preferences of a candidate motif. (2) ability to estimate the correct TF motif length (with possible gaps within) and (3) using importance sampling for efficiency while still able to estimate the EM parameters unbiasedly. As a result, we showed that SEME is substantially better, both in terms of accuracy and efficiency, compared to the existing motif finding programs.

Moreover, in the task of finding co-TF motif in the ChIP-Seq data, SEME not only report more accurate co-TF motifs than other programs but also correctly estimated the position and sequence rank distribution of each co-TF's motif. We showed that such information provides useful insights on the interaction between the ChIPed TF and the predicted co-TFs. SEME does have a few limitations. Firstly, it assumes that the target motif contains conserved 5-mer region. In cases without such 5-mer, SEME also allows user to provide custom seeds. Secondly, SEME is more suitable for large scale input ($\geq$ 100 sequences) since it needs enough samples to determine whether we should do extension (EEM) or include additional binding preferences (REM).

References

1. Ashburner, M.: Gene ontology: Tool for the unification of biology. Nature Genetics 25, 25–29 (2000)
2. Bailey, T.L.: Dreme: Motif discovery in transcription factor chip-seq data. Bioinformatics 27(12), 1653 (2011)
3. Bailey, T.L., Elkan, C.: Fitting a mixture model by expectation maximization to discover motifs in biopolymers. In: Proc. Int. Conf. Intell. Syst. Mol. Biol., vol. 2, pp. 28–36 (1994)
4. Berger, M.F., Bulyk, M.L.: Protein binding microarrays (pbms) for rapid, high-throughput characterization of the sequence specificities of dna binding proteins. Methods in Molecular Biology-Clifton then Totowa 338, 245 (2006)
5. Chen, X., Hughes, T.R., Morris, Q.: Rankmotif++: a motif-search algorithm that accounts for relative ranks of k-mers in binding transcription factors. Bioinformatics 23(13), i72 (2007)
6. Chen, X., Xu, H., Yuan, P., Fang, F., Huss, M., Vega, V.B., Wong, E., Orlov, Y.L., Zhang, W., Jiang, J., et al.: Integration of external signaling pathways with the core transcriptional network in embryonic stem cells. Cell 133(6), 1106–1117 (2008)
7. Ettwiller, L., Paten, B., Ramialison, M., Birney, E., Wittbrodt, J.: Trawler: de novo regulatory motif discovery pipeline for chromatin immunoprecipitation. Nature Methods 4(7), 563–565 (2007)
8. Euskirchen, G.M., Rozowsky, J.S., Wei, C.L., Lee, W.H., Zhang, Z.D., Hartman, S., Emanuelsson, O., Stolc, V., Weissman, S., Gerstein, M.B., et al.: Mapping of transcription factor binding regions in mammalian cells by chip: comparison of array-and sequencing-based technologies. Genome Research 17(6), 898 (2007)
9. Frith, M.C., Hansen, U., Spouge, J.L., Weng, Z.: Finding functional sequence elements by multiple local alignment. Nucleic Acids Research 32(1), 189 (2004)
10. Gao, N., Zhang, J., Rao, M.A., Case, T.C., Mirosevich, J., Wang, Y., Jin, R., Gupta, A., Rennie, P.S., Matusik, R.J.: The role of hepatocyte nuclear factor-3α (forkhead box a1) and androgen receptor in transcriptional regulation of prostatic genes. Molecular Endocrinology 17(8), 1484 (2003)
11. Glynn, P.W., Iglehart, D.L.: Importance sampling for stochastic simulations. Management Science, 1367–1392 (1989)
12. Hu, M., Yu, J., Taylor, J.M.G., Chinnaiyan, A.M., Qin, Z.S.: On the detection and refinement of transcription factor binding sites using chip-seq data. Nucleic Acids Research 38(7), 2154 (2010)
13. Keilwagen, J., Grau, J., Paponov, I.A., Posch, S., Strickert, M., Grosse, I.: De-novo discovery of differentially abundant transcription factor binding sites including their positional preference. PLoS Computational Biology 7(2), e1001070 (2011)
14. Kong, S.L., Li, G., Loh, S.L., Sung, W.K., Liu, E.T.: Cellular reprogramming by the conjoint action of erα, foxa1, and gata3 to a ligand-inducible growth state. Molecular Systems Biology 7(1) (2011)
15. Kulakovskiy, I.V., Boeva, V.A., Favorov, A.V., Makeev, V.J.: Deep and wide digging for binding motifs in chip-seq data. Bioinformatics 26(20), 2622 (2010)
16. Lam, T.W., Sadakane, K., Sung, W.K., Yiu, S.M.: A space and time efficient algorithm for constructing compressed suffix arrays. Computing and Combinatorics, 21–26 (2002)
17. Linhart, C., Halperin, Y., Shamir, R.: Transcription factor and microRNA motif discovery: The Amadeus platform and a compendium of metazoan target sets. Genome Research 18(7), 1180 (2008)

18. Liu, X.S., Brutlag, D.L., Liu, J.S.: An algorithm for finding protein–dna binding sites with applications to chromatin-immunoprecipitation microarray experiments. Nature Biotechnology 20(8), 835–839 (2002)

19. Liu, Y., Schmidt, B., Liu, W., Maskell, D.L.: CUDA-MEME: Accelerating motif discovery in biological sequences using CUDA-enabled graphics processing units. Pattern Recognition Letters (2009)

20. Mahony, S., Auron, P.E., Benos, P.V.: Dna familial binding profiles made easy: comparison of various motif alignment and clustering strategies. PLoS Computational Biology 3(3), e61 (2007)

21. Narang, V., Mittal, A., Sung, W.K.: Localized motif discovery in gene regulatory sequences. Bioinformatics 26(9), 1152 (2010)

22. Pavesi, G., Mauri, G., Pesole, G.: An algorithm for finding signals of unknown length in DNA sequences. Bioinformatics 17(suppl. 1), 207–214 (2001)

23. Raphael, B., Liu, L.T., Varghese, G.: A uniform projection method for motif discovery in dna sequences. IEEE Transactions on Computational biology and Bioinformatics, 91–94 (2004)

24. Reid, J.E., Wernisch, L.: Steme: efficient em to find motifs in large data sets. Nucleic Acids Research 39(18), e126–e126 (2011)

25. Roth1JT, F.P., Hughes, J.D., Estep, P.W., Church, G.M.: Finding DNA regulatory motifs within unaligned noncoding sequences clustered by whole-genome mRNA quantitation. Nature Biotechnology 16, 939 (1998)

26. Sahu, B., Laakso, M., Ovaska, K., Mirtti, T., Lundin, J., Rannikko, A., Sankila, A., Turunen, J.P., Lundin, M., Konsti, J., et al.: Dual role of foxa1 in androgen receptor binding to chromatin, androgen signalling and prostate cancer. The EMBO Journal 30(19), 3962–3976 (2011)

27. Sharov, A.A., Ko, M.S.H.: Exhaustive Search for Over-represented DNA Sequence Motifs with CisFinder. DNA Research (2009)

28. Sinha, S.: On counting position weight matrix matches in a sequence, with application to discriminative motif finding. Bioinformatics 22(14) (2006)

29. Sinha, S., Tompa, M.: A statistical method for finding transcription factor binding sites. In: Proceedings of the Eighth International Conference on Intelligent Systems for Molecular Biology, pp. 344–354 (2000)

30. Valouev, A., Johnson, D.S., Sundquist, A., Medina, C., Anton, E., Batzoglou, S., Myers, R.M., Sidow, A.: Genome-wide analysis of transcription factor binding sites based on chip-seq data. Nature Methods 5(9), 829 (2008)

31. Wasserman, W.W., Sandelin, A.: Applied bioinformatics for the identification of regulatory elements. Nature Reviews Genetics 5(4), 276–287 (2004)

32. Wu, Q., Ng, H.H.: Mark the transition: chromatin modifications and cell fate decision. Cell Research (2011)

33. Zhang, Z., Chang, C.W., Goh, W.L., Sung, W.K., Cheung, E.: Centdist: discovery of co-associated factors by motif distribution. Nucleic Acids Research 39(suppl. 2), W391 (2011)

Author Index

Printed in the United States
By Bookmasters